D0946889

PRINCIPLES OF ELECTRONICS
A User-Friendly Approach

ELLIS HORWOOD SERIES IN *ELECTRICAL AND ELECTRONIC ENGINEERING*

Series Editor: PETER BRANDON,
Emeritus Professor of Electrical and Electronic Engineering, University of Cambridge

P.R. Adby	APPLIED CIRCUIT THEORY: Matrix and Computer Methods
J. Beynon	PRINCIPLES OF ELECTRONICS: A User-Friendly Approach
M. J. Buckingham	NOISE IN ELECTRONIC DEVICES AND SYSTEMS
A.G. Butkovskiy & L.M. Pustylnikov	THE MOBILE CONTROL OF DISTRIBUTED SYSTEMS
S.J. Cahill	DIGITAL AND MICROPROCESSOR ENGINEERING
R. Chatterjee	ADVANCED MICROWAVE ENGINEERING: Special Advanced Topics
R. Chatterjee	ELEMENTS OF MICROWAVE ENGINEERING
R.H. Clarke	DIFFRACTION THEORY AND ANTENNAS
P.G. Ducksbury	PARALLEL ARRAY PROCESSING
J.-F. Eloy	POWER LASERS
D.A. Fraser	INTRODUCTION TO MICROCOMPUTER ENGINEERING
J. Jordan, P. Bishop & B. Kiani	CORRELATION-BASED MEASUREMENT SYSTEMS
F. Kouril & K. Vrba	THEORY OF NON-LINEAR AND PARAMETRIC CIRCUITS
S.S. Lawson	WAVE DIGITAL FILTERS
P.G. Mclaren	ELEMENTARY ELECTRIC POWER AND MACHINES
J.L. Min & J.J. Schrage	DESIGNING ANALOG AND DIGITAL CONTROL SYSTEMS
P. Naish & P. Bishop	DESIGNING ASICS
M. Ramamoorty	COMPUTER-AIDED DESIGN OF ELECTRICAL EQUIPMENT
J. Richardson & G. Reader	ANALOGUE ELECTRONICS CIRCUIT ANALYSIS
J. Richardson & G. Reader	DIGITAL ELECTRONICS CIRCUIT ANALYSIS
J. Richardson & G. Reader	ELECTRICAL CIRCUIT ANALYSIS
J. Seidler	PRINCIPLES OF COMPUTER COMMUNICATION NETWORK DESIGN
P. Sinha	MICROPROCESSORS FOR ENGINEERS: Interfacing for Real Time Applications
E. Thornton	ELECTRICAL INTERFERENCE AND PROTECTION
L.A. Trinogga, K.Z. Guo & T.H.Edgar	MICROWAVE CIRCUIT DESIGN
C. Vannes	LASERS AND INDUSTRIES OF TRANSFORMATION
R.E. Webb	ELECTRONICS FOR SCIENTISTS
A.M. Zikic	DIGITAL CONTROL

ELECTRONIC AND COMMUNICATION ENGINEERING

R.L. Brewster	TELECOMMUNICATIONS TECHNOLOGY
R.L. Brewster	COMMUNICATION SYSTEMS AND COMPUTER NETWORKS
J.N. Slater	CABLE TELEVISION TECHNOLOGY
J.N. Slater	SATELLITE BROADCASTING SYSTEMS: Planning and Design
J.G. Wade	SIGNAL CODING AND PROCESSING: An Introduction Based on Video Systems

PRINCIPLES OF ELECTRONICS
A User-Friendly Approach

JOHN BEYNON B.Sc., Ph.D., M.Inst.P., C.Phys.
Department of Physics
Brunel University, Uxbridge, Middlesex

ELLIS HORWOOD
NEW YORK LONDON TORONTO SYDNEY TOKYO SINGAPORE

First published in 1990 by
ELLIS HORWOOD LIMITED
Market Cross House, Cooper Street,
Chichester, West Sussex, PO19 1EB, England

A division of
Simon & Schuster International Group

Typeset in Times by Ellis Horwood Limited
Printed and bound in Great Britain
by Hartnolls, Bodmin, Cornwall

British Library Cataloguing in Publication Data

Beynon, John
Principles of electronics.
I. Title
537.5
ISBN 0-13-717117-X

Library of Congress Cataloging-in-Publication Data available

Table of Contents

Dedicated
to my
father's
memory

Preface

This textbook is designed, particularly, to be of help to students of materials science, chemistry, and biology who, through no special desire of their own, have found themselves having to study a course in electronics. Although the student may have no intense liking of electronics, the need to obtain some basic knowledge of the subject is vitally important not only for his/her general education but also because a wide variety of measurement techniques are electronics-based. The textbook should also assist those physics students at school, college, and university who need a firm theoretical framework on which to support an understanding of more complicated electronics systems.

The book takes the student on a journey through the world of digital and analogue electronics: from number codes, logic gates, and flip-flops to decoders and multiplexers; from definitions of current and electric field to semiconductor diodes and transistors and operational amplifiers. This book should give the student an understanding of some of the basic concepts which underpin elementary electronics, and on which he/she can build detailed explanations of more complicated circuits and systems.

Problem-solving exercises play an important role in seven of the chapters. They have a multifunctional purpose: to gain practice in employing and manipulating some fundamental relationship; to reinforce a particular concept or construct; to enable further information about a given device or circuit to be extracted. In short, the worked examples form an integral part of the book and should not be omitted if you hope to gain a coherent understanding of a topic.

When all is said and done, electronics is a practically-oriented subject, and a one hundred percent appreciation of the subject matter in this textbook will not compensate for not sitting down at a workbench and *doing* electronics. Some of the devices mentioned in this book can be bought fairly cheaply nowadays, so there is no real excuse for not performing a few of the experiments referred to. This book, then, tells only half a story. The other half is for you to script.

ACKNOWLEDGEMENTS

My thanks to the Department of Physics, Brunel University, for allowing me to use a few of the ideas contained in some undergraduate laboratory scripts.

1

General introduction with definitions

Objectives

(i) To identify the properties of a linear circuit
(ii) The importance of the principle of superposition in network theory
(iii) To distinguish between electric current, charge, and current density
(iv) To distinguish between potential difference, elctromotive force, and electric field
(v) To obtain the laws for adding resistances in series and parallel
(vi) To define Ohm's law and verify it experimentally
(vii) To identify differences in the electrical behaviour of the resistor, inductor, and capacitor
(viii) To define the voltage and the current source (to be used throughout the book).

1.1 CURRENT FLOW

Electrostatics is the study of electrical charges at rest, whereas *electrodynamics* is about electric charges in motion. In this book we shall be concerned only with current flow, so it is the latter topic of electrodynamics which will be of interest to us. You will find that many textbooks treat electrodynamics by first setting-up sophisticated equations of motion for electrons (and, in some instances, positive holes). We shall definitely not be taking this route. Instead, we will use an operational approach based on *cause-and-effect*.

What are we meant to understand by the phrase *cause-and-effect*? Well, if you put your hand into a flame — obviously it will get hot and eventually burn if left there too long. The only sensible course of action for you to take is to remove your hand from the flame. Withdrawing your hand is said to be a predictable *response* to the *stimulus* of heat. This is a pretty straightforward example. However, there are situations in which the response (effect) of a particular stimulus (cause) cannot always be anticipated — weather forcasting is a case in point because there are many unpredictable factors involved. In electronics, the stimulus might be either a voltage or current source, and the response might be either the current flowing through a component or the potential difference across it.

In this book we are going to concentrate on a class of circuits and devices described as *linear*, and following the more usual terminology in electrical work, we shall call the stimulus the *input* and the response the *output*.

Linear devices and circuits have the following special properties:

(i) The output Y produced by a given input X is reproducible, and does not depend on when the measurements are carried out.

(ii) If X can be broken down into a number of inputs $X_1, X_2, \ldots$, etc., then Y can be determined by studying the effect of each input separately.

The output Y of the linear circuit in Fig. 1.1 is directly proportional to the input X;

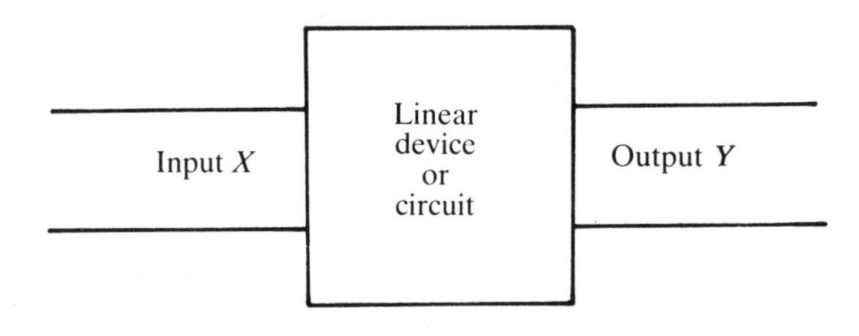

Fig. 1.1.

doubling the input doubles the output. If the input X is equal to the sum of the inputs X_1 and X_2 then by (ii), using an arrow to represent the phrase *leads to*, we have

$$X_1 \rightarrow Y_1$$
$$X_2 \rightarrow Y_2 \; . \tag{1.1}$$

Therefore

$$X_1 + X_2 \rightarrow Y_1 + Y_2 \; . \tag{1.2}$$

Relation (1.2) is a statement of the *Principle of Superposition*. It can be extended to the case where the input is the sum of N individual inputs, $X_1, X_2, X_3, \ldots X_N$ and the output is the sum of N individual outputs $Y_1, Y_2, Y_3, \ldots Y_N$. Then, using the summation sign,

$$\sum_{r=1}^{N} X_r \rightarrow \sum_{r=1}^{N} Y_r \; .$$

1.2 BASIC DEFINITIONS

1.2.1 Electric current

A current results whenever there is a flow of charge. Its formal definition is:

$$\text{CURRENT} = \text{RATE OF FLOW OF ELECTRICAL CHARGE} \tag{1.3}$$

UNITS: coulombs per second ($C s^{-1}$) or ampere (A)

Usually, the charges are electrons and the direction of conventional current flow (as indicated by the arrow placed on circuit diagrams) is taken to be the opposite direction to this.

Fig. 1.2 shows a wire cut so that the cross-section is at right angles to the length of

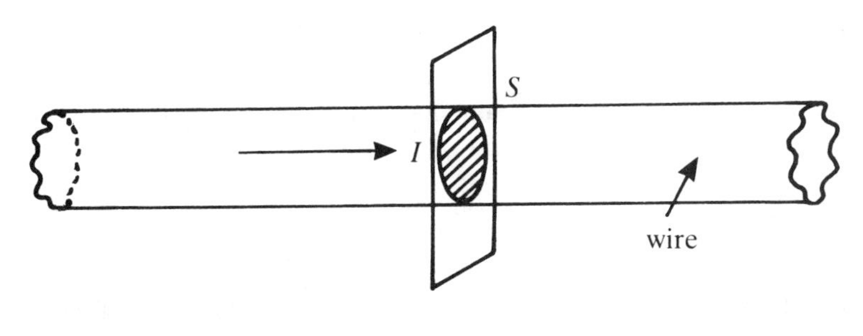

Fig. 1.2.

the wire. Thus

$$\text{CURRENT DENSITY } j = \frac{\text{CURRENT } I}{\text{CROSS-SECTIONAL AREA } S} \tag{1.4}$$

UNITS: ampere per unit area ($A m^{-2}$)

For those students familiar with vectors the relation between current and current density is

$$\mathbf{I} = \mathbf{j} \cdot \mathbf{S} \ . \tag{1.5}$$

One restriction that must be placed on this definition is that the flow of charge must be the same through every point on the cross-section. Then the current density will be uniform. If this is not the case a slightly more mathematical procedure must be adopted in which the wire is divided into a large number of filaments, each having a tiny cross-sectional area which we shall call ΔS, see Fig. 1.3.

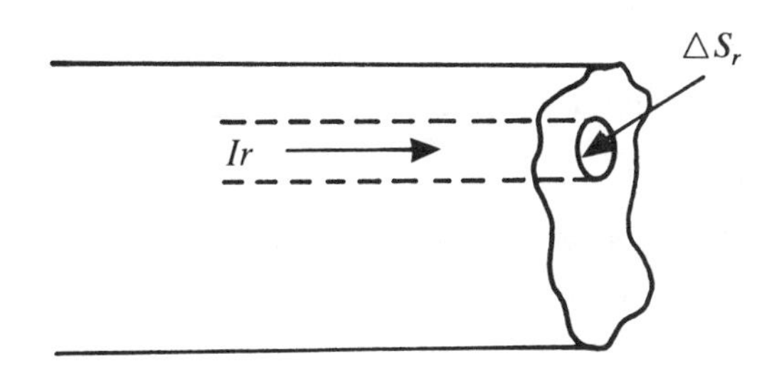

Fig. 1.3.

The current density j_r, for the rth filament, can be expressed in terms of the current I_r flowing through it. That is,

$$I_r = j_r \Delta S_r \ . \tag{1.6}$$

Then the total current I flowing through the wire is obtained by summing (1.6) for all the filaments composing the wire. Thus

$$I = \sum_{r=1}^{N} j_r \Delta S_r \tag{1.7}$$

If the cross-sectional area of each filament is infinitesimally small, then the summation sign can be replaced by an integral sign, as in

$$I = \int j_r \, \mathrm{d}S_r \ . \tag{1.8}$$

Suppose that the charges flowing through the wire have an average speed $\bar{v}$. The number passing through the cross-section per unit time is the number contained within a cylinder of length $\bar{v}$, constructed perpendicular to the cross-section — see Fig. 1.4. If there are n charges per unit volume, then the total number in the cylinder

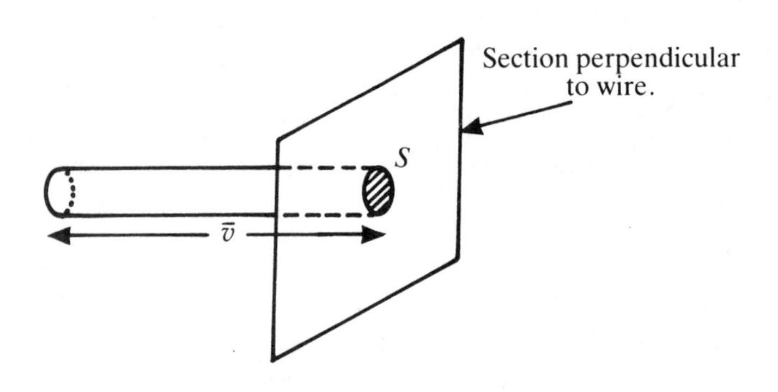

Fig. 1.4.

is $n\bar{v}S$. This is also the number of charges which pass through the cross-section in unit time. Multiplying this number by the charge q will give us the current I, that is

$$I = nq\bar{v}S \ . \tag{1.9}$$

$\bar{v}$ is usually called the *drift* velocity of the charge carriers, and is about equal to 10^4 to $10^5\,\mathrm{cm\,s^{-1}}$.

In some devices the current may arise from the movement of positive charges as well as electrons. The semiconductor diode and bipolar transistor are examples of such devices.

☆ ☆

AN ASIDE: As electrons do not travel with infinite speed the current is not the same at two points in a circuit. However, because the current can be measured only at the terminals of a component it is usual to ignore the finite speed of the charge carriers and assume that the current flowing into one terminal is equal to the current flowing out of the other. This is shown in Fig. 1.5.

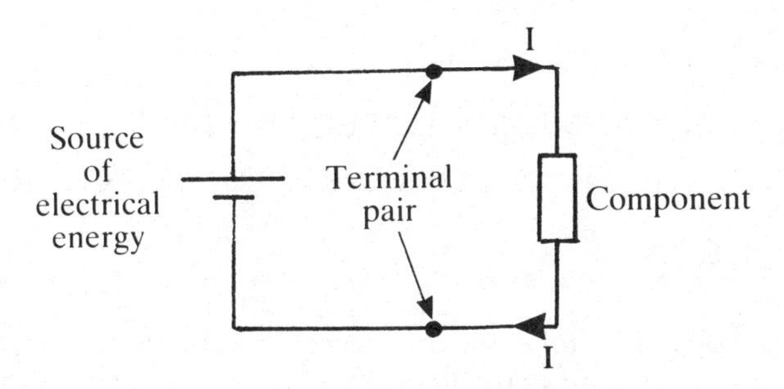

Fig. 1.5.

☆ ☆

1.2.2 Electric field and potential difference

Just as water will flow through a pipe only if there is a pressure difference between its ends, so current will flow in a conductor only if there is a potential difference across it.

Suppose that a small charge q moves through a potential difference V. The work done W results in the charge gaining an amount of kinetic energy equal to the product of the charge q and the potential difference V. A formal definition of potential difference is

$$\text{POTENTIAL DIFFERENCE} = \frac{\text{WORK DONE}}{\text{CHARGE}} = \frac{\text{GAIN IN KINETIC ENERGY}}{\text{CHARGE}} \tag{1.10}$$

Thus

$$V = \frac{W}{q} \; . \tag{1.11}$$

UNITS: joules per coulomb (JC^{-1}) or volt (V).

If 1 C of charge gains 1 J of kinetic energy, then it is said to move through a potential difference of 1 V.

Work done is the product of a force F and distance L. So let us rewrite (1.11) as

$$V = \frac{(F \times L)}{q} \; .$$

Here, L is identical to the distance through which the charge is accelerated. Now transpose terms to give

$$\frac{V}{L}=\frac{F}{q},$$

As V/L is the magnitude of the electric field strength E, we determine

$$F=qE \tag{1.12}$$

(1.12) allows us to define E in the following way:

ELECTRIC FIELD STRENGTH is NUMERICALLY EQUAL TO THE FORCE ACTING ON A UNIT POSITIVE CHARGE (1.13)

☆ ☆

AN ASIDE: Let there be a potential difference δV between the ends of a section of wire of length δL. Then the electric field may also be defined by the relation

$$E=-\frac{\delta V}{\delta L}\ . \tag{1.14}$$

Fig. 1.6 illustrates the origin of the minus sign. Let the increase in potential occur along the positive X-axis. Then as the electric field is always directed from high to low potential, this means that it points in along the negative X-axis — the direction of decreasing potential.

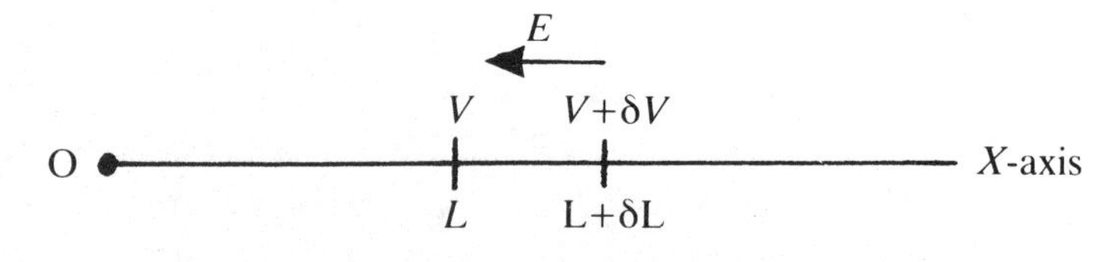

Fig. 1.6.

☆ ☆

1.3 OHM'S LAW

DEFINITION: In a conductor held at a *fixed* temperature the current I is directly proportional to the potential between its ends. That is

$$I \alpha V\Big|_{T=\text{const}} \tag{1.15}$$

The constant of proportionality in (1.15) is the *conductance* G or the *resistance* R, depending on which way round we write Ohm's law. So we can either have

$$V=IR \quad \text{or} \quad I=G\ V\ . \tag{1.16}$$

At any other temperature, Ohm's law is still obeyed but the constant of proportionality must have a different value; in metals, for example, the resistance R

increases as the temperature T increases. This latter result is normally written in terms of the temperature coefficient of resistance (TCR) α of the material. To obtain α, measure the resistance R_0 at some reference temperature T_0, such as room temperature or the temperature of melting ice, say, and also the resistance R_T at some other temperature T. Then as

$$R_T = R_0\{1 + \alpha(T - T_0)\} \tag{1.17}$$

the TCR is defined as

$$\alpha = \frac{\text{CHANGE OF RESISTANCE PER UNIT CHANGE IN TEMPERATURE}}{\text{REFERENCE RESISTANCE}}$$

UNITS: $°C^{-1}$ or K^{-1}.

Table 1.1 gives some values of the TCR for some common materials between 0° and 100°C.

Table 1.1 — Resistivity and TCR of some materials

Material	ρ/Ω.m (273 K)	TCR/$°C^{-1}$
Aluminium	2.8×10^{-8}	3.9×10^{-3}
Copper	1.7×10^{-8}	3.9×10^{-3}
Mercury	94.0×10^{-8}	—
Platinum	10.0×10^{-8}	3.9×10^{-3}
Silver	1.6×10^{-8}	3.8×10^{-3}
Constantan	44.0×10^{-8}	1.0×10^{-5}
Manganin	44.0×10^{-8}	1.0×10^{-5}
Nichrome	100×10^{-8}	1.0×10^{-4}
Glass (soda lime)	5×10^{11}	
Mica (sheet)	10^{11}–10^{15}	
Quartz	$>5 \times 10^{16}$	
Silicon	$\sim 1 \times 10^{0}$–1×10^{-5}	
Sulphur	1×10^{15} (depends on impurity concentration)	

1.3.1 Electrical resistivity

We wish now to investigate how the length and cross-sectional area of a conductor affect its resistance. Let us begin by assuming that the resistor is a cylinder of length L and cross-sectional area S, as in Fig. 1.7a. We can increase its length by connecting two cylinders in series, as in Fig. 1.7b. If there is a potential difference V across each of them, then, because the same current I flows through both, I must be given by

$$I = \frac{V}{R} = \frac{2V}{R_s}$$

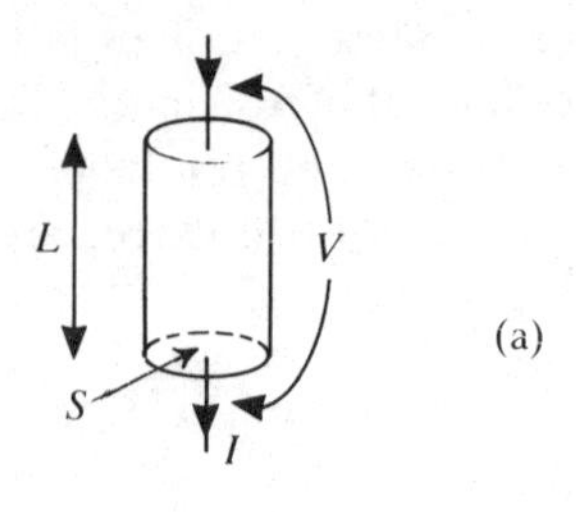

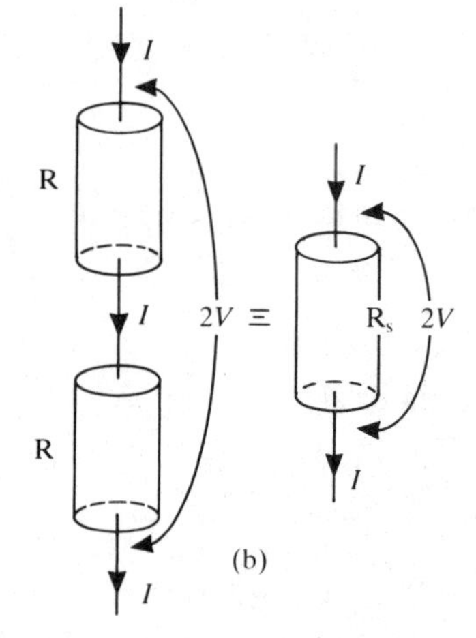

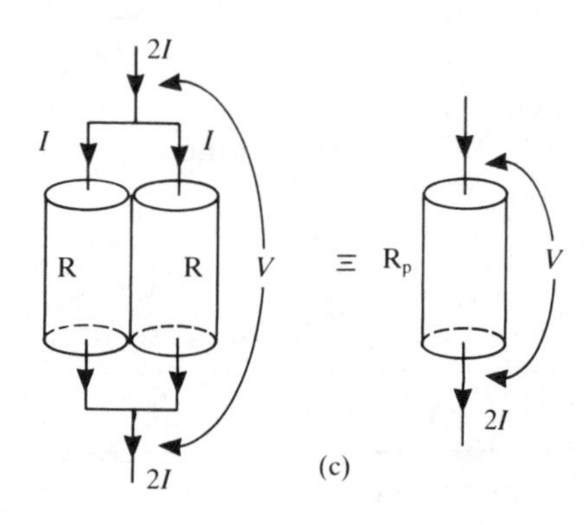

Fig. 1.7.

where R_s is the resistance of the combination. It is easy to work out that

$$R_s = 2R \ .$$

This result tells us that doubling the length of the conductor doubles the resistance. We can go on increasing the resistance of the conductor by connecting more cylinders together (say N of them), when we should find that, in general,

$$R_s = NR \tag{1.18}$$

Relation (1.8) tells us that

$$\text{RESISTANCE is } \begin{matrix} \text{DIRECTLY PROPORTIONAL to} \\ \text{the LENGTH of the conductor} \end{matrix} \tag{1.19}$$

Now let us increase the cross-sectional area of the conductor by placing two cylinders in contact with each other, as in Fig. 1.7c. The current I flowing through each of them is the same, as is the potential difference V across them. So we obtain

$$I = \frac{V}{R}$$

for each cylinder, and

$$2I = \frac{V}{R_p}$$

for the combination, where R_p is the effective resistance. Hence

$$\frac{V}{R_p} = \frac{2V}{R}$$

giving

$$R_p = \frac{R}{2} \ .$$

So doubling the cross-sectional area of the conductor halves the resistance. It should not be difficult to see that if the cross-sectional area is increased by a factor of N times then

$$R_p = \frac{R}{N} \ . \quad (1.20)$$

(1.20) tells us that

RESISTANCE is INVERSELY PROPORTIONAL TO CROSS-SECTIONAL AREA (1.21)

(1.19) and (1.21) can be combined into one expression, viz.

$$R \propto \frac{L}{S}$$

$$= \frac{\rho L}{S} \quad (1.22)$$

The Greek letter ρ (rho) is usually used for the constant of proportionality. It is called the *resistivity* of the conductor.

UNITS: ohm metre (ohm.m). The ohm has the symbol Ω (Greek capital omega).

The important point to remember about the resistivity is that it is a property of the conductor, itself, and does not depend on the geometrical shape.

The reciprocal of the resistivity is the conductivity (symbol: sigma σ). Thus

$$\rho = \frac{1}{\sigma} \quad (1.23)$$

UNITS: ohm^{-1} metre^{-1} (ohm^{-1} m^{-1}) or siemens metre^{-1} (S m^{-1}) .

Table 1.1 includes values of the resistivity of some common metals, alloys, and insulators.

In short, for resistors of unequal resistance in series,

$$R_s = R_1 + R_2 + R_3 + \ldots + R_N \tag{1.24}$$

and for resistors in parallel

$$\frac{1}{R_p} = \frac{1}{R_1} + \frac{1}{R_2} + \frac{1}{R_3} \cdots + \frac{1}{R_N} \; . \tag{1.25}$$

1.3.2 Experiments to verify the series and parallel resistor rules

1.3.2.1 *Series resistance rule*

Connect a low-voltage power supply and two resistance boxes (R_1 and R_2) in series. Measure the potential difference (pd) across the resistance boxes and the current through them.

These instructions appear fairly inoccuous until you attempt to sketch an appropriate circuit diagram. Why? Because there are apparently two ways of connecting the current meter and the voltmeter. Fig. 1.8a shows one way and Fig.

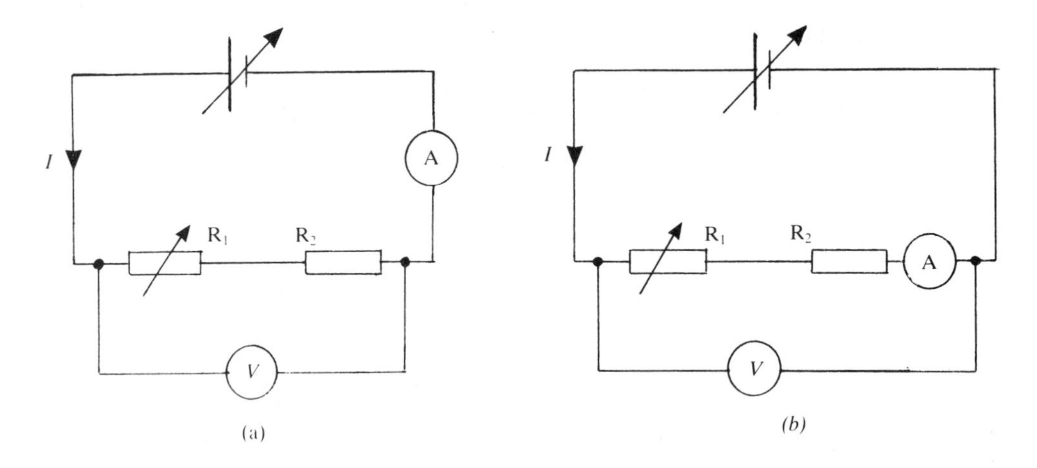

Fig. 1.8.

1.8b the other. So, which circuit diagram is correct. The answer hinges on the resistances of two meters. If the resistance of the current meter is large compared with R_1 and R_2 then there will be a substantial pd across this meter — so Fig. 1.8b cannot be used. However, another problem rears its head with Fig. 1.8a! Does the current meter measure the current through R_1 and R_2 only? It will, if the resistance of the voltmeter is very large compared with the sum of R_1 and R_2. A digital voltmeter (DVM) is useful because it has a resistance of about 10 megohms (MΩ).

Set the power supply to 9 V, say. Fix the value of R_2 to 1000 Ω and measure the pd V and the current I for various values of R_1 from 100 Ω to 2000 Ω in 200 Ω steps.

As

$$V = I(R_1 + R_2)$$

$$\frac{V}{I} = R_1 + R_2 \tag{1.26}$$

If the series resistance rule is correct, a graph of V/I vs R_1 should be linear with the slope of 45° and an intercept equal to the value of R_2.

1.3.2.2 *Parallel resistance rule*
Connect R_1 and R_2 in parallel and fix R_2 at 2000 Ω. Vary R_1 from about 6000 Ω to 300 Ω and measure V and I in each case.

Now

$$R_{\mathrm{TOT}} = (R_1 + R_2)^{-1}$$
$$= \frac{R_1 \times R_2}{(R_1 + R_2)}$$

and

$$V = \frac{IR_1 \times R_2}{(R_1 + R_2)}$$

or

$$\frac{I}{V} = \frac{1}{R_1} + \frac{1}{R_2} \tag{1.27}$$

A graph of I/V vs $1/R_1$ will be linear if the parallel rule holds. The slope of the line is 45° and the intercept is $1/R_2$.

1.4 THE INDUCTOR

This circuit element stores electrical energy when current passes through it. In its simplest form it consists of a helix, or coil, of wire. Often the inductor consists of a coil wound around a core of magnetic material — this design allows more energy to be stored by constraining the lines of magnetic induction to the immediate region near the coil.

Suppose that current from a signal generator produces a potential difference across an inductor, as shown in Fig. 1.9. Both the current i and the potential difference v may be displayed on an oscilloscope, and Fig. 1.10 should be obtained. As the signal generator delivers alternating current (a.c.) the waveforms have a sinusoidal dependence on time. The v-waveform *leads* the i-waveform by 90°.

NB. Small (lower case) letters are always used for time-varying quantities, in contrast to the use of capital (upper case) symbols for d.c. quantities - LOOK BACK OVER THE PREVIOUS SECTIONS.

Two important points to note about Fig. 1.10 are:

(i) the v- and i-waveforms are displaced from each other by 90°;
(ii) v is not related to i but to the slope $\mathrm{d}i/\mathrm{d}t$.

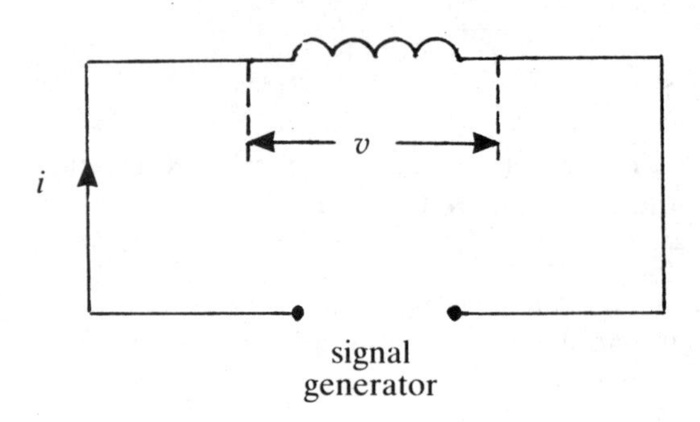

Fig. 1.9.

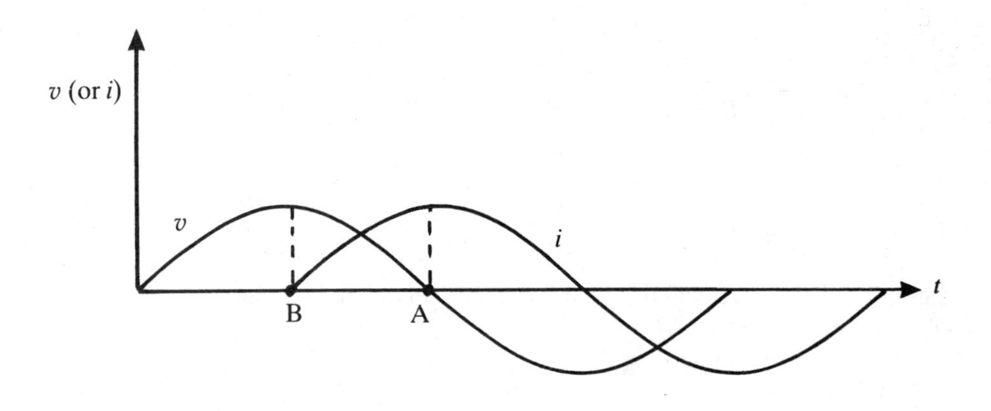

Fig. 1.10.

For (ii), we can say that at A: $v=0$ and $di/dt=0$, and at B: $v=$ maximum and $di/dt=$ maximum.

So, in this way we see that the potential difference across an inductor is directly proportional to the rate of change of current. That is

$$v \propto \frac{\mathrm{d}i}{\mathrm{d}t}$$

or

$$v = L\frac{\mathrm{d}i}{\mathrm{d}t} \tag{1.28}$$

where L, the constant of proportionality, is called the *inductance*.

UNITS: Henry (H).

1.5 THE CAPACITOR

The capacitor also stores electrical energy. However, it stores energy only when there is a potential difference across it. The stored energy does not depend on the current but only on the amount of charge on the capacitor plates. As with the inductor, the i-waveform and the v-waveform can be displayed on an oscilloscope. You should notice, in particular, that the v-waveform *lags* the i-waveform by 90°.

As the charge q on the capacitor plates builds up, the potential difference between the plates increases. So we can write

$$q \alpha v$$

or

$$q = Cv \tag{1.29}$$

where the constant of proportionality is known as the capacitance.

UNITS: farads (F).

We said in section 1.1 that current is the rate of flow of charge. In calculus terms this means that

$$i = dq/dt\ .$$

On differentiating (1.29) with respect to time we obtain

$$i = C\frac{\mathrm{d}v}{\mathrm{d}t}\ . \tag{1.30}$$

The capacitance is assumed to be a constant. The capacitance can vary in value owing, for example, to changes in temperature (it has a negative temperature coefficient (TCC)). If this is the case then (1.30) should be written in full as

$$i = \frac{\mathrm{d}(Cv)}{\mathrm{d}t}\ . \tag{1.31}$$

Why does a current appear to flow through the capacitor? The arrival of an electron on one plate of the capacitor increases the amount of negative charge there. To restore charge equilibrium this 'extra' electron repels an electron from the other plate, making it, in turn, more positive. The flow of electrons to one plate and the flow away from the other gives rise to a current. It is specifically known as the *displacement* current. No current actually flows *through* the capacitor!

1.6 POWER AND ENERGY

We already know that the current I can be expressed in terms of the rate of flow of charge by (1.3), viz.

$$I = \frac{dq}{dt}$$

If this quantity of charge is accelerated through a potential difference V then its kinetic energy will increase by an amount dW, which is given by

$$dW = V\,dq$$

$$dW = VI\,dt \ .$$

UNITS: joule (J).
This energy is gained in time dt.

If you wish to find the kinetic energy gained by the charge over a longer period t, then you can proceed as follows: Find how many intervals of duration Δt there are in the time t. Clearly, this is $t/\Delta t$. Call this number N. Then the total kinetic energy gained will be

$$W = NVI\Delta t$$

There is another way of writing this, using the mathematical summation sign Σ. This is

$$W = \Sigma\, VI\Delta t \ .$$

For those students who have a knowledge of the calculus we can go a step further. If the interval of time Δt is infinitesimally small, then the summation sign can be replaced by an integral sign, when

$$W = \int VI\,dt \ . \qquad (1.33)$$

Power is the rate of doing work *on* the charge. This is expressed as

$$P = \frac{dW}{dt} = VI \ . \qquad (1.34)$$

UNITS: watts (W).

Electric fires are quoted as, say, 1kW heaters and electric light bulbs as, say, 60 W. What exactly does this statement mean? For each electric fire, it says that a dynamo in a power station is doing work on the electrons in the heater coil. The amount of work that the dynamo does every second is measured as 1000 J of energy. Every second the power station loses this amount of energy whilst the electrons in the electric fire gain this amount. Similarly with the light bulb, the work done by the power station every second is measured as 60 J.

We shall not concern ourselves here with the actual processes by which electrical energy is transferred. (It comes under the name *joule heating* or *resistive heating*.)

However, it is important to recognise that 1000 J is not dissipated or consumed in the heating process. Energy cannot be 'eaten up' in this way. Rather, you should always think of energy as being a measure of the work done by one system on another.

One important point arises out of the definition of power in (1.34). It is that energy cannot change *instantaneously*, in the way depicted in Fig. 1.11, for this

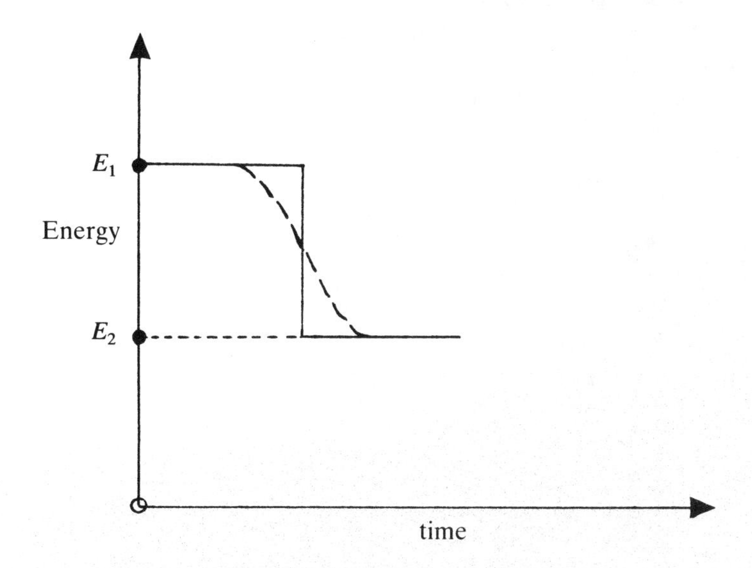

Fig. 1.11.

implies that the power is infinite. This conclusion is contrary to our understanding of the real world. Therefore energy must change gradually over a finite time (dotted curve in Fig. 1.11). A similar theme occurs in mechanics also. The velocity of a body does not change instantaneously—rather, its velocity just *after* a force is applied to it is equal to the velocity which it had just *before* the force was applied. This kind of continuity condition will be used later to analyse switching circuits (section 3.10).

1.7 SOURCES OF ELECTRICAL ENERGY

The *electromotive force* (emf) can be defined as the potential difference between the terminals of a cell when it is on open circuit. This is a satisfactory definition only so long as the concept of potential difference is understood. An alternative way of looking at the emf is to say that if a source, such as an accumulator, is able to supply energy to an electric current then it has an emf associated with it. Then

$$\text{EMF} = \frac{\text{POWER SUPPLIED}}{\text{CURRENT}} \tag{1.35}$$

UNITS: watts per ampere (WA^{-1}) or volt (V).

It is important to recognize, also, that the potential difference is reversed on reversing the direction of the current but the emf is not because the rate at which energy is dissipated cannot be reversed.

1.7.1 The voltage source

DEFINITION: If the *electromotive force* (emf) generated by a source is relatively independent of the circuit to which it is connected then it is called a voltage source.

Fig. 1.12a illustrates that the emf stays approximately constant at 10 V as currents

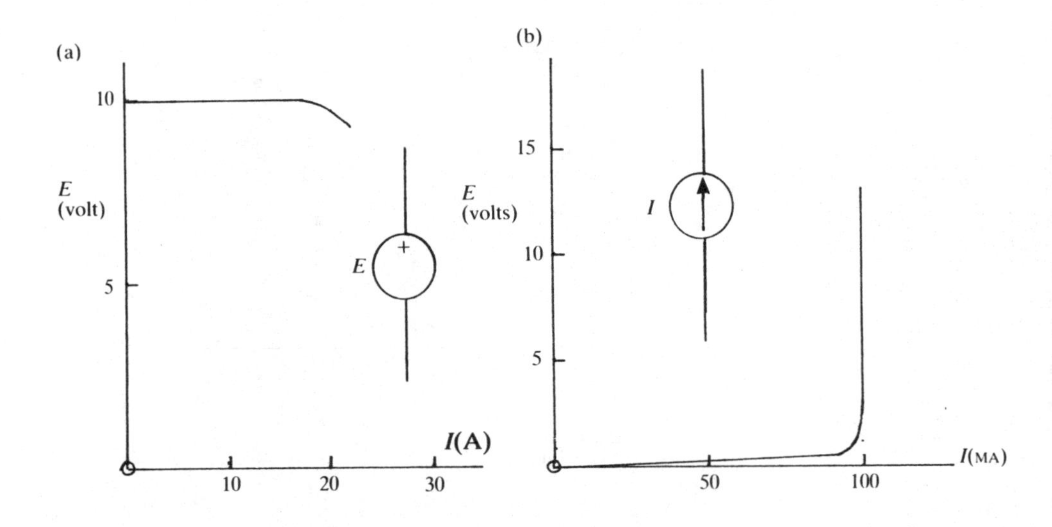

Fig. 1.12.

up to 20 A are drawn from the source, but decreases thereafter. The symbol for the voltage source is also shown. An example of a voltage source is the *storage battery*.

1.7.2 The current source

DEFINITION: If the *output current* is relatively independent of the complexity of the circuit to which the source is connected, then the source is a current source.

Fig. 1.12b illustrates that above 1.0 V the current is constant at about 100 mA. It is only in this region that the source behaves as a current source. The symbol is also shown. An example is the *bipolar transistor*.

2

Digital electronics

Objectives

(i) To differentiate between digital and analogue signals
(ii) How to count in binary, decimal, and hexadecimal
(iii) The importance of codes
(iv) To list the characteristics of a typical logic circuit
(v) How to formulate logic statements by using truth tables
(vi) To introduce Boolean algebra as a mathematical tool for analysing logic statements
(vii) To introduce the Venn diagram as a geometrical tool for analysing logic statements
(viii) To introduce the basic gates used in combinational logic
(ix) To design the simplest form of combinational logic circuits by using the Karnaugh map
(x) To discuss the operation of the different types of flip-flops used in sequential logic
(xi) How to use the JK flip-flop in counters and shift registers
(xii) How to program logic functions with a multiplexer.

2.1 INTRODUCTION

The brain makes sense of the world about us by receiving information from it in the form of *visual*, *tactile*, *acoustical*, *olfactoral*, and *taste* data. The five sense *organs* — the eyes, skin, ears, nose, and tongue — act like input ports. The *sense* data are converted into electrical signals, which are then transmitted to the brain *via* the *central nervous system*. There, the data are processed and interpreted. However, if there is a malfunction, for example, through a severe hit on the head, our perception of the world becomes distorted because the brain cannot do the job properly.

The brain behaves like the central processing unit of a highly complex computer. (Undoubtedly, it does more than this, but this analogy is a useful one to follow up.) It accepts data *via* the input ports (mentioned above) and the data bus, which is the complex system of synapses and ganglions that make up the central nervous system. It then performs a variety of logical and arithmetic operations at an extremely rapid rate. The processed data are sent along the data bus, the central nervous system, and finally outputted in the form of a visual display or through one of the other senses.

This very brief review of the brain's functions obviously does not do justice to the actual way that the brain behaves. This whole book would be unable to do justice to the brain's activities because although much is known about how the brain enables us to perceive the external world, and, as a result, permits us to survive, there is much which is still a mystery. It is extremely unlikely that the brain will reveal all its secrets. It is more likely that as one layer of knowledge is revealed another layer will present itself for further investigation.

2.2 DIGITAL vs ANALOGUE

In this section we wish to obtain a clear distinction between digital and analogue information. We shall attempt to do this through some examples.

On scanning across the primary or secondary rainbows we observe a gradual change in the colours; the colours seem to vary in intensity as well as merge into one another. The visual data that the eye receives through this scanning process change in a *continuous* way. This is the essential characteristic of an analogue signal — the change from one value to another is continuous, and all values, within given limits, are accepted. However, suppose that instead of using our eyes we use an optical instrument to record only those colours which have an intensity lying *above* a specific level. In other words the instrument behaves like a *filter* accepting some information and rejecting the rest. There is now a clear analogy with a toggle switch, which is either *off* or *on*. The instrument makes *discrete* or *digital* measurements.

As another example, consider a beaker being filled with water from a tap. If the tap is dripping then the volume of water in the beaker increases in a series of steps, as we see in Fig. 2.1a. Further, if all the drips have the same radius, then the sizes of the steps will be equal; the volume of water increases in a staircase fashion. Obviously, if the drips have different radii then the staircase will be irregular. When the tap runs continuously the volume of water in the beaker also increases continuously (Fig. 2.1b). The slope of the straight line graph gives the rate of change of the volume of water.

All the electrical signals grouped together in Fig. 2.2a demonstrate a continuous change in the current or potential difference with time. These are analogue signals. Those in Fig. 2.2b are digital signals because the y-ordinate changes abruptly from one value to another.

2.2.1 Some digital and analogue devices

Voltmeters can be of either the analogue or the digital variety. The DVM (digital voltmeter) is a very familiar abbreviation frequently heard in *lab* classes nowadays!

The speedometer of a car is an analogue device because it records the continuous change in the car's speed with time. On the other hand, the odometer, which records miles (or kilometres) is digital because it records changes in units of one-tenth of a mile (or kilometre). A room light switch may be of either kind: the toggle switch, as we have already mentioned, is digital because it is either ON or OFF, whereas the

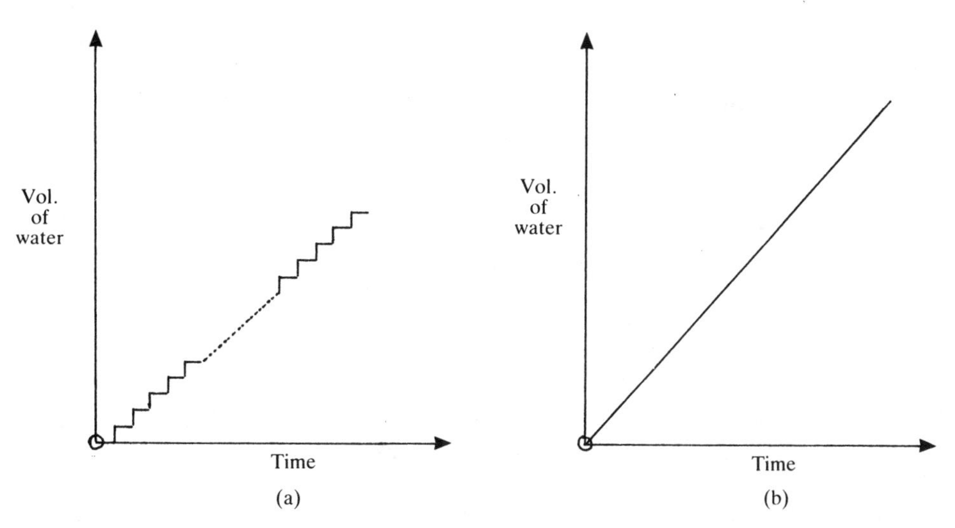
Vol.
of
water
Time
(a)
Vol.
of
water
Time
(b)

Fig. 2.1.

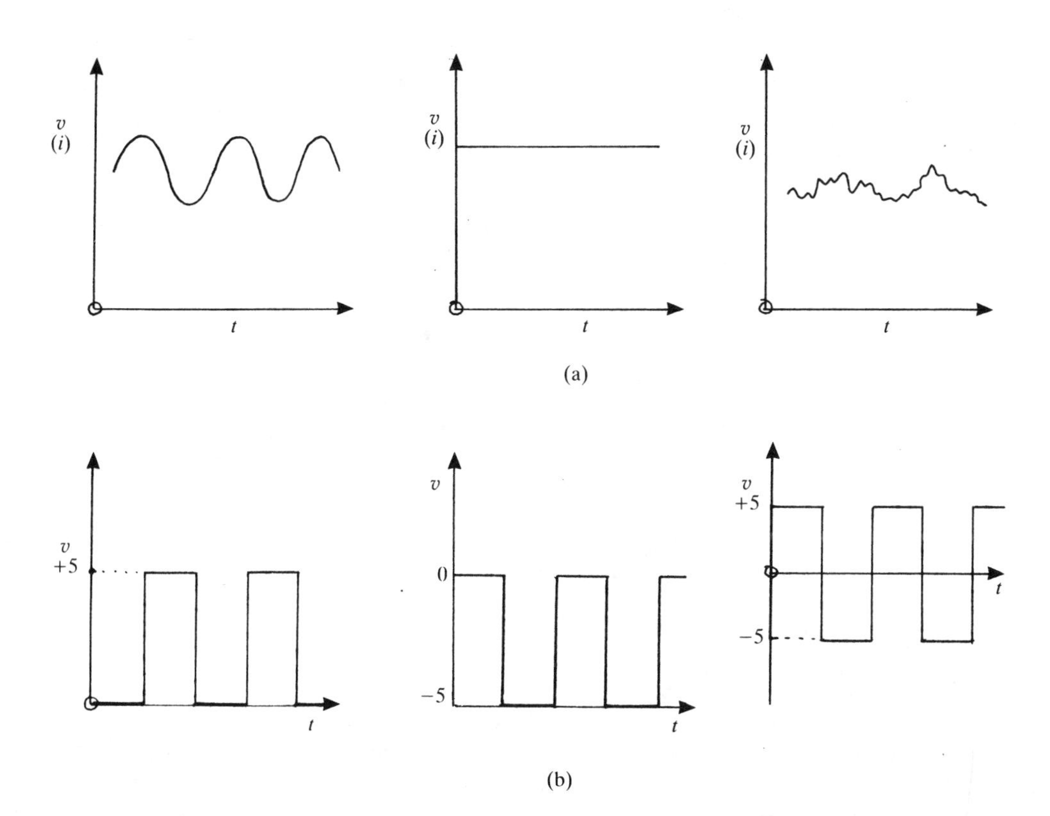
v
(i)
t
v
(i)
t
v
(i)
t
(a)
v
+5
t
v
0
−5
t
v
+5
t
−5
(b)

Fig. 2.2.

dimmer switch is analogue because it allows the light intensity to be varied in a continuous way.

Digital techniques are widely used in communication systems because they are less susceptible to noise. This is vitally important if information is not to be corrupted. Digital circuits have increased in popularity partly for this reason and partly because of their lower cost. Nowadays, the number of digital elements on each square millimetre of a silicon chip has increased to more than 400. The following abbreviations are in vogue with electronics engineers.

(i) SSI — small-scale integration. There are fewer than 50 elements mm^{-2}.
(ii) MSI — medium-scale integration. There are between 50 and 100 elements mm^{-2}.
(iii) LSI — large-scale integration. There are between 100 and 500 elements mm^{-2}.
(iv) VLSI — very large-scale integration. Over 500 elements mm^{-2}.

So let us begin our exploration into the basics of digital electronics by considering number systems and codes.

2.3 NUMBER SYSTEMS

2.3.1 Decimal (base 10)

If we include zero, there are ten digits: 0, 1, 2, . . ., 9. The change from one digit to the next is called a *unit* or *quantum*. A decimal number is the *weighted* sum of the powers of the base. For example, 576 is written:

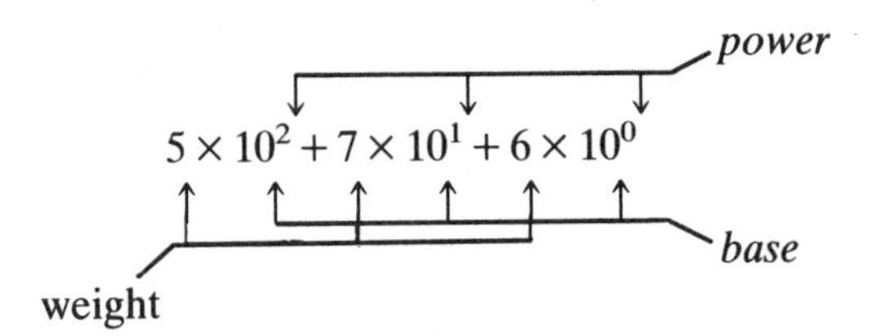

In general, any decimal number, including fractional parts, may be written

$$N_{10} = d_i \,.\, 10^i + d_{i-1} \,.\, 10^{i-1} + d_{i-2} \,.\, 10^{i-2} + \ldots + d_o \,.\, 10^0 + \underbrace{d_{-1} \,.\, 10^{-1} + \ldots}_{\text{fraction}} \quad (2.1)$$

d_i is one of the digits 0 to 9. Or, in set language, $d_i \in (0, 1, \ldots, 9)$.

2.3.2 Binary (base 2)

Pure binary numbers have two digits only: 0 and 1. These are usually referred to as *bits*. The term *pure* is used because, as we shall see in section 2.4, there are other ways of representing binary numbers. The subscript '2' at the end of the binary number indicates that it is pure binary. In the same way the subscript '10' indicates decimal and the subscript '16' hexadecimal.

A pure binary number can be written in a similar way to the decimal number in (2.1). For example, 1101_2 is:

$$1 \times 2^3 + 1 \times 2^2 + 0 \times 2^1 + 1 \times 2^0$$

which is 13 in decimal. In general, binary numbers are written:

$$b_i b_{i-1} \ldots b_1 b_0$$

where the symbol b stands for *bit*; b_i can only be 0 or 1. Bit b_0 is the *least significant bit*.

2.3.2.1 Conversion from decimal to binary

A simple division technique can be used. The decimal number (N) is divided by 2 to obtain a new decimal number (NEW) and a remainder (R). Even if there is no remainder it is essential to include it (R 0). The process is continued until NEW is 0. Then only the remainders are used to write down the binary number. An example will make this clear.

Suppose that we wish to convert 163_{10} to binary. Set out the division sum as follows:

2	163	
2	81	R 1
2	40	R 1
2	20	R 0
2	10	R 0
2	5	R 0
2	2	R 1
2	1	R 0
	0	R 1

The division process stops when the quotient is 0 (NEW = 0). The arrow indicates how the remainders must be read to obtain the corresponding binary number. It is 10100011_2.

2.3.2.2 Binary number sizes

Binary numbers are also called binary words. A 4-bit word is known as a *nibble* and an 8-bit word is known as a *byte*. In the case of a nibble there are 16 possible binary words. Some of these are: 0000, 0001, 0010, 0011, 0100, . . ., 1111. Each binary word is also known as a *state*. Once you know the number of bits in a binary word, it is easy to calculate the number of states. In the present case there are 4 bits. So the number of states equals 2^4. With a byte, there are 2^8 states, that is 256 in all.

The largest decimal number L that may be represented by the 4-bit word is given by

$$2^4 - 1 \text{ or } 15$$

With a 6-bit word it is

$$2^6 - 1 \text{ or } 63$$

and with an 8-bit word

$$2^8 - 1 \text{ or } 255$$

and so on, until, in general, with an n-bit word it is given by

$$2^n - 1 \qquad (2.2)$$

On the other hand, if you know the largest decimal number that you will be dealing with, then you can work backwards from the general result in (2.2) to determine the maximum number of bits in the binary word. That is,

$$n = 3.32 \log_{10}(L + 1) \qquad (2.3)$$

The number 1 in the parentheses can be neglected if L is quite large. For example, let L be 500, then

$$\begin{aligned} n &= 3.32 \log_{10}500 \\ &= 3.32 \times 2.70 \\ &= 8.96 \end{aligned}$$

As n must always be integral, this means that 9 bits must be used in the binary word: 111110100_2.

2.4 BINARY CODES

Can you convert the pure binary number 1010011_2 to decimal? How long did it take you? Although the conversion process is not difficult it is rather time consuming. It is for this reason that other ways of representing a binary number were looked for. One of these is *binary-coded-decimal*. It is more *user-friendly*. Binary-coded-decimal (BCD) represents each digit in a decimal number by a nibble. The most common BCD code is the 8–4–2–1 weighted code. What does this mean? The weight of each bit in the nibble is two times greater than the previous bit. For example, 83_{10} in BCD is found by writing 8 as 1000 and 3 as 0011. Then the complete binary-coded number is 1000 0011.

There are a few disadvantages in using BCD. These are:

(i) More bits are needed than in pure binary; 83_{10} in pure binary is 1010011_2, which has one bit less than BCD;
(ii) Digital circuitry is more complicated for BCD than for pure binary because more logic elements are required;
(iii) Arithmetic operations are more time consuming.

So a trade-off is necessary between the user-friendliness of BCD in providing a more efficient communication link between the digital equipment and the user, and the above disadvantages.

There are many other kinds of binary codes, such as excess-3 (XS3), that have been developed. These will not be discussed here, but interested students can read about them in some of the books listed in the bibliography at the end of this book.

2.5 HEXADECIMAL CODE

This system uses base 16. It was devised to meet the need for expressing binary numbers more concisely. It is ideal for representing the contents of memories and registers which might typically contain 8, 16, 24, or 32 bits. It uses the decimal numbers 0 to 9 and then the alphabetic letters A, B, C, D, and E for the numbers 10 to 15. As a nibble covers the decimal range 0 to 15, all pure binary numbers are divided into nibbles. For example, to convert 101100101100_2 to hexadecimal, divide the binary number into three nibbles:

1011 0010 1100

These represent decimal 11, 2, and 12 which is hexadecimal are B, 2, and C. So we end up with $B2C_{16}$.

To convert from decimal to hexadecimal use the same division procedure as in section 2.3.2.1, but with the divisor 16. For example, 163 in hexadecimal is obtained as follows:

```
16 |163
16 |_10  R 3  ↑
      0  R A  |
```

So reading upwards in the direction of the arrow we have $A3_{16}$.

To convert from hexadecimal to decimal, adopt the kind of power expansion used in (2.1):

$$N_{16} = h_i \times 16^i + h_{i-1} \times 16^{i-1} + \ldots + h_0 \times 16^0 \qquad (2.4)$$

Then, for example, 113_{16} is

$$1 \times 16^2 + 1 \times 16^1 + 3 \times 16^0$$

or 275_{10}.

2.6 THE GRAY CODE

The Gray code is called an *error-minimizing* code because it reduces the possibility of an ambiguity occurring within the computer's electronic circuitry when a binary word is changed. For this reason it is often used in control systems in which the position of an object is required as, for example, in secondary-surveillance radar systems.

Suppose that we wish to change from the binary word 0010_2 to 0001_2. To do this two bits must change at the same time: b_0 from $0 \rightarrow 1$ and b_1 from $1 \rightarrow 0$. The likelihood of this happening simultaneously is practically zero. Instead, we may have either $0010 \rightarrow 0011$ occurring first or $0010 \rightarrow 0000$. So, to avoid this ambiguity binary words are converted to Gray code.

Consider the pure binary number 1011_2. To convert to Gray code, first write down bit b_3 – the 1. Then add together neighbouring pairs of bits, but remembering to throw away the 'carry' bit: $1+0=1; 0+1=1; 1+1=0$. The corresponding Gray code is 1110_G. Once again the subscript 'G' indicates the fact that we are using Gray code. Let us show this again —

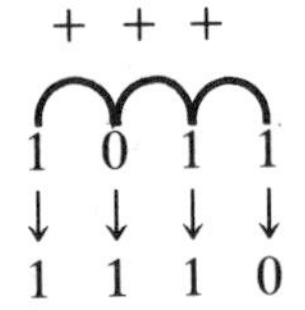

Now let us convert 15_{10} to Gray code.

First, we must convert 15_{10} to pure binary. Referring to section 2.3.2.1, we find that this is 1111_2. Now we can use the above procedure to enable us to arrive at the Gray code.

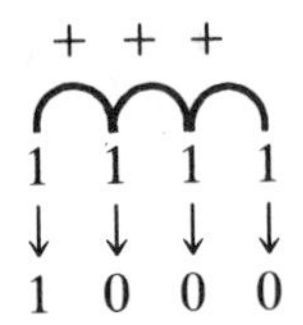

In each of the pair of additions, the carry bit, a '1', is discarded.

Table 2.1 lists some of the pure binary and Gray numbers from 1_{10} to 15_{10}. Space is left for you to fill in the remainder. Note that only one bit changes at a time in the Gray code, whereas this is not the case in binary.

The disadvantage of the Gray code is that arithmetic is difficult to do — so it is usual to work with binary. We shall use Gray code for labelling Karnaugh maps (see section 2.13.2.2).

It is relatively simple to convert from Gray to binary. Consider 1000_G. Write down bit b_3—the '1'. Now add it to bit b_2 in the Gray-coded number; $1+0=1$; write down '1'. Add this result to bit b_1 in the Gray number: $1+0=1$; write down a '1'.

Table 2.1 — Conversion of decimal and pure binary numbers to Gray code

Decimal	*Pure binary*	*Gray*
0	0000	0000
1	0001	0001
2	0010	0011
3	0011	0010
4	....	....
5	....	....
6	....	....
7	0111	0100
8	1000	1100
9	....	....
10	1010	1111
11	....	....
12	....	....
13	....	....
14	1110	1001
15	1111	1000

Add this result to bit b_0 in the Gray number: $1+0=1$; write down a '1'. The final result is 1111_2. You can check that this is correct by looking at Table 2.1. The following schema would make the procedure clear.

$$\begin{array}{cccc} 1 & 0 & 0 & 0_G \\ \downarrow + & + & + & \\ 1 & 1 & 1 & 1_2 \end{array}$$

2.7 THE ASCII CODE

ASCII stands for American Standard Code for Information Interchange. In the early days of computers, manufacturers often used their own alpha-numeric code for getting data into and out of the computer. You can imagine the confusion this created if a peripheral, such as a printer, made by another manufacturer was desired to be purchased! So to avoid this serious problem the ASCII code was devised.

ASCII code has seven bits. This means that there are 2^7 or 128 characters and functions that can be represented through it — see Table 2.2. There you will see that the character L is given by 100 1100. So every time you hit the L key on the keyboard this seven bit word is generated. The space bar is represented by 010 0000.

2.8 LOGIC CIRCUIT CHARACTERISTICS

There are essentially three types of digital integrated circuits on the market. These use transistor–transistor logic (TTL), metal-oxide semiconductor logic (MOS), and emitter-coupled logic (ECL). Each of these families has its own capabilities and

Table 2.2 — The ASCII code

	BITS					
	$b_6 b_5 b_4$					
$b_3 b_2 b_1 b_9$	010	011	100	101	110	111
0000	SP	O	@	P		p
0001	!	1	A	Q	a	q
0010	"	2	B	R	b	r
0011	#	3	C	S	c	s
0100	$	4	D	T	d	t
0101	%	5	E	U	e	u
0110	&	6	F	V	f	v
0111	'	7	G	W	g	w
1000	(	8	H	X	h	x
1001	)	9	I	Y	i	y
1010	*	:	J	Z	j	z
1011	+	;	K		k	{
1100	,	<	L		l	
1101	-	=	M		m	}
1110	.	>	N		n	
1111	/	?	O	—	0	DEL

limitations. It is not the intention to discuss these digital circuits at this stage, but, rather, to concentrate on the general characteristics which all of them possess. The characteristics are: logic levels, propagation delay, power dissipation, noise immunity, fan-out.

2.8.1 Logic levels

These are the voltage values assigned to binary '1' and binary '0'. In *positive* logic, binary '1' corresponds to a *high* voltage and binary '0' to a *low* voltage. The opposite is true for *negative* logic. It is proposed to deal entirely with positive logic elements in this chapter.

There is no sharp voltage cut-off below which you have binary '0' and above which you have binary '1'. Such a cut-off is unacceptable because factors like temperature and power supply variations, component tolerances, etc. could upset the circuit logic; a designed binary '1' state could become a binary '0' state, and vice versa. So, in TTL, for example, binary '0' corresponds to any voltage below about 0.4 V and the binary '1' state to voltages above about 2.0 V (Fig. 2.3). This kind of

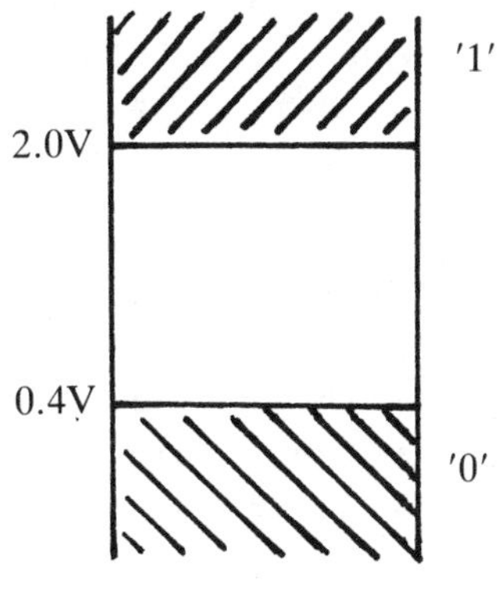

Fig. 2.3.

information is given in the manufacturer's specifications, and you are urged to read the data-sheets in order to familiarise yourselves with them.

2.8.2 Propagation delay

The propagation delay gives a measure of the speed of operation of a logic element or circuit. It is the time taken for the output to react to a change in the input. Fig. 2.4

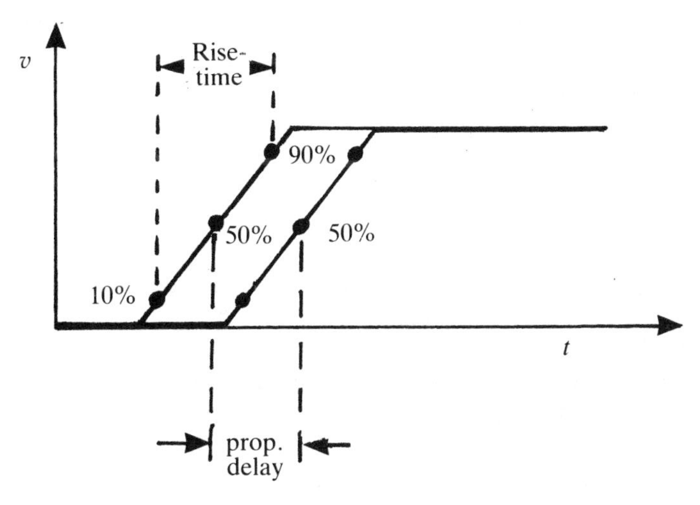

Fig. 2.4.

depicts the change in the voltage levels on the *leading edge* of a pulse at the input and output of a digital device. The rise-time is measured between the 10% and 90% points, whereas the propagation delay is measured between the 50% points. A digital circuit normally consists of a number of logic gates, each one of which will have its own propagation delay. Hence, the total propagation delay for the circuit is the sum of the individual delays. Typical values range from 1 ns to several ns.

2.8.3 Power dissipation

Power dissipation is quoted in microwatts or milliwatts per logic element. It is an average value because the power consumption will be different for the binary '0' and binary '1' states; for the former very little current flows, whereas for the latter a larger current flows. MOS logic elements have a power dissipation $\simeq \mu$W and ECL logic elements $\simeq$ 100 mW.

There is a direct relationship between speed of operation of a logic circuit and power dissipation. To achieve high speed you have to pay the penalty of high power dissipation. A compromise has to be reached. MOS circuits have low power ratings partly because their input *impedance* is very high. Switching times are low because they have a capacitive structure, and charge has to be removed before the logic levels may be altered. These circuits are useful when dry batteries are to be used and low speed is not a disadvantage. ECL circuits are very fast but have a high power rating.

2.8.4 Noise immunity

The manufacturer's data sheet may say that a logic circuit has a noise immunity of 1 V. What does this mean? Noise signals with an amplitude below about 1 V will not disturb the logic levels of the circuit, but above this value the circuit cannot be guaranteed to operate correctly.

2.8.5 Fan-out

Fan-out indicates how much of a load can be connected to the output of a digital circuit. For example, a fan-out of 10 means that 10 identical standard loads can be attached to the output of the digital circuit without upsetting the operation of the circuit.

There are two ways of connecting a load to the output of the digital circuit: (i) between the output and earth; (ii) between the output and the high voltage supply rail. (i) is used in *current source* logic and (ii) in *current sink* logic. We need to look at these briefly in order to gain a better appreciation of fan-out.

2.8.5.1 Current source logic

Fig. 2.5 consists of a basic bipolar transistor logic element — the *inverter* with two identical loads (each being a base resistor R and the emitter–base junction) connected to it. Although we have not yet studied the operation of a bipolar transistor we can still understand how this circuit works if we use two simple rules:

A bipolar transistor conducts current (is ON) if a HIGH voltage (>2 V, say) is applied to its base; the transistor does not conduct (is OFF) if a LOW voltage (<0.4 V) is applied to the base.

When T_1 conducts, point X is pulled down towards earth potential. The base

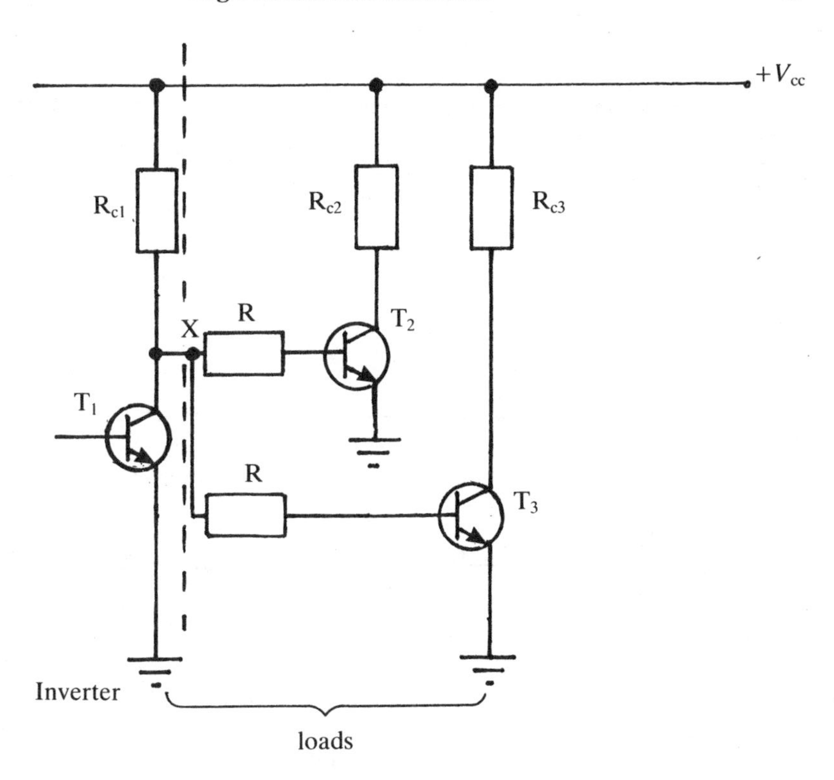

Fig. 2.5.

potentials on T_2 and T_3 are LOW and so they do not conduct. When T_1 does not conduct, no current flows through R_{c_1} and X is pulled up towards the potential of the supply rail (which lies above the HIGH logic level — 2 V with TTL). Now T_2 and T_3 are switched ON and current is able to flow through R_{c_1} and then through each load.

The overall load resistance is obtained by using the parallel-resistor rule (1.25). It is less than the resistance of each individual load. As more loads are connected to X, the overall load resistance falls still further — approaching zero — with the result that the potential at X is pulled down towards earth potential. If it falls below 2 V, T_2 and T_3 switch OFF, and the inverting action fails. Fan-out is the maximum number of loads which may be connected to X without disturbing the logical operation of the circuit.

2.8.5.2 Current sink logic

Fig. 2.6 shows two loads connected between the high voltage supply rail and point X. This time they consist of a diode/resistor combination.

When T_1 is OFF, the potential at X is pulled up to V_{cc} and no current flows through the diodes because they are *reverse-biased*. When T_1 is ON, the potential at X is nearly earth potential, and now current is able to flow through each load. Adding loads decreases the overall load resistance with the result that the potential at X is

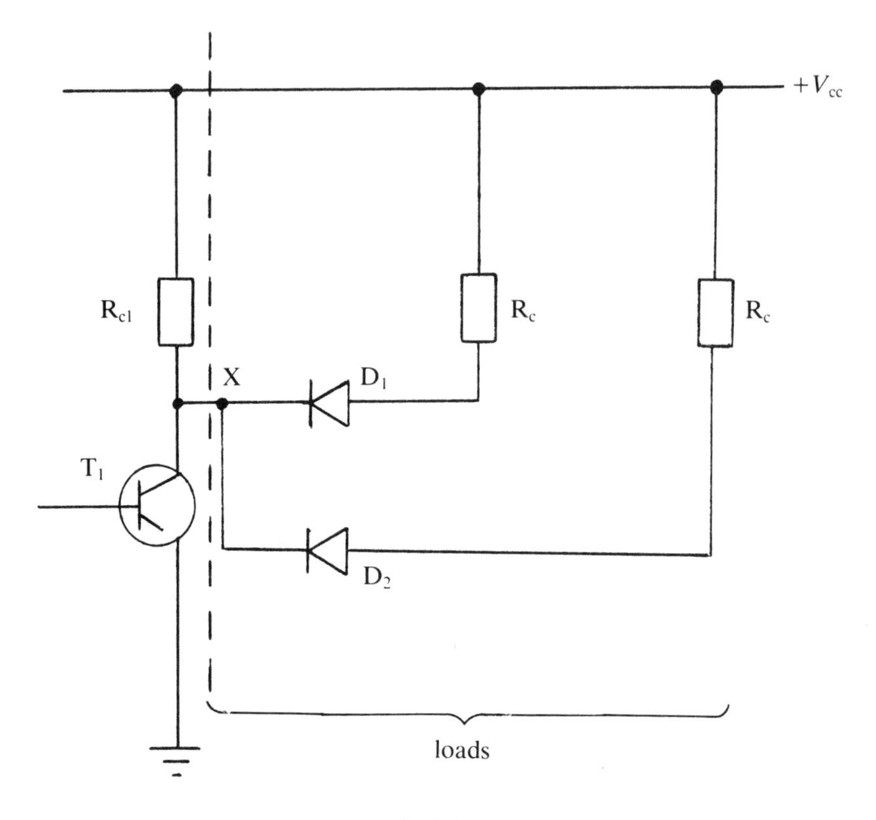

Fig. 2.6.

pulled towards V_{cc}. If V_x becomes larger than about 0.4 V, then T_1 will stop conducting. Once again, the fan-out is the maximum number of identical loads that can be connected without the circuit logic becoming disturbed.

2.9 COMBINATIONAL vs SEQUENTIAL LOGIC CIRCUITS

In a combinational logic circuit the output always has the same value for the same set of inputs. An example is the *adder*: $2_{10} + 8_{10}$ is always equal to 10_{10}. Other examples are the decoder, the encoder, and the multiplexer. A combinational circuit is often referred to as a *decision-making* circuit. The basic unit is called a *gate*.

In a sequential logic circuit the output at a given instant depends not only on the inputs at that time but also on all the previous inputs. An example is the *accumulator*: Inputs 2, 5, and 7 (see Table 2.3) are entered into the accumulator at set time intervals and the sum (the output) is calculated after each entry.

Other examples are counters and shift registers. The basic unit is the flip-flop.

The way to distinguish between a combinational and a sequential logic circuit is through the time-dependence of its output.

Table 2.3 – To illustrate the accumulator as a sequential circuit

Time/arbitrary units	*Input*	*Output*
1	2	2
2	5	7
3	7	14

2.10 LOGIC STATEMENTS

Consider the following statement:
The circuit will be successful if the soldering is good AND the mains switch is on. *Good soldering* and *the mains switch is on* may be referred to as *input variables*, and *the circuit will be successful* is the *output variable*. These may be either TRUE or FALSE. For simplicity, let us use the symbols G, M, and S for the three variables. Then we can say that

S is TRUE if G is TRUE *AND* M is TRUE

If only one of the input variables is TRUE, or both are FALSE, then S will be FALSE. We can set up a TRUTH TABLE which will indicate all the possible combination of input variables, see Table 2.4a. We can further simplify the truth

Table 2.4 — Truth tables for the logical AND function

G	*M*	*S*	*G*	*M*	*S*
FALSE	FALSE	FALSE	0	0	0
FALSE	TRUE	FALSE	0	1	0
TRUE	FALSE	FALSE	1	0	0
TRUE	TRUE	TRUE	1	1	1
	(a)			(b)	

table by letting binary '0' stand for FALSE and binary '1' for TRUE, as we have done in Table 2.4b.

We are now in a position to write down the logic statement in a mathematical way, viz.

S = G *AND* M

or

$$S = G \, . \, M \; . \tag{2.5}$$

Where the dot stands for the logical *and* function.

The logic *OR* statement can be understood in a similar way. Consider the statement:

I will pass the examination if I am lucky OR I work hard.
The input variables are: *I am lucky* (L); *I work hard* (W).
The output variable is: *I will pass the examination* (S).

S is TRUE if either L is TRUE *OR* W is TRUE *OR* both are TRUE. S is FALSE only if both L and W are FALSE. Hence the truth table becomes (Table 2.5).

Table 2.5—Truth table for the logical OR function

L	*W*	*S*
0	0	0
0	1	1
1	0	1
1	1	1

Mathematically, the OR function is written

$$L + W = S \tag{2.6}$$

The + sign represents the logical *OR* function.

The logic *NOT* statement is easier than the *AND* and *OR* statements to understand. Consider the statement:

Mercury is a liquid at room temperature.

If this statement is represented by the variable A, then the negation of this statement is

Mercury is NOT a liquid at room temperature.

It is represented by the variable $\overline{A}$, pronounced the *complement* of A. Since A is TRUE (1), $\overline{A}$ is NOT TRUE (0).

2.11 THE BINARY CONNECTIVES

Fig. 2.7 shows two inputs, labelled A and B, applied to a logic system with the output

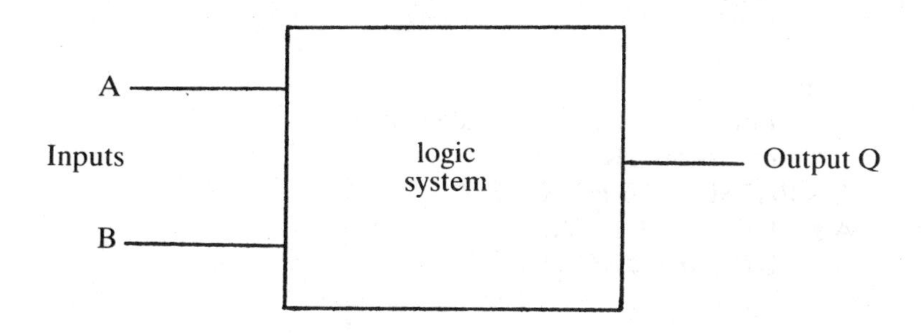

Fig. 2.7.

Q. The logic system is placed inside a black box because we are not interested in it, other than through its inputs and output. We shall try different combinations of AB and look at what output we get from each one.

AB can be 00, 01, 10, or 11 (where, again, 0 stands for FALSE and 1 for TRUE). There are 16 different sequences of outputs for this set of inputs, as Table 2.6

Table 2.6 — Binary connectives for two input variables

	A	0	1	0	1		
	B	0	0	1	1		
						Function	*Symbol*
	0	0	0	0	0	Always false	O
	1	0	0	0	1	AND	A.B
	2	0	0	1	0	—	—
	3	0	0	1	1	Variable B	B
	4	0	1	0	0	—	—
	5	0	1	0	1	Variable A	A
	6	0	1	1	0	exclusive OR	$A \oplus B$
Outputs	7	0	1	1	1	OR	A + B
	8	1	0	0	0	NOR	$\overline{A + B}$
	9	1	0	0	1	coincidence	$\overline{A \oplus B}$
	10	1	0	1	0	NOT A	$\bar{A}$
	11	1	0	1	1	—	—
	12	1	1	0	0	NOT B	$\bar{B}$
	13	1	1	0	1	—	—
	14	1	1	1	0	NOT AND (NAND)	$\overline{A.B}$
	15	1	1	1	1	Always true	1

illustrates. Each sequence is called a *connective*. Connective 3 is simply input B, connective 5 is input A, connective 1 is the AND statement, connective 7 the OR statement, and so on. The inputs AB, taken with the outputs on each line, are truth tables for specific logic functions.

Worked examples 2.1

Q1. Express connective 13 in mathematical form.
This says that output Q is TRUE if

A is FALSE *and* B is FALSE
or A is TRUE *and* B is FALSE
or A is TRUE *and* B is TRUE.

Using the variable itself to represent a TRUE statement and the complement for a FALSE statment, we have

$$Q = \overline{A} . \overline{B} + A . \overline{B} + A . B$$

Q2. Connective 6 is the Exclusive–Or (XOR) function. Show that Connective 9 is its complement.

The truth table for the XOR function is:

A	*B*	*Q*
0	0	0
1	0	1
0	1	1
1	1	0

It should be noticed that Q is TRUE only if *one* of the inputs is TRUE. The mathematical symbol for the XOR function is $A \oplus B$. The truth table for conenctive 9 is

A	*B*	*Q*
0	0	1
1	0	0
0	1	0
1	1	1

Hence it is easy to see that Q is the complement of the output of the XOR function. It is the coincidence or equivalence function, written $\overline{A \oplus B}$. It is used in memory searches.

2.12 BOOLEAN ALGEBRA

Boolean algebra is a mathematical technique for manipulating logic statements. We shall not go into great mathematical detail other than to introduce the topic. There are basically three kinds of theorems which are helpful for simplifying logic statements. These are concerned with;

(i) logical operations on constants
(ii) logical operations on *one* variable.
(iii) Logical operations on *two or more* variables.

Let us draw up truth tables to illustrate AND, OR, and NOT statements.
For (i),

AND	*OR*	*NOT*
$0.0=0$	$0+0=0$	$\bar{0}=1$
$0.1=0$	$1+0=1$	$\bar{1}=0$
$1.0=0$	$0+1=1$	
$1.1=1$	$1+1=1$	

These are sometimes referred to as Huntington's postulates. We came across these logical operations in section 2.11, as they are three of the binary connectives.
For (ii), you must remember that the variable can be either TRUE or FALSE —

AND	*OR*	*NOT*
$A.0=0$	$A+0=0$	
$A.1=A$	$A+0=1$	$A=\bar{\bar{A}}$
$A.A=A$	$A+A=A$	
$A.\bar{A}=0$	$A+\bar{A}=1$	

The double bar in the NOT statement means the complement of the complement of A. which is A itself.
For (iii), it is no longer sensible to confine ourselves to the basic AND, OR, and NOT statements, because the theorems deal with combinations of the three. Let us list the essential ones. All can be proved by using truth tables.

(a)

$$\left.\begin{array}{l} A+B=B+A \\ A.B=B.A \end{array}\right\} \text{(\textit{commutative}; the order of the variables is unimportant).} \qquad \begin{array}{r}(2.7)\\(2.8)\end{array}$$

(b)

$$\left.\begin{aligned} A + A.B &= A \\ A.(A + B) &= A \end{aligned}\right\} \textit{(absorptive)} \qquad \begin{aligned}(2.9)\\(2.10)\end{aligned}$$

(c)

$$\left.\begin{aligned} A + (B + C) &= (A + B) + C = (A + C) + B \\ A.(B.C) &= (A.B).C = (A.C).B \end{aligned}\right\} \textit{(associative)} \qquad \begin{aligned}(2.11)\\(2.12)\end{aligned}$$

(d)

$$\left.\begin{aligned} A.(B.C) &= A.B + A.C \\ A + (B.C) &= (A + B).(A + C) \end{aligned}\right\} \textit{(distributive)} \qquad \begin{aligned}(2.13)\\(2.14)\end{aligned}$$

(e)

$$\left.\begin{aligned} \overline{A + B} &= \bar{A}.\bar{B} \\ \overline{A.B} &= \bar{A} + \bar{B} \end{aligned}\right\} \textit{(de Morgan's theorems)} \qquad \begin{aligned}(2.15)\\(2.16)\end{aligned}$$

Worked examples 2.2

Q1. Set up a truth table which will verify de Morgan's first theorem.

de Morgan's first theorem is stated mathematically in (2.15). We shall need 7 columns in the truth table. Each variable and logic function in each row is treated seperately in order that the mechanics of constructing any truth table will, we hope, become apparent.

A	B	A + B	$\bar{A}$	$\bar{B}$	$\overline{A + B}$	$\bar{A}.\bar{B}$
0	0	0	1	1	1	1
1	0	1	0	1	0	0
0	1	1	1	0	0	0
1	1	1	0	0	0	0

The last two columns are identical. This proves the validity of de Morgan's first theorem. The second theorem can be approached in an identical manner. As experience is gained in constructing truth tables, the number of columns can be reduced to include only the absolutely essential ones.

Q2. Prove that: $A + \bar{A}.B = A + B$, by three different methods.

The three methods referred to are: (i) truth tables; (ii) Venn diagrams; (iii) Boolean algebra.

(i) Six columns are essential for the truth table.

A	B	$\bar{A}$	$\bar{A}.B$	$A+\bar{A}.B$	$A+B$
0	0	1	0	0	0
1	0	0	0	1	1
0	1	1	1	1	1
1	1	0	0	1	1

(ii) The Venn diagram is probably familiar to you if you have studied SET theory. True statements, that is those represented by the variables A and B, will be depicted by the shaded region within circles in a universe, ε (a rectangle), which includes all possible logical operations on the input variables. hence

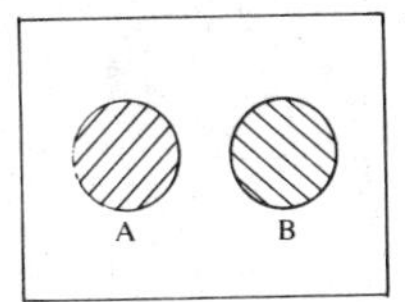

A . B is represented by the heavily-shaded region

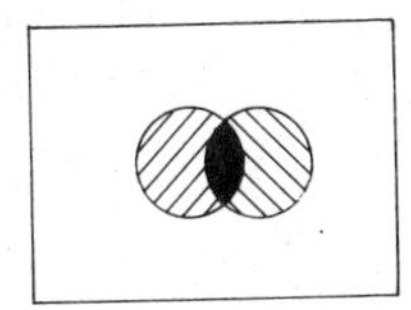

Since it is only within this shaded region that both A and B are TRUE. A + B is

since within this shaded region either A is TRUE only OR B is TRUE only OR A AND B is TRUE.

$\bar{A}$ is

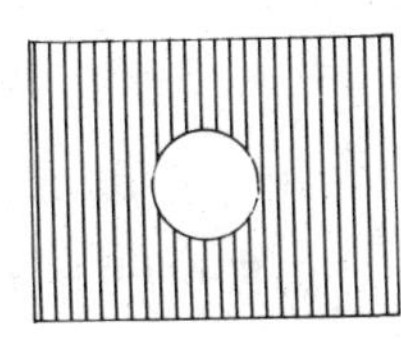

The hatching indicates the region in which A is FALSE.
$\overline{A}\,.\,B$ is the cross-hatched region outlined in black.

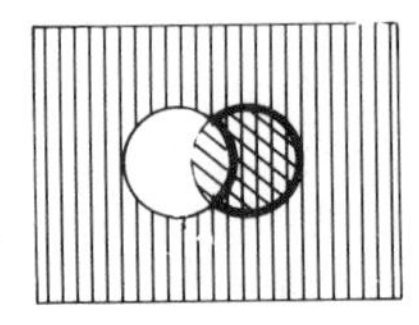

since only in this region is $\overline{A}$ AND B TRUE.
Therefore, $A + \overline{A}.B$ is

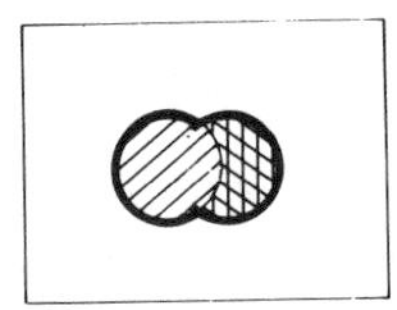

which is simply $A + B$.
(iii) The proof is more difficult by Boolean algebra because the route to take is not always obvious. Such is the case here. Only after much practice (and, sometimes, much patience) does the method become obvious. Like most things which come into this category, the method is obvious once it is shown to you!

$$A + \overline{A}\,.\,B = A\,.\,(B + 1) + \overline{A}\,.\,B \quad \text{(because } B + 1 = 1 \text{ and } A\,.\,1 = A\text{)}$$
$$= A\,.\,B + A\,.\,1 + \overline{A}\,.\,B$$
$$= A + (A + \overline{A})\,.\,B$$
$$= A + 1\,.\,B$$
$$= A + B$$

Q3. Using Venn diagrams, prove that: $A.\,(\overline{A} + B) = A.B$
With the help of the Venn diagrams used in Q.2, $\overline{A} + B$ is the region outside the black crescent in the left-hand diagram below.

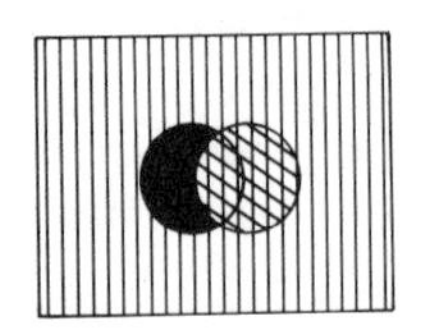

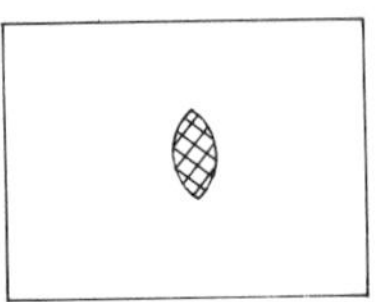

Thus $A.(\overline{A} + B)$ is the cross-hatched region in the right-hand diagram, which is A.B — the intersection of A AND B.

The Boolean expressions in Q2 and Q3 belong to a set of *minimization* theorems.

2.13 COMBINATIONAL LOGIC

It was stated previously that we shall be concerned only with circuits and systems which use positive logic. Let us remind ourselves what this means: binary '0' is represented by a low voltage level, <0.4 V, say, and binary '1' by a high voltage, >2.0 V. No sharp cut-off divides the binary states, for the reasons given in section 2.8.1.

Combinational logic circuits are constructed from logic gates. The most common gates are the AND, OR, NOT (or INVERTER), NAND, and NOR. These are illustrated in Fig. 2.8. Except for the NOT gate, they all have two inputs although multi-input gates are possible.

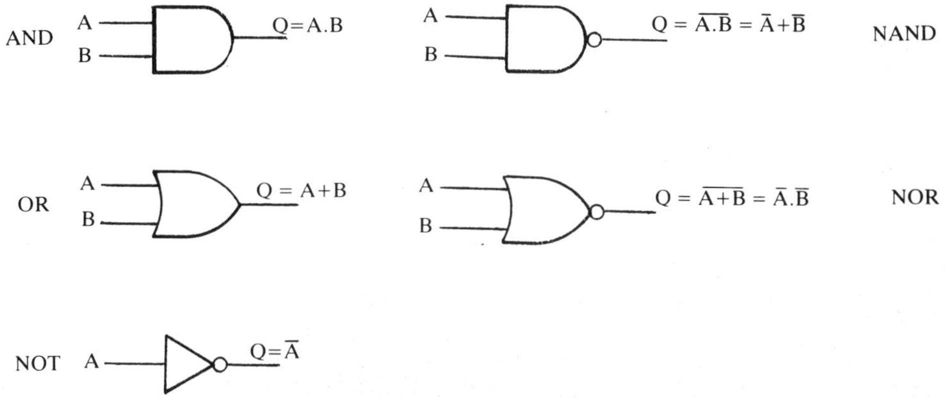

AND A B $Q = A.B$

NAND A B $Q = \overline{A.B} = \bar{A} + \bar{B}$

OR A B $Q = A + B$

NOR A B $Q = \overline{A+B} = \bar{A}.\bar{B}$

NOT A $Q = \bar{A}$

Fig. 2.8.

The NAND gate is a combination of an AND gate and a NOT gate, as can be inferred from the Boolean expression. The bubble indicates that the polarity of the output of the AND gate is inverted.

2.13.1 NAND logic

All logic gates can be constructed with NAND gates. NAND technology is very important because it enables logic circuits to be fabricated easier, cheaper, with a higher package density and with lower power consumption. The 7400 TTL series of integrated circuits is one such example. Table 2.7 lists some of the standard gates. The terms *quad*, *triple*, *dual*, and *single*, respectively, refers to the number of such gates to be found on a chip, and <number–input> indicates the maximum number of input variables.

Fig. 2.9 indicates how the function of the AND, OR, and NOT gates can be simulated by using NAND gates.

Table 2.7 — Standard TTL logic gates

Type	*Quad 2-input*	*Triple 3-input*	*Dual 4-input*	*Single 8-input*
NAND	7400	7410	7420	7430
NOR	7402	7427	7425	
AND	7408	7411	7421	
OR	7432			

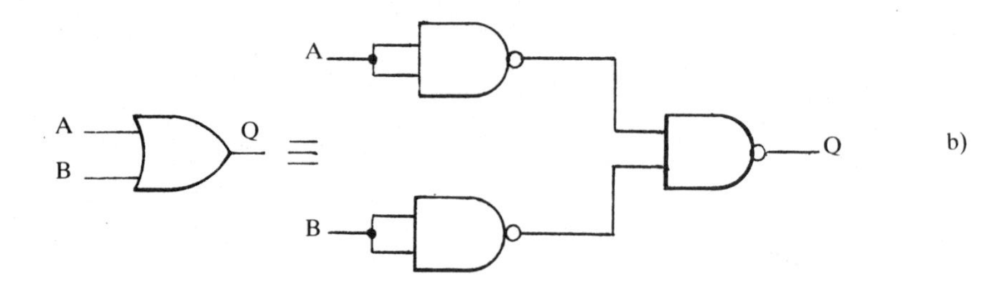

Fig. 2.9.

Worked examples 2.3

Q1. Construct the logic circuit for the function

$$Q = A . \overline{B} + B . C$$

from: (i) conventional AND, OR, and NOT gates; (ii) NAND gates

(i) There are two AND gates, a NOT gate, and an OR gate. The circuit is drawn below.

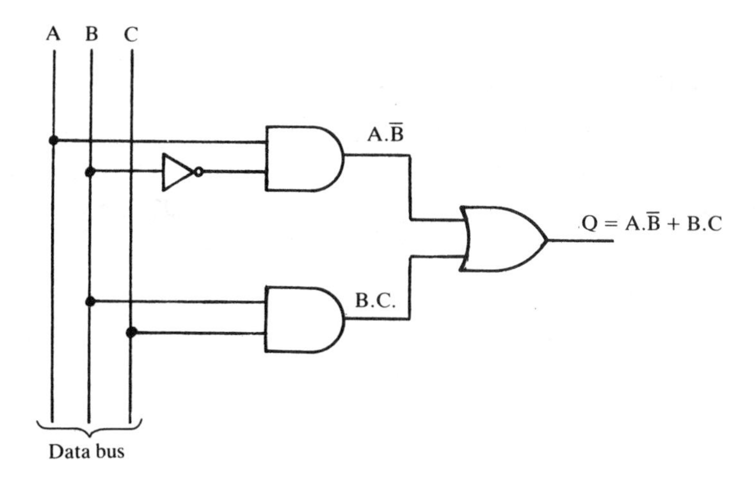

The set of parallel leads is called a *data bus*. It transfers information about the input variables A, B, and C around the logic system — a microcomputer, perhaps.

The circuit is not well-designed, because the NOT gate connected to variable B produces a propagation delay. We will not dwell on this problem here.

(ii) We can make use of Fig. 2.9 and replace the conventional gates by their NAND counterparts to give

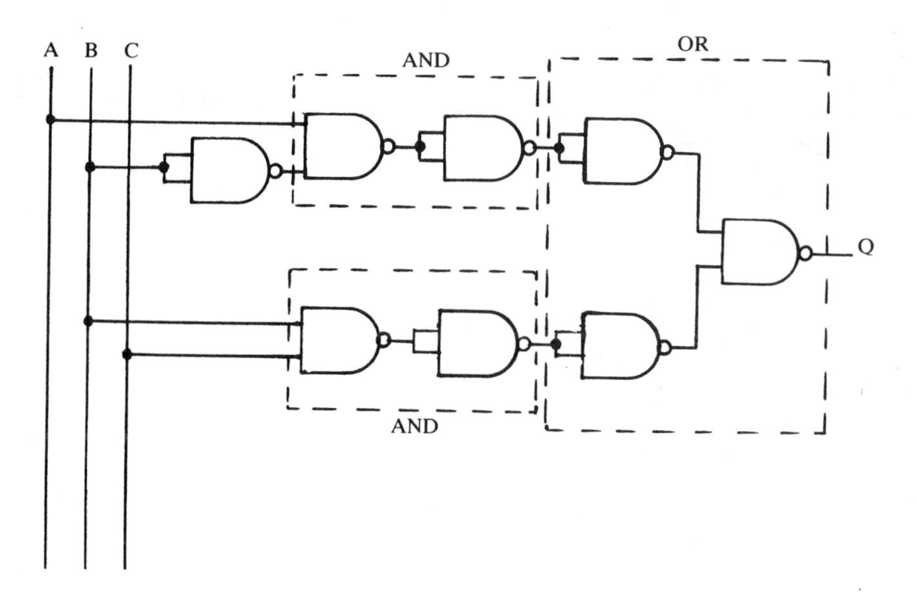

The circuit can be simplified by noticing that the effect of the two NOT gates in series is redundant because

$$\bar{\bar{A}} = A \quad \text{and} \quad \bar{\bar{B}} = B \; .$$

Hence a better designed circuit is

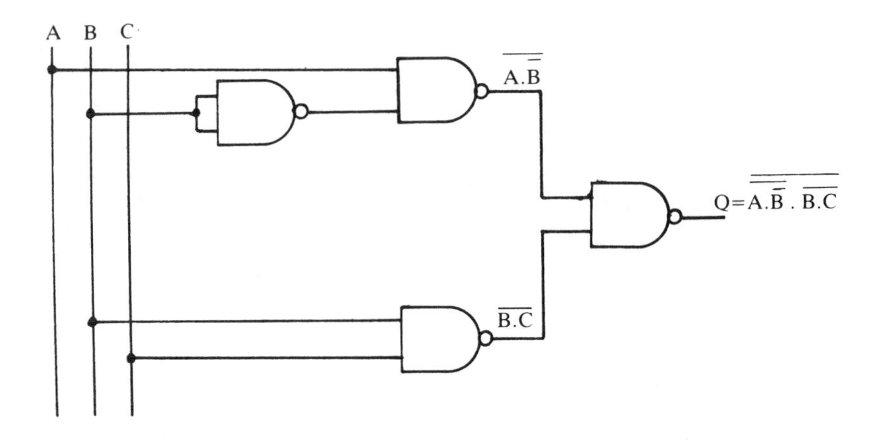

Mathematically, NAND logic gives an output Q expressed as

$$Q = \overline{\overline{A\,.\,\bar{B}}\,.\,\overline{B\,.\,C}}$$

$$= \overline{\overline{A\,.\,\bar{B}}} + \overline{\overline{B\,.\,C}} \qquad \text{(de Morgan's theorem)}$$

$$= A\,.\,\bar{B} + B\,.\,C \qquad \text{(de Morgan's theroem)}$$

What is interesting about (ii) is that its structure is identical to (i). The AND and OR gates are replaced by NAND gates and the inverter by its NAND equivalent.

Q2. Draw the logic circuit for:

$$Q = A + B\,.\,C$$

using (i) AND and OR gates; (ii) NAND gates.

(i)

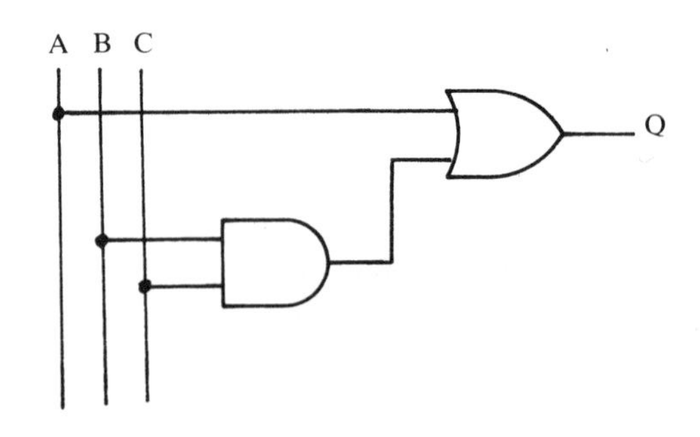

(ii)

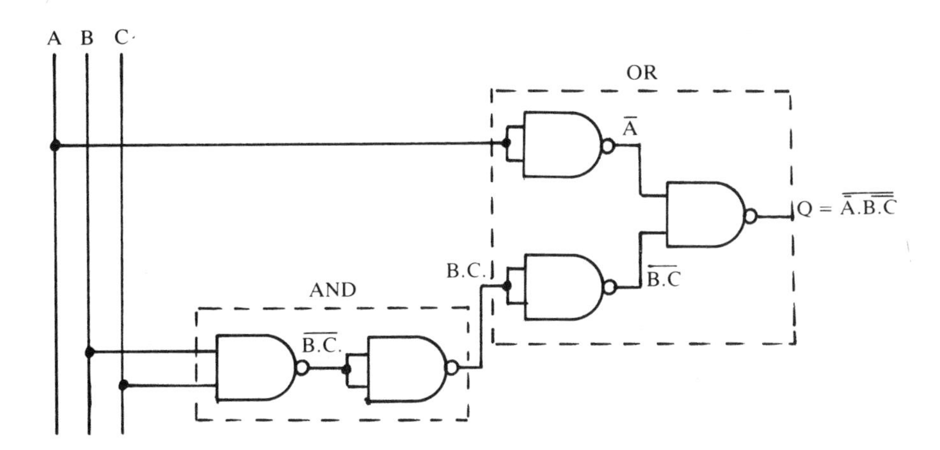

Once again, the two NOT gates in series are redundant and can be removed from the circuit. The 'final' circuit is:

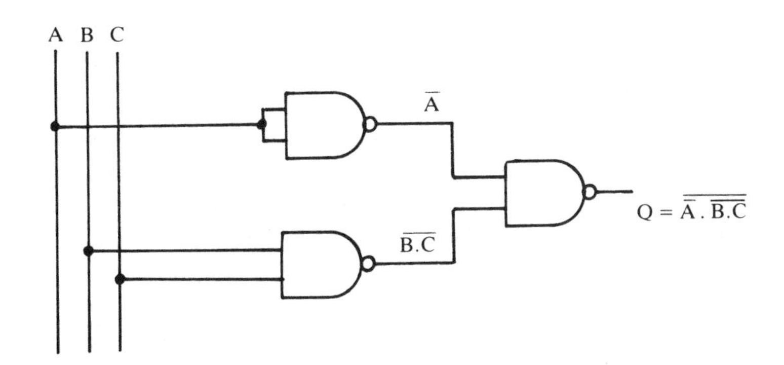

This circuit gives us another piece of valuable information: Whenever an input variable is connected directly to the final OR gate, as in (i), a NOT gate must be inserted in this line in NAND technology. Otherwise the structure of the NAND circuit is identical to the circuit made from conventional gates.

2.13.2 The first canonical form (sum-of-products) of a logic function

Suppose that a hypothetical truth table has been drawn up for a particular application. It is given in Table 2.8

Following the procedure discussed in Q.1 of worked examples 2.1, we note tha the truth table says that:

$$Q = 1 \textit{ if } A = 0 \text{ AND } B = 0 \text{ AND } C = 1, \text{ OR}$$

$$A = 0 \text{ AND } B = 1 \text{ AND } C = 1, \text{ OR}$$

$$A = 1 \text{ AND } B = 0 \text{ AND } C = 1$$

Table 2.8 — Hypothetical truth table

A	*B*	*C*	*Q*
0	0	0	0
0	0	1	1
0	1	0	0
0	1	1	1
1	0	0	0
1	0	1	1
1	1	0	0
1	1	1	0

This logic statement as a Boolean equation is

$$Q=\overline{A}\,.\,\overline{B}\,.\,C+\overline{A}\,.\,B\,.\,C+A\,.\,\overline{B}\,.\,C \tag{2.17}$$

This is the first canonical form of a logic statement obtained directly from the truth table. It consists of a set of three *minterms*, each one formed by ANDing the input variables. Once again, the variable itself is used whenever its binary value is '1' and the complement whenever the binary value is '0'. The ANDed outputs are then ORed together to give the final output. It should be relatively straightforward to construct the logic circuit.

2.13.2.1 Shorthand notation

Replace the variables in each minterm in (2.17) by their corresponding binary values, and then convert these to decimal. The result is:

$$\overline{A}\overline{B}C \equiv 001_2 \equiv 1_{10}$$

$$\overline{A}BC \equiv 011_2 \equiv 3_{10}$$

$$A\overline{B}C \equiv 101_2 \equiv 5_{10}$$

Then the shorthand form of (2.17) is written:

$$Q(ABC)=\Sigma\ (1,3,5) \tag{2.18}$$

The Σ sign is *not* a summation. It indicates that we are considering the first canonical form of a logic statement.

Worked examples 2.4

Q1. Determine the Boolean equation represented by

$$Q = \Sigma\ (2,4,7)\ .$$

The number of input variables is unknown. In (2.18) we have this information because the variables are included in parentheses after the output Q. This is not always the case, as suggested here. So, how do we determine the number of input variables?

RULE: Convert the largest decimal number *first* to the corresponding binary word and then to the minterm. The number of bits in this binary word will indicate the number of input variables.

Decimal	*Binary*	*Minterm*
7	111	ABC
4	100	$A\overline{B}\,\overline{C}$
2	010	$\overline{A}B\overline{C}$

Hence

$$Q(ABC) = \overline{A}\,.\,B\,.\,\overline{C} + A\,.\,\overline{B}\,.\,\overline{C} + A\,.\,B\,.\,C$$

The logic circuit can also be constructed as in:

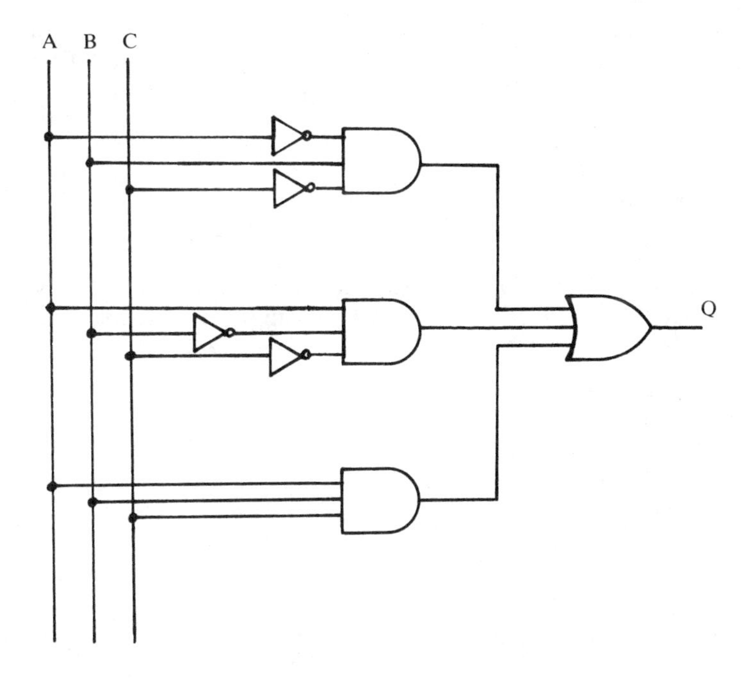

Note that all the gates have three inputs.

Q2. Determine the first canonical form of the logic statement represented by

$$Q = \Sigma(2,4,6,8,12) \ .$$

As in Q.1 we shall set up a table for converting the decimal numbers to minterms, starting with the largest number — 12.

Decimal	*Binary*	*Minterm*
12	1100	$AB\overline{C}\overline{D}$
8	1000	$A\overline{B}\overline{C}\overline{D}$
6	0110	$\overline{A}BC\overline{D}$
4	0100	$\overline{A}B\overline{C}\overline{D}$
2	0010	$\overline{A}\overline{B}C\overline{D}$

Hence

$$Q(ABCD) = \overline{A}\,.\,\overline{B}\,.\,C\,.\,\overline{D} + \overline{A}\,.\,B\,.\,\overline{C}\,.\,\overline{D} + \overline{A}\,.\,B\,.\,C\,.\,\overline{D} + A\,.\,\overline{B}\,.\,\overline{C}\,.\,\overline{D} + A\,.\,B\,.\,\overline{C}\,.\,\overline{D}$$

There is a second canonical form for a logical expression which consists of ANDing *maxterms* formed by ORing the input variables. We shall not discuss it here. However, interested students will discover details in some of the books listed in the bibliography.

2.13.2.2 The Karnaugh (K –) map

An important question that must be now asked is: How can we be sure that the logic circuit is being constructed with the minimum number of gates? From a technological, economical, and aesthetic standpoint this is a vital question to ponder.

Consider

$$Q = B\,.\,C + B\,.\,\overline{C} \ . \qquad (2.19)$$

At first glance, it seems that four logic gates are necessary to simulate this function: 2 AND, 1 NOT, and 1 OR. However, (2.19) can be rewritten as

$$Q = B(C + \overline{C})$$

which reduces to

$$Q = B \ .$$

This tells us that the output is identical with input variable B and that variable C plays no part in the logic process. This example is very basic, and does not demand any special minimizing technique. It is not obvious, however, whether (2.17) is in its minimal form. This is where the Karnaugh map is useful. It is a graphical technique which embraces the Venn diagram and the minimization theorems.

Divide the universe in a Venn diagram vertically into two halves. The right-hand half represents variable A and the other half its complement $\overline{A}$, see Fig. 2.10a. Next,

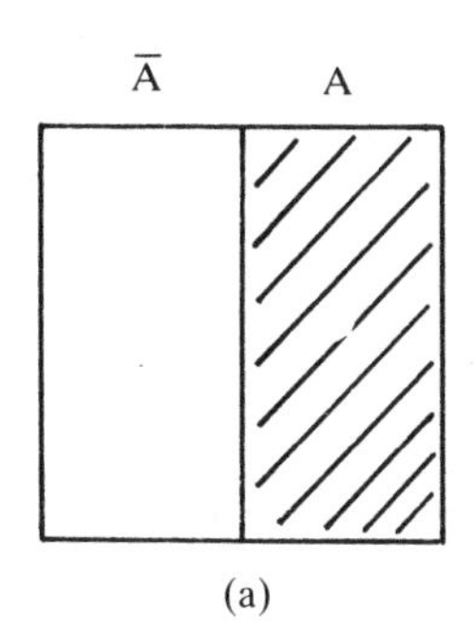

(a)

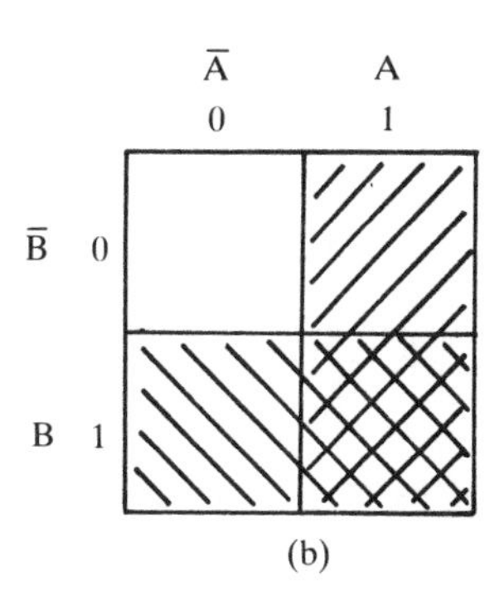

(b)

Fig. 2.10.

divide the universe horizontally so that the lower half represents variable B and the upper half $\overline{B}$, as in Fig. 2.10b. The K-map consists of four cells.

It is straightforward to represent the OR function formed from two variables with a K-map. This has been done in Fig. 2.11.

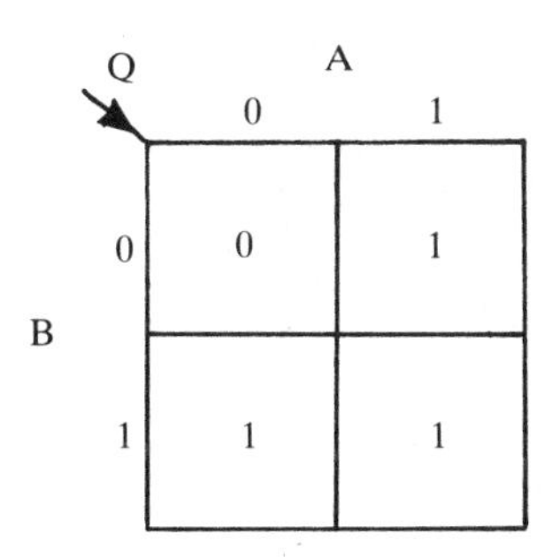

Fig. 2.11.

STEPS:
(i) Write down the truth table — see Table 2.5;
(ii) In cell (A = 0, B = 0 write '0';
(iii) In cell (A = 1, B = 0) write '1';
(iv) In cell (A = 0, B = 1) write '1';

(v) In cell (A = 1, B = 1) write '1'.

With three variables the horizontal side of the universe is labelled AB and the vertical side C (Fig. 2.12a). Note that Gray code has been used for AB. This has the

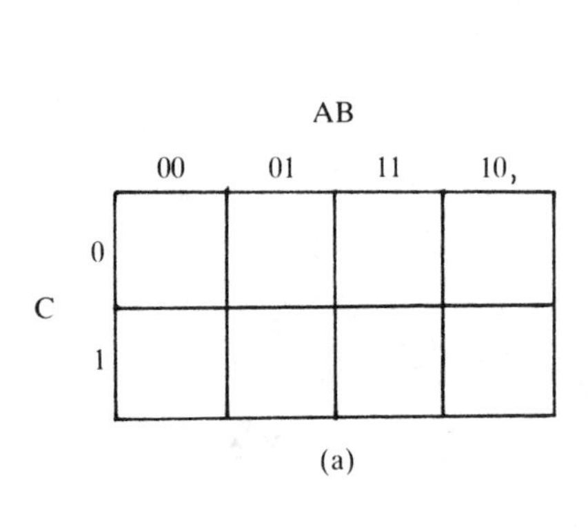

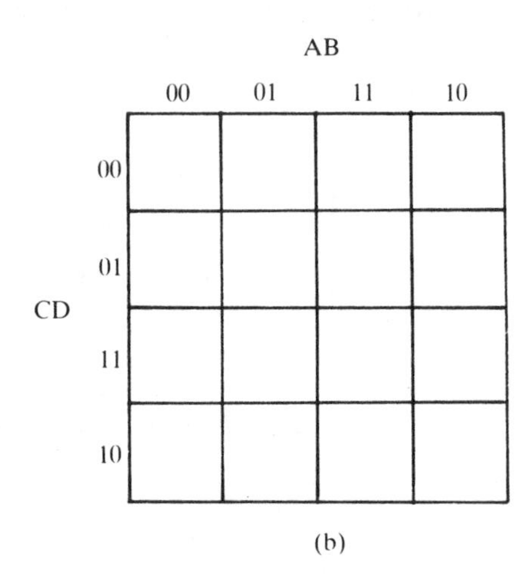

Fig. 2.12.

advantage that in going from one cell to the next only one variable changes at a time. With four variables, the vertical side of the universe is labelled CD, making a total of sixteen cells in all (Fig. 2.12b).

The K-map is two-dimensional for up to four variables but needs a third dimension for five and six variables. It is not constructed on a plane surface but on the surface of a torus (which is doughnut shaped). This means that cells in the top and bottom rows are pulled around until they are adjacent to one another. The same is true for cells in the first and fourth columns.

2.13.2.2.1 *Minimizing procedure*

Once again there are a set of rules for achieving minimization.

RULES:

(i) Write '1' in cells for which Q = 1, and '0' elsewhere.
(ii) Look for groups of 1, 2, 4, 8 . . . adjacent cells containing '1's and draw a loop around them. (The larger the group the simpler the logic).
(iii) Loops may overlap provided that they contain at least one other cell.
(iv) Ignore any loop that lies entirely within another loop.
(v) Edges of the K-map are adjacent.

Worked examples 2.5

Q1. Find the minimal first canonical form of

$$Q = \Sigma\ (0,2,3,6,7,8,10,12)\ .$$

As in the previous examples, begin with the largest decimal number in order to determine the number of bits in the binary words.

Decimal	*Binary*	*Minterm*
12	1100	$A.B.\overline{C}.\overline{D}$
10	1010	$A.\overline{B}.C.\overline{D}$
8	1000	$A.\overline{B}.\overline{C}.\overline{D}$
7	0111	$\overline{A}.B.C.D$
6	0110	$\overline{A}.B.C.\overline{D}$
3	0011	$\overline{A}.\overline{B}.C.D$
2	0010	$\overline{A}.\overline{B}.C.\overline{D}$
0	0000	$\overline{A}.\overline{B}.\overline{C}.\overline{D}$

The K-map must be like Fig. 2.12b.

Following RULE (i) of section 2.13.2.2.1, '1's and '0's can be inserted on to the K-map and then RULES (ii) and (iii) invoked to draw loops — The result is:

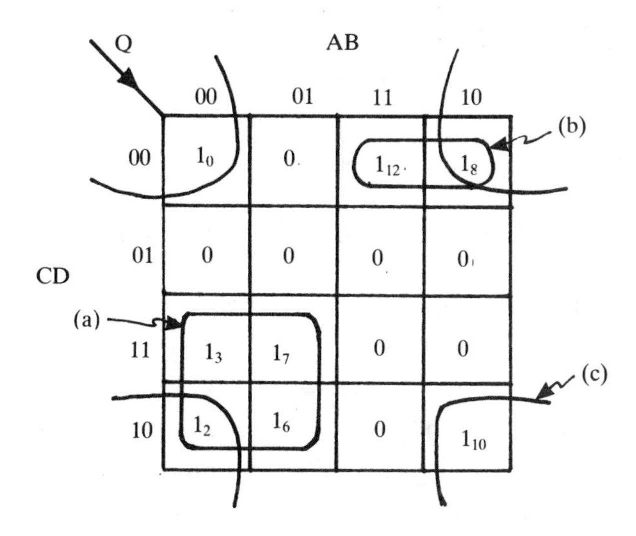

Loop (c) is interesting. Owing to the toroidal nature of the K-map four corner cells are adjacent. The idea now is to look for variables that stay constant in moving from one cell to another within each loop, and AND them together.

For loop (a): the '1's correspond to the logic function:

$$Q = \overline{A}.C\ .$$

For loop (b):

$$Q = A \,.\, \overline{C} \,.\, \overline{D}$$

For loop (c):

$$Q = \overline{B} \,.\, \overline{D}$$

Now the minimal first canonical form is obtained by ORing these Q's. That is

$$Q = \overline{A} \,.\, C + \overline{B} \,.\, \overline{D} + A \,.\, \overline{C} \,.\, \overline{D} \;.$$

☆ ☆

ASIDE: To understand why loop (a) leads to the result $Q = \overline{A} \,.\, C$, write out the logic function in detail. This is:

$$\begin{aligned} Q = 1 \text{ if } & AB = 00 \text{ AND } CD = 11, \text{ OR} \\ & AB = 01 \text{ AND } CD = 11, \text{ OR} \\ & AB = 00 \text{ AND } CD = 10, \text{ OR} \\ & AB = 01 \text{ AND } CD = 10 \;. \end{aligned}$$

That is

$$Q = \overline{A} \,.\, \overline{B} \,.\, C \,.\, D + \overline{A} \,.\, B \,.\, C \,.\, D + \overline{A} \,.\, \overline{B} \,.\, C \,.\, \overline{D} + \overline{A} \,.\, B \,.\, C \,.\, \overline{D} \;.$$

Collect terms together, as follows

$$\begin{aligned} Q &= \overline{A} \,.\, (B + \overline{B}) \,.\, C \,.\, D + \overline{A} \,.\, (B + \overline{B}) \,.\, C \,.\, \overline{D} \\ &= \overline{A} \,.\, C \,.\, D + \overline{A} \,.\, C \,.\, \overline{D} \\ &= \overline{A}.C\ (D + \overline{D}) \\ &= \overline{A} \,.\, C \end{aligned}$$

which is the same result obtained directly from the K-map.

☆ ☆

Occasionally, the techniques of Boolean algebra can be used directly to reduce a logical expression to its simplest form. Neither is it always easy to do, or always necessary. Once the rules for minimization are understood fully, the K-map is the most sensible approach to take.

Q2. Minimize:

$Q = \Sigma\ (1,2,3,6)$

Decimal	*Binary*	*Minterm*
6	110	$A.B.\overline{C}$
3	011	$\overline{A}.B.C$
2	010	$\overline{A}.B.\overline{C}$
1	001	$\overline{A}.\overline{B}.C$

The K-map has three variables, so it resembles Fig. 2.12a. Its completed form is

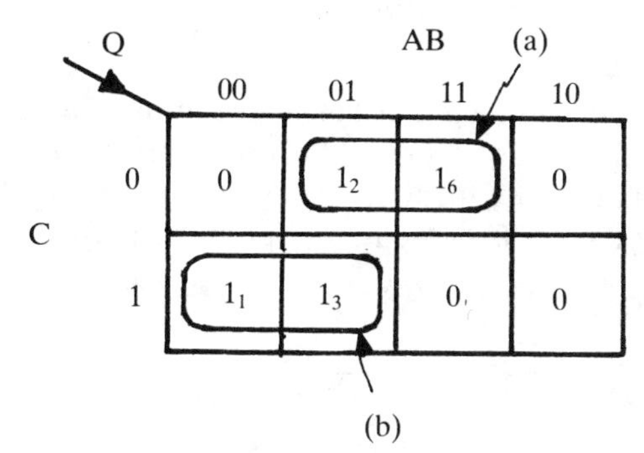

1_2 and 1_3 need not be looped together, because they have already been looped.
For loop (a):

$$Q = B.\overline{C}\ .$$

For loop (b):

$$Q = \overline{A}.C\ .$$

Hence the minimal first canbonical form is

$$Q = \overline{A}.C + B.\overline{C}\ .$$

Q3. Each light emitting bar in the seven-segment display of a digital wristwatch is connected to a logic circuit. The bars light up when their respective outputs are at level '1'. Design the logic circuitry which will cause bar g in the following diagram to light-up.

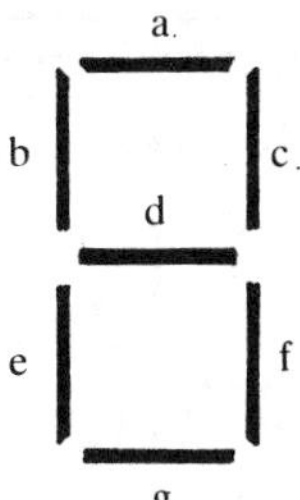

Bar g will light up for the following decimal numbers: $0_{10}, 2_{10}, 3_{10}, 5_{10}, 6_{10}, 8_{10}, 9_{10}$

— but not for 1_{10}, 4_{10}, 7_{10}. We now require to convert these decimal numbers to their logical equivalents. That is

Decimal	*Binary*	*Minterm*
9	1001	$A\,\bar{B}\,\bar{C}\,D$
8	1000	$A\,\bar{B}\,\bar{C}\,\bar{D}$
6	0100	$\bar{A}\,B\,C\,\bar{D}$
5	0101	$\bar{A}\,B\,\bar{C}\,D$
3	0010	$\bar{A}\,\bar{B}\,C\,D$
2	0010	$\bar{A}\,\bar{B}\,C\,\bar{D}$
0	0000	$\bar{A}\,\bar{B}\,\bar{C}\,\bar{D}$

The binary word has 4 bits, although certain combinations corresponding to the decimal numbers 10_{10} to 15_{10} are not needed. We 'don't care' what outputs light up these numbers, but we will indicate them on the K-map by writing the letter 'd' in their cells. 1_{10}, 4_{10} and 7_{10} have a '0' in their appropriate cell. Thus we have

Q: CD \ AB	00	01	11	10
00	1_0	0	d	1_8
01	0	1_s	d	1_9
11	1_3	0	d	d
10	1_2	1_6	d	d

RULE: Put d = 1 if it will help to increase the size of a loop.
Otherwise, assume that d = 0. We arrive at

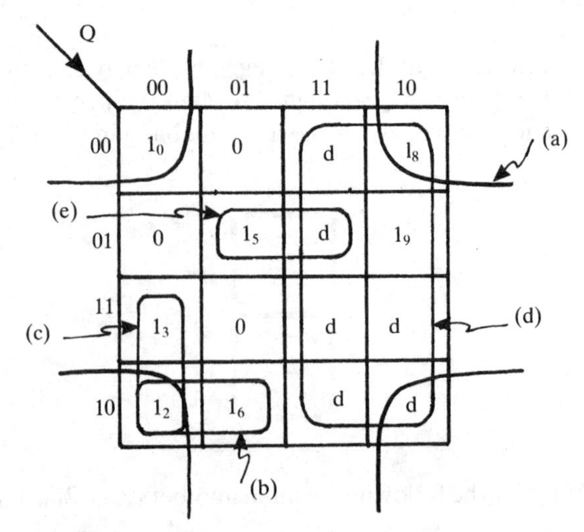

Loop (a) gives $Q = \bar{B} . \bar{D}$.
Loop (b) gives $Q = \bar{A} . C . \bar{D}$.
Loop (c) gives $Q = \bar{A} . \bar{B} . C$.
Loop (d) gives $Q = A . \bar{D}$.
Loop (e) gives $Q = B . \bar{C} . D$.

Hence the minimal first canonical form is

$$Q = \bar{A} . \bar{B} . C + \bar{A} . C . \bar{D} + A . \bar{D} + \bar{B} . \bar{D} + B . \bar{C} . D \ .$$

Now the logic circuit can be constructed with the smallest number of gates. This is left as an exercise for the student.

2.14 ARITHMETIC CIRCUITS

What kind of logic circuit is required in order to add together two binary digits A and B? We can represent the binary addition by

$$\begin{array}{cc} & A \\ + & B \\ \hline C & S \\ \hline \end{array}$$

where S is the sum bit and C is the carry bit. Now let us take various values of A and B and tabulate the results (Table 2.9).

Table 2.9 — Binary addition of two numbers

A	*B*	*S*	*C*
0	0	0	0
0	1	1	0
1	0	1	0
1	1	0	1

The *Carry* column is reminiscent of the output to an AND gate, whilst the S column bears some resemblance to the output of an OR gate. In fact, there is a gate which has this output. It is called an EXCLUSIVE OR gate (XOR, for short). The output to this gate will be at level '1' so long as A and B are not equal. It has the symbol shown in Fig. 2.13. The XOR gate can have more than two inputs; the output will only be at '1' so long as an odd number of inputs are at binary '1'.

So why not use an XOR gate to generate the sum bit and an AND gate to generate the carry bit, as in Fig. 2.14?

The circuit is called the *half-adder*.

The next task that presents itself is to add the carry bit to two more bits in the next column. As an illustation, consider the following addition sum:

		1	
		+	
		0	
	C = 0	S = 1	
		C_p = 1	← carry bit from previous column
new carry bit →	C′ = 1	S′ = 0	← final sum bit
final carry bit →	C″ = 1		

The initial addition, 1 + 0, can be carried out with a half-adder and generates a sum bit S and a carry bit C. Next, S is added to the carry bit C_p from the previous column, using a half-adder to generate a final sum bit S′ and a carry bit C′. Then C and C′ are added together to give the final carry bit C″. This is achieved by using an OR gate. The answer is C″S′ = 10_2. The logic circuit, therefore, consists of two half-adders and an OR gate. It is called the full-adder. It is depicted in Fig. 2.15.

Before leaving this section, let us give one more illustration of binary addition:

		1	
		+	½-adder
		1	
	C = 1	S = 0	½-adder
OR gate		C_p = 1	
	C′ = 0	S′ = 1	
		C′ = 1	*Answer* is C″S′ = 11_2

2.15 SEQUENTIAL LOGIC

Section 2.9 stated that a sequential circuit has an output which depends not only on the present state of the inputs but on all previous inputs in a given time sequence.

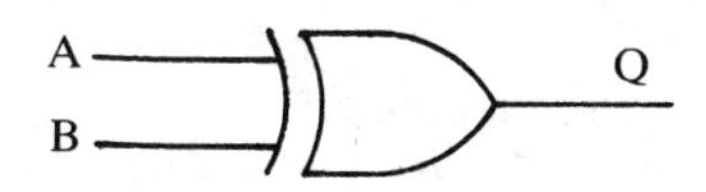

Fig. 2.13.

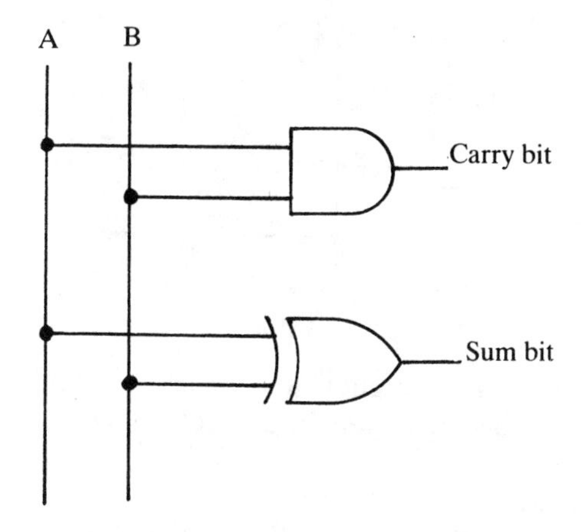

Fig. 2.14.

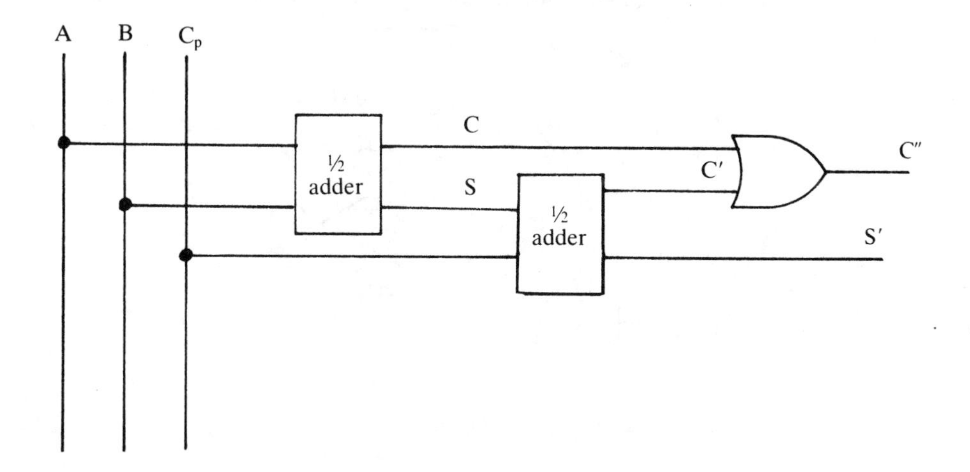

Fig. 2.15.

This means that the same inputs applied to a sequential circuit at times t and t' may generate different outputs.

The basic element of the sequential circuit is the *flip-flop*. There are a number of different kinds: SR-, D-, and JK-. We shall look at the basic characteristics of each type in turn.

2.15.1 SR flip-flop (SRFF)

The symbol is illustrated in Fig. 2.16. S stands for *SET* and R for *RESET*. There are

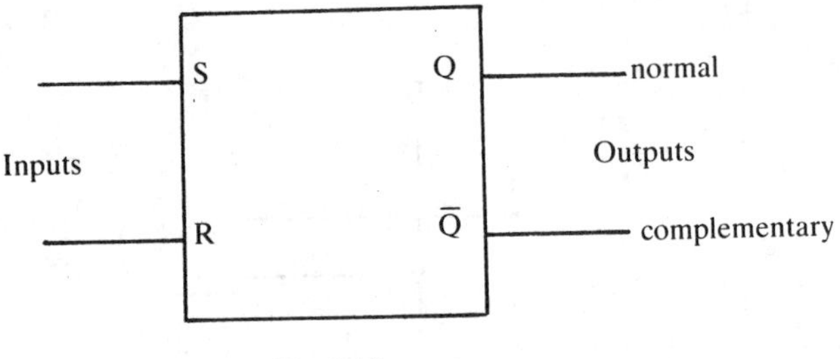

Fig. 2.16.

two outputs: the normal output Q and the complementary output $\bar{Q}$. The operation of the SR flip-flop can be understood by following some basic rules.

RULES:

(i) If $S=1$ and $R=0$ then $Q=1$. This is the SET condition.
(ii) If $S=0$ and $R=1$ the $Q=0$. This is the RESET condition.
(iii) If $S=0$ and $R=0$ then Q stays at its previous value. It is a *highly stable* input condition.
(iv) $S=1$ and $R=1$ are NOT ALLOWED because $Q\bar{Q}=11$; $\bar{Q}$ is no longer complementary to Q. This is an unacceptable output state.

Fig. 2.17 shows the construction of the SRFF in terms of NAND gates.

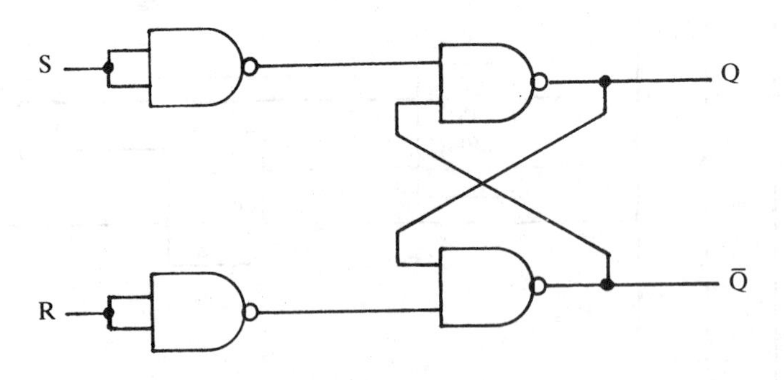

Fig. 2.17.

An interesting application of the SRFF is as a *buffer* in overcoming *contact bounce* in mechanical switches. These switches, of the *toggle* type, may be used to change the logic state in a circuit. They suffer from a major problem, however, in that their contacts do not close immediately but continue to make and break for some time after. This is an undesirable state of affairs because it causes the logic state of the circuit to fluctuate. An SRFF placed between the switch and the circuit overcomes this problem, see Fig. 2.18.

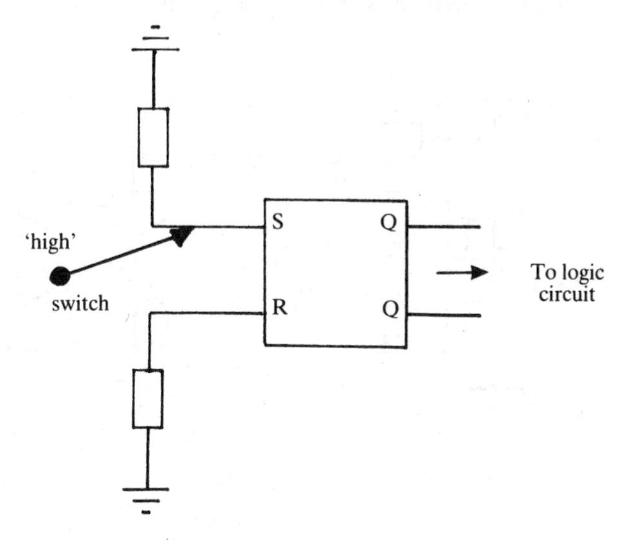

Fig. 2.18.

OPERATION:
When the switch makes, S becomes *high* and SR = 10. The flip-flop SETS with Q = 1. If the switch now *breaks*, S goes *low* and SR = 00. Q stays at 1. When the switch *makes* again, SR = 10 and the flip-flop SETS. But because Q is already at 1, it retains this state. Thus we see that the switch can no longer influence the logic state of the circuit in an adverse way.

2.15.2 D-flip-flop (or latch)

The 'D' stands for *delay*. Fig. 2.19 shows that the presence of an inverter avoids the input condition SR = 11. SR can either be 01 or 10.

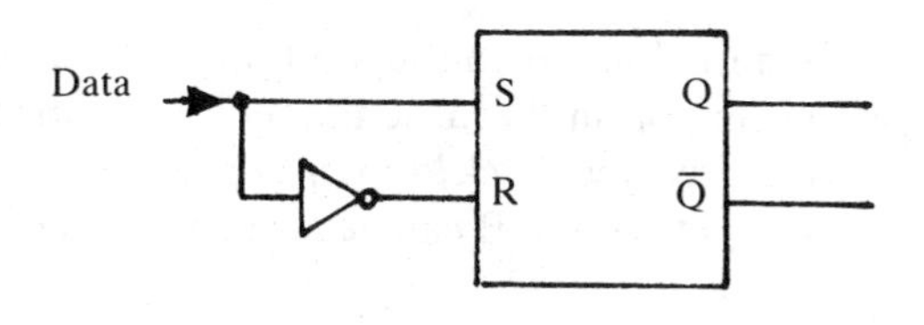

Fig. 2.19.

OPERATION:
For SR = 10, the flip-flop SETs ($Q\bar{Q}$ = 10) whereas for SR = 01 it RESETs ($Q\bar{Q}$ = 01).

Notice that the output replicates the input state but with a delay equal to the time for information to propagate through the flip-flop.

For the latch to operate in a predetermined way the SRFF is modified to allow a stream of timing pulses to be fed to it from a clock operating at a well-defined frequency. The basic circuitry is shown in Fig. 2.20.

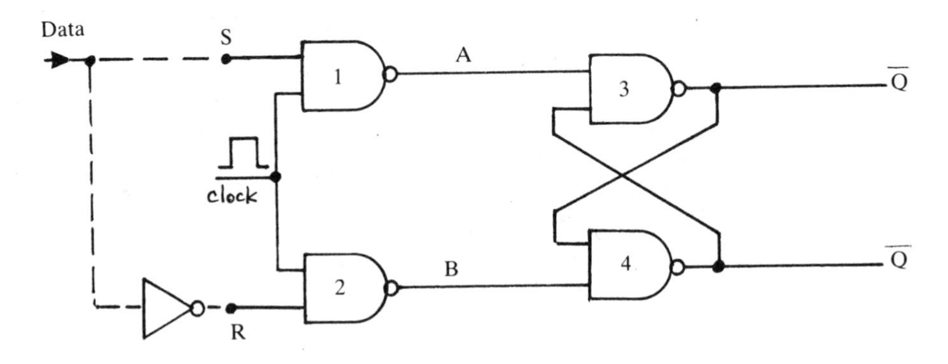

Fig. 2.20.

OPERATION:
Only when a clock pulse is *high* will gates 1 and 2 allow data to be inputted to gates 3 and 4; gates 1 and 2 are said to be *enabled*. If D = 1, SR = 10, AB = 01 and $Q\overline{Q}$ = 10. This is the SET condition. If D = 0, SR = 01, AB = 10 and $Q\overline{Q}$ = 01. This is the RESET condition.

When a clock pulse is *low*, AB = 11 because gates 1 and 2 are *disabled* or *inhibited*. No data can enter the SRFF. This is an obvious disadvantage because the flip-flop can *latch* on to data only during periods when the clock is high. There is another problem that can be classed as a disadvantage, viz. that the output will change if the data vary when the clock is high. It is for this latter reason that this flip-flop is sometimes called the *transparent* latch.

2.15.3 JK flip-flop

In Fig. 2.21, J and K are external inputs to the flip-flop; the letters bear no special significance. This flip-flop is important because the input condition JK = 11 is now acceptable. This is because the two feedback loops produce complementary inputs at gates 1 and 2 so that the oputputs A and B can never be the same.

OPERATION:
Once again, gates 1 and 2 are enabled only when the clock pulse is *high*. Four separate cases can be considered.

(i) JK = 00
If $Q\overline{Q}$ = 10, AB = 00 = SR and Q stays at 1.

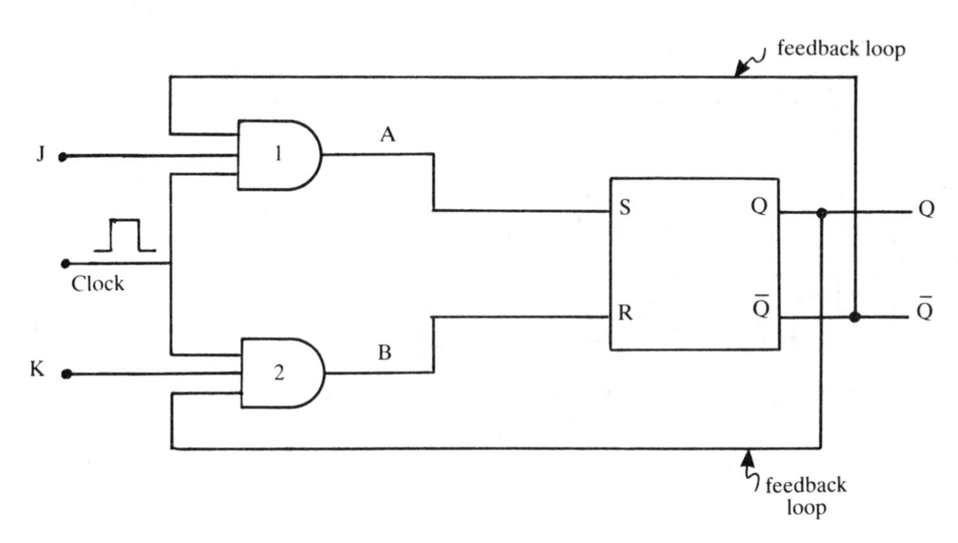

Fig. 2.21.

If $Q\bar{Q} = 01$, $AB = 00 = SR$ and Q stays at 1.

(ii) JK = 10

If $Q\bar{Q} = 01$, $AB = 10 = SR$ and the flip-flop SETs, i.e. Q goes to 1. Now $Q\bar{Q} = 10$, $AB = 00 = SR$ and Q stays at 1.

(iii) JK = 01

If $Q\bar{Q} = 10$, $AB = 01 = SR$ and the flip-flop RESETs, i.e. Q goes to 0.
If $Q\bar{Q} = 01$, $AB = 00 = SR$ and Q stays at 0.

(iv) JK = 11

If $Q\bar{Q} = 10$, $AB = 01 = SR$ and the flip-flop RESETs. Now $Q\bar{Q} = 01$, $AB = 10 = SR$ and the flip-flop SETs. This means that we return to $Q\bar{Q} = 10$ . . . The output *oscillates* continuously between 1 and 0 so long as the clock is *high*.

Case (iv) presents a problem. It may be solved by making the clock pulses of extremely short duration; but, then, the circuit loses its versatility. So what is normally done is to modify the JKFF further by using a second SRFF, when it is called the *master–slave* JKFF.

2.15.4 Master–slave JKFF

Fig. 2.22 shows the final form of the JKFF. It is well to point out here that all commercially-available JKFFs are of the master–slave type. Some JKFFs change their output when the clock pulse goes from *low* to *high*, i.e. on the *leading-edge*. We shall only discuss the operation of JKFFs which change their output on the *falling edge* of the clock pulse.

OPERATION:

(i) When the clock is *high*, gates 3 and 4 are disabled owing to the presence of the INVERTER. So $S_2R_2 = 00$ and Q_2 stays the same. Gates 1 and 2 are enabled so data can enter the master and Q_1 is free to change its state.

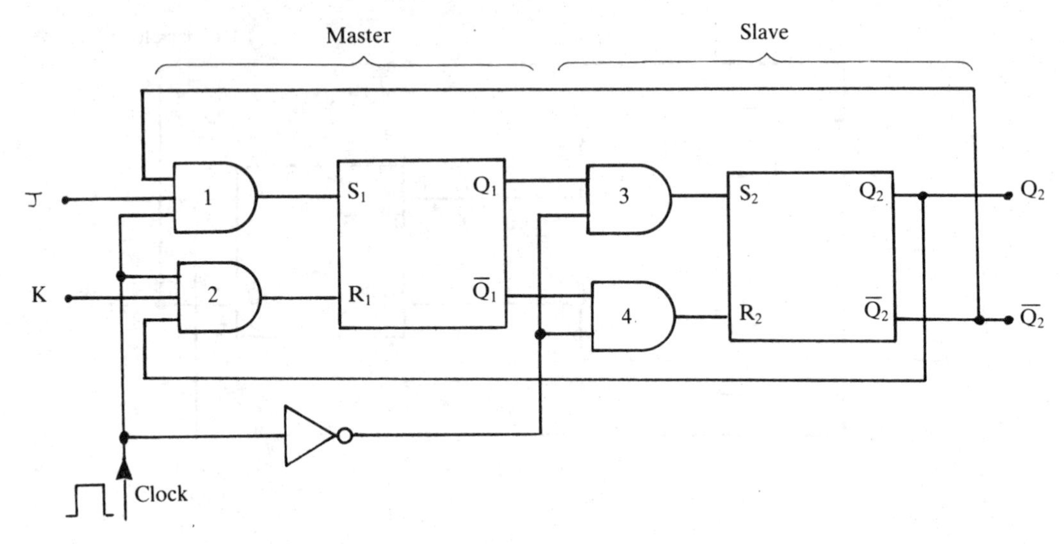

Fig. 2.22.

(ii) When the clock is *low*, gates 1 and 2 are *disabled*. This means that $S_1R_1 = 00$ and Q_1 stays the same.
Gates 3 and 4 are *enabled*, data enters the slave from the master, and Q_2 may change its value.

In short, data enter the JKFF when the clock is *high* but the output may alter only when the clock goes *low*.

Table 2.10 indicates the state of the normal output Q after the passage of one

Table 2.10 — Truth table for the JKFF

J	K	Q_t	Q_{t+1}	
0	0	0	0	
0	0	1	1	
1	0	0	1	*SET*...
1	0	1	1	
0	1	0	0	
0	1	1	0	*REST*...
1	1	0	1	*TOGGLE*...
1	1	1	0	*TOGGLE*...

clock pulse for various combinations of the inputs JK. The arrows indicate those input combinations which generate the same set of outputs. For example, $Q_t \rightarrow Q_{t+1} = 0 \rightarrow 0$ occurs for JK = 00 and 01. This means that it is essential that J stays at 0 whereas we *don't care* what value K has. Similarly, $Q_t \rightarrow Q_{t+1} = 0 \rightarrow 1$ occurs for JK = 10 and 11. J must stay at 1 but we *don't care* what value K has. A *reduced* truth table or *transition* table can be set up which should be used for determining the output state after each clock pulse (see Table 2.11). The letter 'd' is used to indicate a *don't care* condition.

Table 2.11 — Transition table for the JKFF

J	K	Q_t	$\rightarrow$	Q_{t+1}
0	d	0	$\rightarrow$	0
d	0	1	$\rightarrow$	1
1	d	0	$\rightarrow$	1
d	1	1	$\rightarrow$	0

We shall now briefly discuss the operation and applications of three sub-species of the JKFF.

2.15.4.1 Asynchronous JKFF

Fig. 2.23 shows that the JK inputs are permanently held *high* and data are inputted

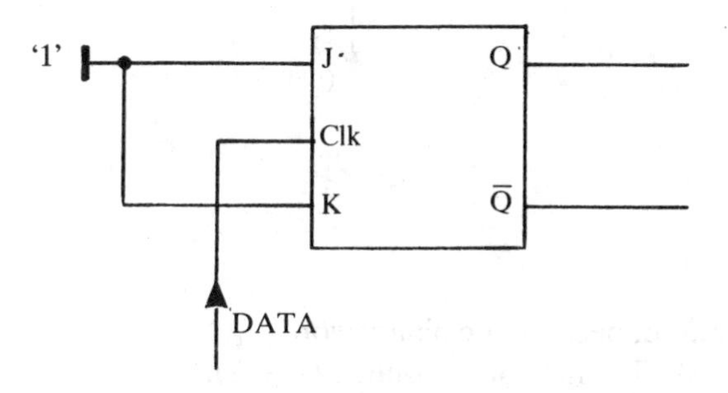

Fig. 2.23.

via the clock line. The flip-flop toggles (see Table 2.10) after each data pulse. Fig. 2.24 is instructive because it depicts the changes in the voltage waveform at the output Q in relation to the data pulses. What is interesting about this diagram is that the frequency of the output is one half of the frequency of the data pulses.

Let us now connect two asynchronous JKFFs in series in the manner of Fig. 2.25 in which the output Q_o of the first JKFF is fed to the clock input of the second. The waveforms are shown in Fig. 2.26. It may be observed that the combination behaves

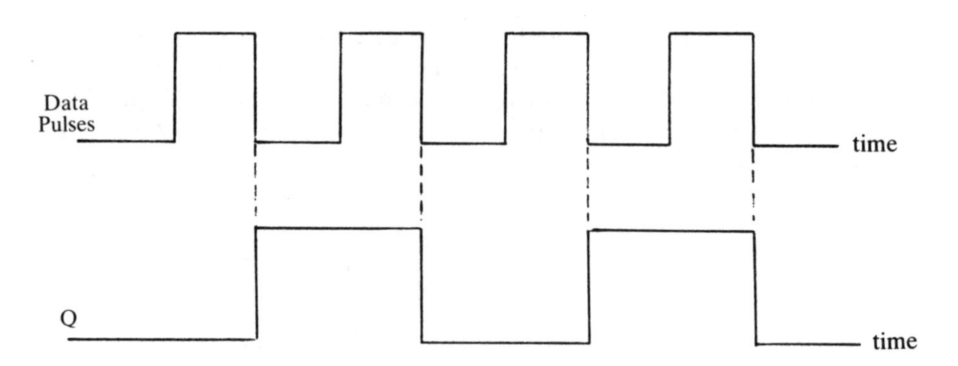

Fig. 2.24.

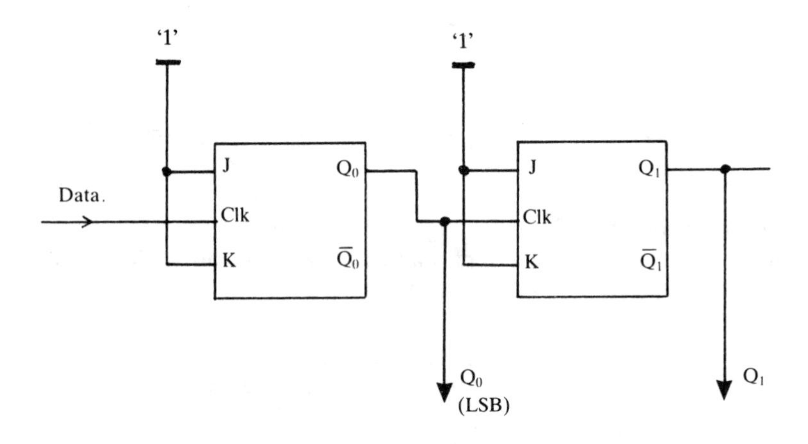

Fig. 2.25.

as a 2-bit ripple counter because the binary word Q_1Q_o goes from $00_2(0_{10})$ to $11_2(3_{10})$ and then resets to 00_2. The maximum count is $(2^3 - 1)$. The term *ripple* is used for this counter because the bits appear to ripple from one JKFF to the next.

2.15.4.2 Synchronous JKFF

Fig. 2.27 shows that in the synchronous JKFF the JK inputs are strapped together so that when data are fed to them they are either at 11 or 00. If the former, then the flip-flop toggles on the falling edge of the clock pulse, whereas if the latter, the output remains unaltered. The voltage waveforms are given in Fig. 2.28.

A limitation of the ripple counter is the rate at which pulses can be counted. The maximum counting rate depends on the propagation delay of the flip-flop; data cannot be fed to the second flip-flop until the first has completed its operation. Ways of overcoming this problem are: (i) use faster logic, such as emitter-coupled logic, or

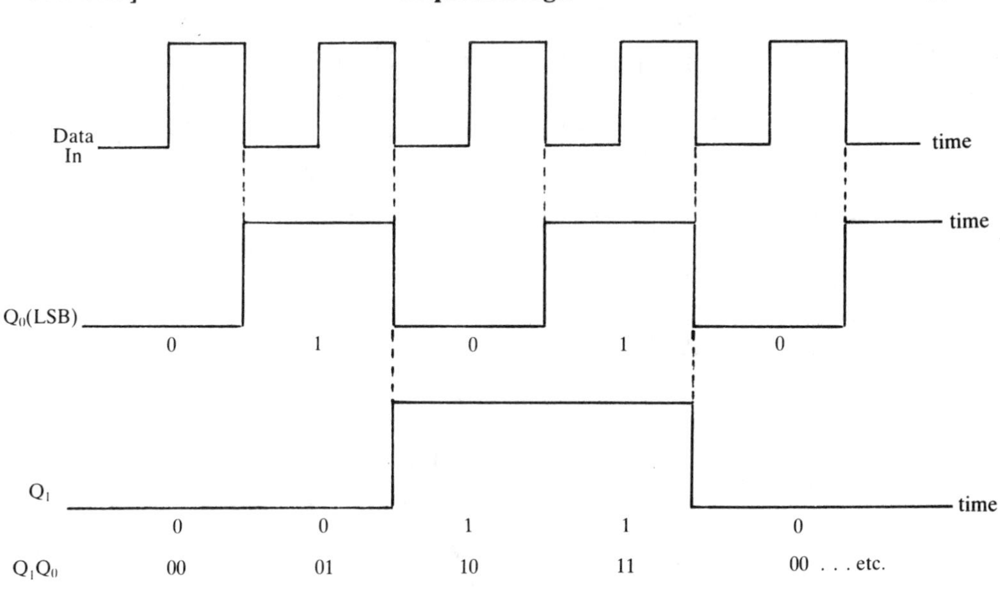

Fig. 2.26.

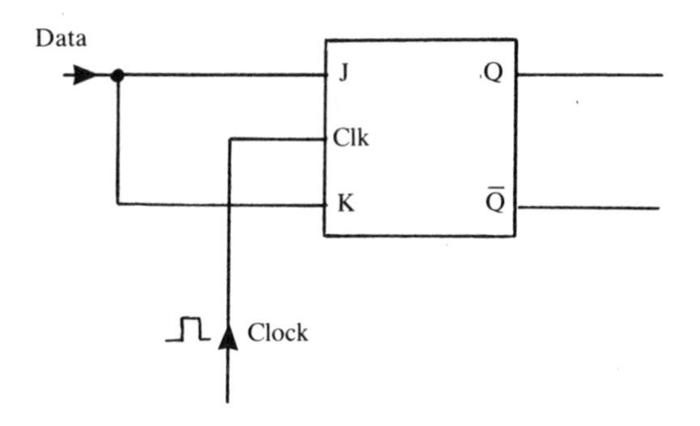

Fig. 2.27.

(ii) have each flip-flop operating at the same time. This is the principle behind the synchronous counter.

Let us take one example — the mod-8 counter, which counts from 0_{10} to 7_{10} before resetting. It is shown in Fig. 2.29. The object is to count the clock pulses. Output A is used as the least significant bit. The operation of the counter can be best examined by tabulating the state of the outputs A, B and C after successive clock pulses (Table 2.12). Thus we see that after 4 clock pulses the counter indicates the binary number 100. The process can be continued up to 7 counts. The diagonal arrows indicate the way in which the inputs of one flip-flop are influenced by the output of the preceding flip-flop.

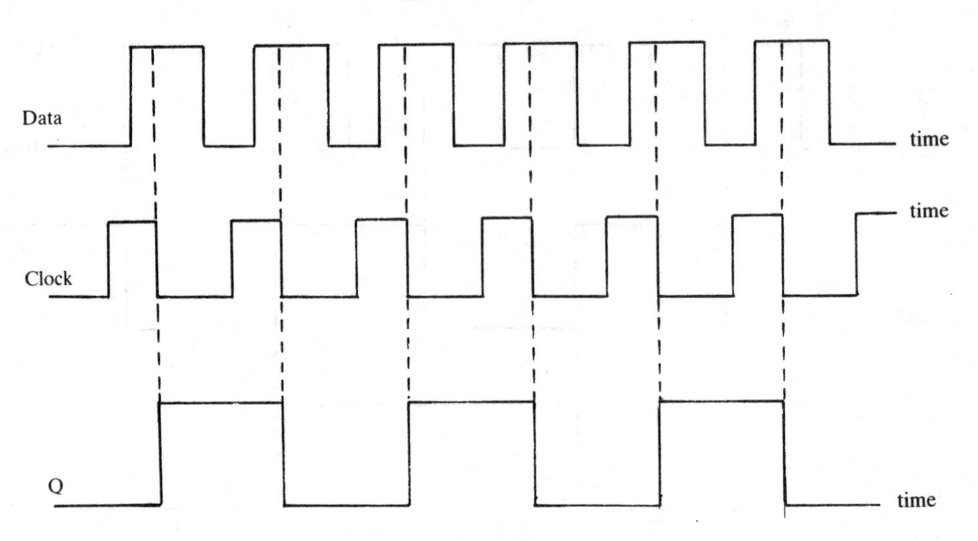

Fig. 2.28.

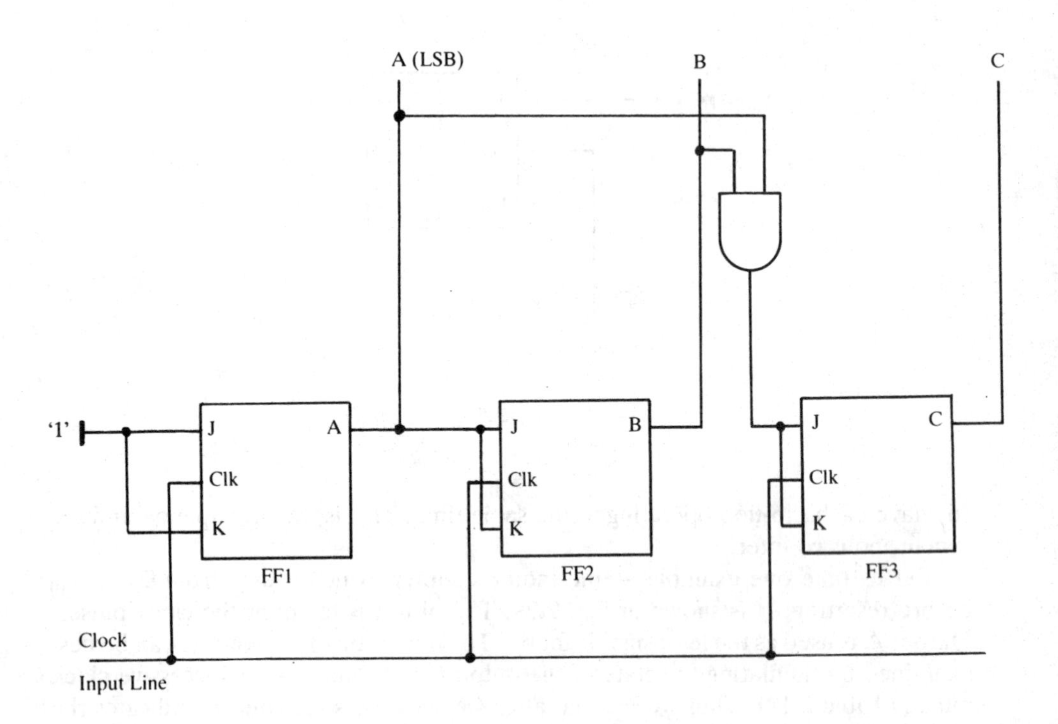

Fig. 2.29.

Table 2.12 — Operation of the mod-8 synchronous counter

Counts	*FF1*	*FF2*	*FF3*	*Comments*
0	A = 0	B = 0	C = 0	Initial count
1	A = 1 ↘	B = 0	C = 0	
		JK = 11		
		↓		
2	A = 0 ↘	B = 1 ↘	C = 0	
		JK = 00	JK = 00	AND gate disabled
		↓	↓	
3	A = 1 ↘	B = 1 ↘	C = 0	
		JK = 11	JK = 11	AND gate enabled
		↓	↓	
4	A = 0	B = 0	C = 1	
	...	....	....	

2.15.4.3 Delay JKFF

This version of the JKFF has complementary inputs so that JK = 10 or 01, as in Fig. 2.30. If JK = 10 the Q = 1, and if JK = 01 then Q = 0. In other words, apart from a propagation delay the state of the output matches the input state.

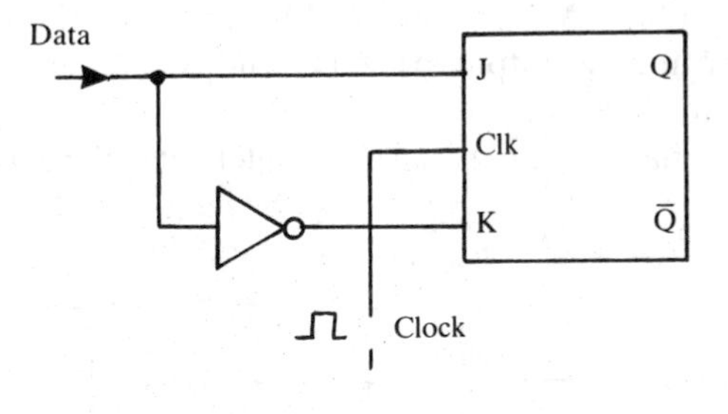

Fig. 2.30.

One important application of the D-JKFF is in the *shift register*. The term *register* applies to any circuit that stores information on a *temporary* basis. Permanent or long-term storage requires magnetic tape or disc. The shift register, then, is able to either store or shift data for arithmetic manipulations. There are a number of types, such as: serial in–serial out; serial in–parallel out; parallel in–parallel out.

OPERATION:
Look at the 3-bit serial in-serial out register shown in Fig. 2.31. On the falling edge of

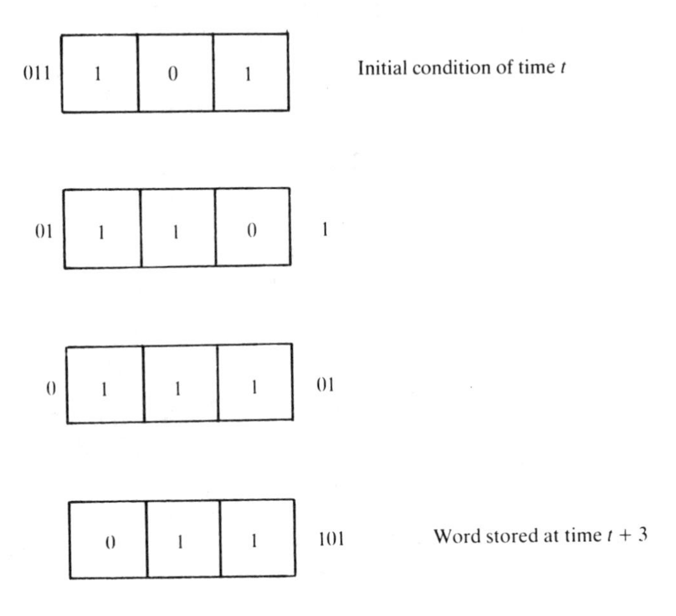

Fig. 2.31.

each clock pulse 1 bit of an external binary word enters the register and 1 bit leaves it. After the third clock pulse the whole of the external word is stored in the register. The data that leave the register do not have to be lost. A feedback loop from output to input could be used to recycle the word. Alternatively, the binary word can be inputted in serial form and the output to each flip-flop examined synchronously. This is the parallel out form of register.

Let us consider the make-up of a serial in-parallel out shift register (see Fig. 2.32).

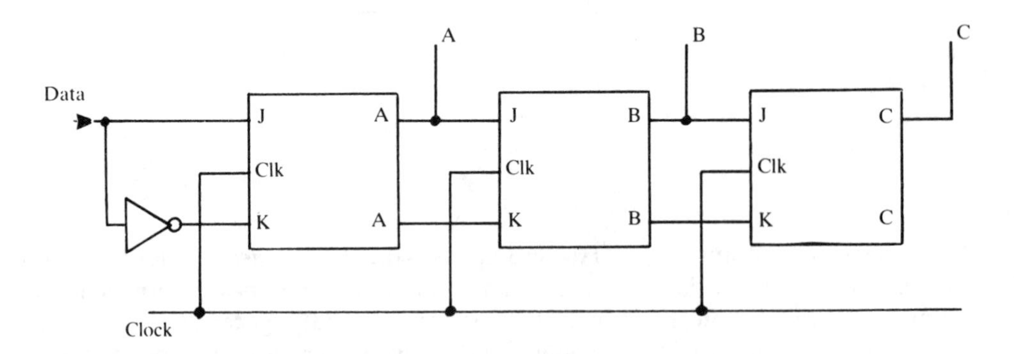

Fig. 2.32.

OPERATION:

(i) The inputs of each flip-flop are JK = 10 or 01. Their outputs SET or RESET accordingly.

(ii) The normal output of each flip-flop provides data for subsequent flip-flops.

Fig 2.33 depicts the voltage waveforms for the clock, serial data, and the three outputs.

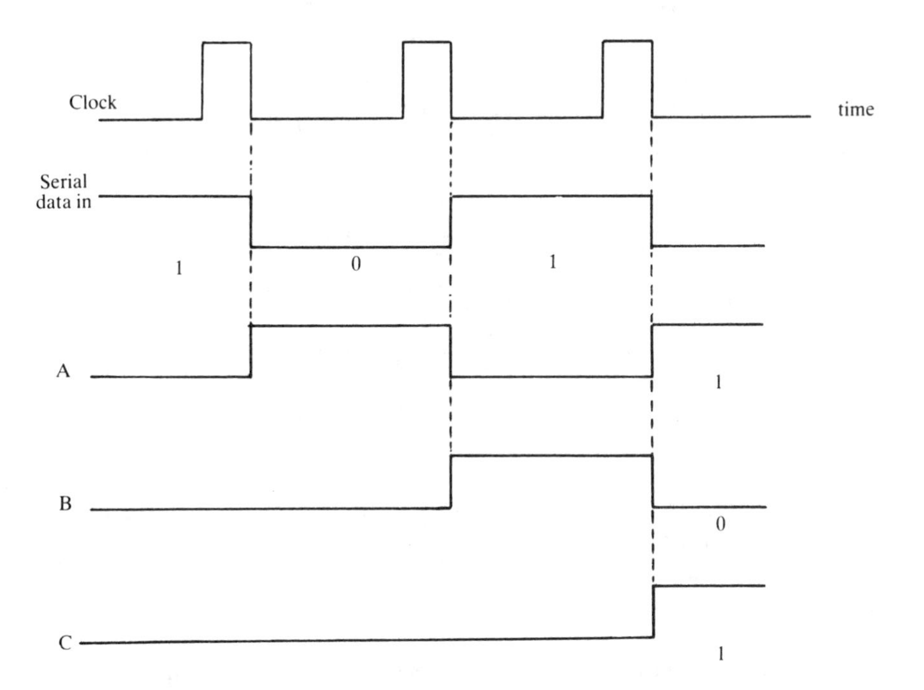

Fig. 2.33.

2.16 THE MULTIPLEXER

The multiplexer, or data selector, is a multi-input, single output device which converts parallel data into serial data. It differs from the parallel in-serial out shift register in that the inputted data can be accessed in any order. Fig. 2.34 shows a 4-input device, the 4:1 multiplexer, with the help of which the principle of operation can be elucidated. The 16:1 multiplexer is another possibility. This has 16 separate inputs.

A and B are called *control* inputs which are used to activate the AND gates. For example, with AB = 00, AND gate 1 is enabled. Data D_1 are outputted to the OR gate. As the outputs of the other AND gates are at 0, the state of the output Q matches D_1. Switching to AB = 11 makes $Q = D_4$, and so on. To output the binary word in serial form: $D_2D_3D_1D_4$, the control inputs need to be selected in the

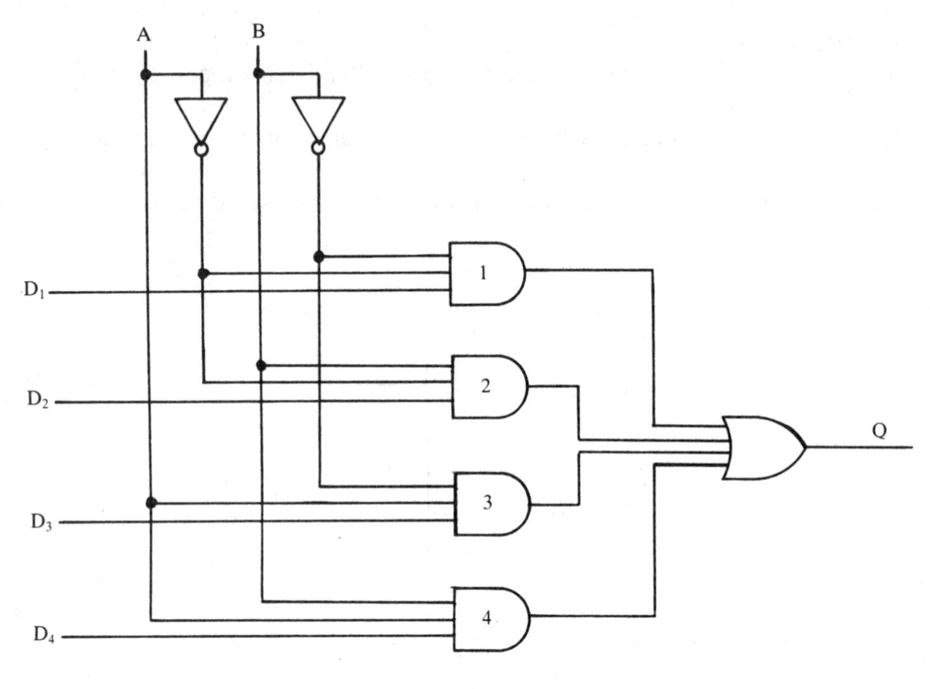

Fig. 2.34.

following order: $\overline{A}$, B, A $\overline{B}$, $\overline{A}$ $\overline{B}$, A B. The control lines, with the inverters, constitute a pure binary-to-decimal decoder. There are other types, such as the BCD-to-decimal decoder.

2.16.1 Programmable logic

So far we have discussed gates and circuits which perform specific logic functions. These are said to be *dedicated* devices because their only task is to give the correct output for a particular set of inputs. The multiplexer, on the other hand, may be used to simulate a variety of logic functions by wiring up the data inputs appropriately.

Worked examples 2.6

Q1. Use the 4:1 multiplexer to simulate the NAND logic function.

The first thing to do is write down the truth table:

A	*B*	*Q*
0	0	1
0	1	1
1	0	1
1	1	0

Now we can relate this truth table to Fig. 2.34. With A and B the control inputs, Q will be *high* if D_1, D_2 and D_3 are *high* and D_4 *low* then the multiplexer can be wired up like

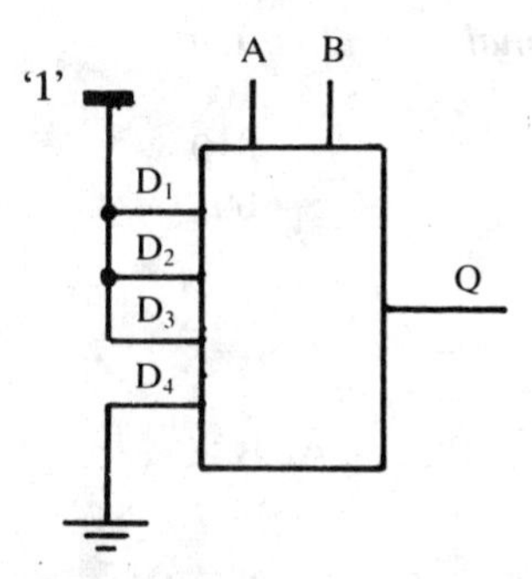

Q2. Use the 4:1 multiplexer to simulate the Exclusive-Or logic function.

The truth table is

A	B	Q
0	0	0
0	1	1
1	0	1
1	1	0

Thus we see that D_2 and D_3 need to be *high* and D_1 and D_4 *low*. So the writing diagram is

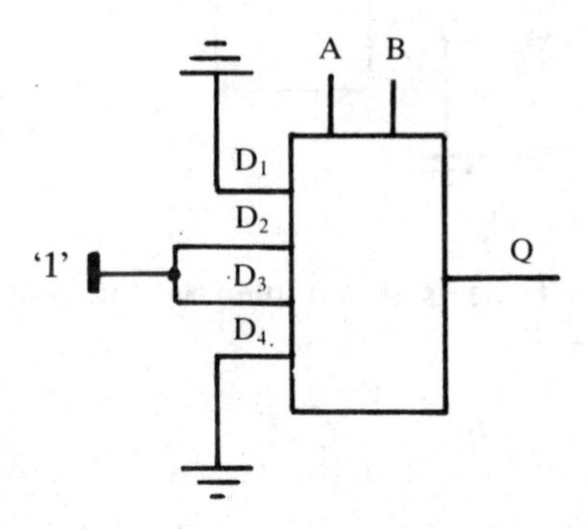

Q3. Use the 4:1 multiplexer to implement the logic function

$$\Sigma\ (1,2,4)\ .$$

First, we need to determine the minterms of the first canonical form, see section 2.13.2.

Decimal	*Binary*	*Minterm*
4	100	$A.\overline{B}.\overline{C}$
2	010	$\overline{A}.B.\overline{C}$
1	001	$\overline{A}.\overline{B}.C$

So

$$Q = A.\overline{B}.\overline{C} + \overline{A}.\overline{B}.C + \overline{A}.B.\overline{C}\ .$$

Now concentrate on the binary terms involving B and C. The minterms include $\overline{B}.\overline{C}$, $\overline{B}.C$, and $B.\overline{C}$. The only omission from the set is $B.C$. However, it can be included if Q is rewritten as

$$Q = A.\overline{B}.\overline{C} + \overline{A}.\overline{B}.C + \overline{A}.B.\overline{C} + (0).B.C\ .$$

So let B and C be the control variables and wire up the multiplexer with D_1 permanently connected to A, D_2 and D_3 connected to $\overline{A}$, and D_4 earthed. Then we have

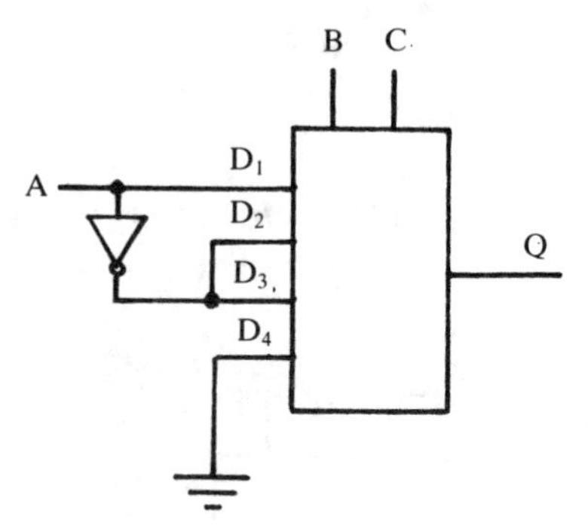

Q4. A 16:1 multiplexer is to be used to implement the logic function described by the following truth table —

A	*B*	*C*	*D*	*Q*
0	0	0	0	1
0	0	0	1	1
0	0	1	0	1
0	0	1	1	0
0	1	0	0	1
0	1	0	1	0
0	1	1	0	1
0	1	1	1	1
1	0	0	0	0
1	0	0	1	0
1	0	1	0	1

1	0	1	1	0
1	1	0	0	0
1	1	0	1	1
1	1	1	0	1
1	1	1	1	0

The outputs Q_r are identical with the data inputs D_r. All *high* data inputs are connected together and all *low* data inputs are connected together. A, B, C, and D act as the control inputs. The wired-up multiplexer looks like

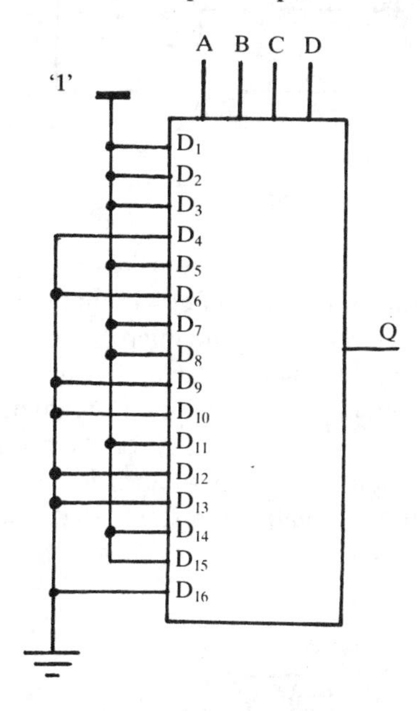

CAUTION:
TTL multiplexers use NAND gates. An example is the 74150 IC. The NAND gates have the effect of inverting the output. Hence, from the truth table above, Q_r is the complement of D_r. So all data inputs which give $Q=0$ should be *high* and those which give $Q=1$ should be *low*. The TTL chips also have a *strobe* input. When the strobe is *high* the chip is disabled, and when it is *low* it is enabled.

2.17 THE DIGITAL BUFFER AND THE TRI-STATE GATE

In many data communication systems information must be sent to two ways: from the user port of a microcomputer, say, to the external world — a robot, perhaps, and from the robot to the microcomputer. Also, the time when the information is to be sent needs to be controlled. The digital buffer allows us to do this. Fig. 2.35 shows one way of doing this.

The R/W line controls the flow of information. When this line is *high*, amplifier 2 is *enabled* and data pass from the robot to the user port. Data are *written* into the

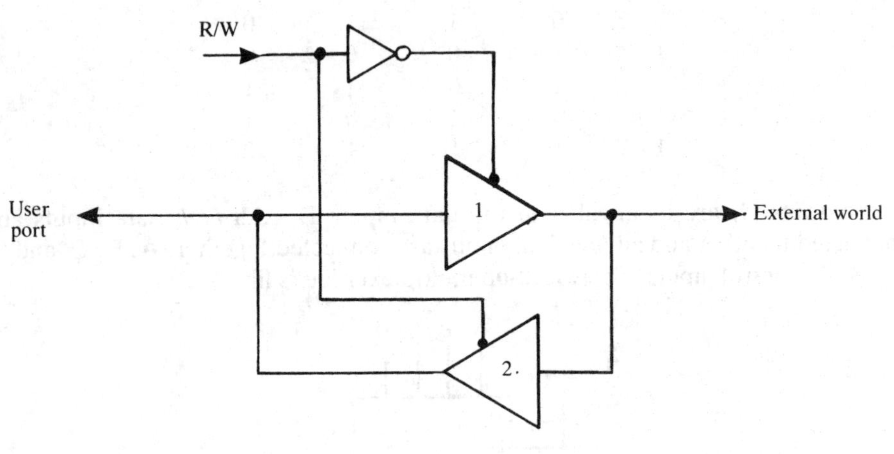

Fig. 2.35.

micro's memory. When the line is *low*, amplifier 1 is enabled and data pass from the user port to the robot. Data are read from the micro's memory.

The tri-state gate is a logic gate which has an additional control input. There are three output states, as the name suggests. As with the digital buffer, when the control input is *high* the output responds to the logic states on the normal inputs. That is, the output can be *high* or *low*. When the control input is *low* the gate's output enters a high impedance state; the output is unable to respond to the normal inputs. Fig. 2.36 shows how the control line is indicated on a few gates.

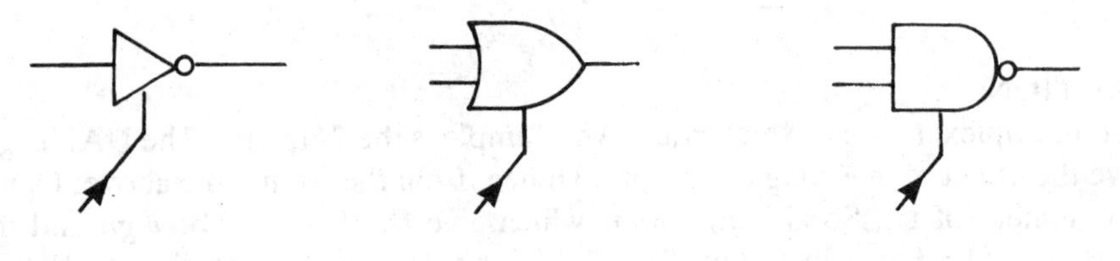

Fig. 2.36.

3

DC circuit analysis

Objectives

(i) Linear circuit computation with Kirchhoff's laws, loop and nodal analysis
(ii) The importance of Thévenin's theorem in dc/ac network analysis
(iii) The concept of impedance-matching
(iv) The potential and current divider rules
(v) The effect of introducing a voltmeter/ammeter into a circuit; loading errors
(vi) How to convert an ammeter into a voltmeter
(vii) The measurement of resistance with the Wheatstone bridge
(viii) The measurement of emf with the potentiometer
(ix) The analysis of R–C, RL, R–C–L circuits with specific initial (boundary) conditions
(x) How to distinguish between under-, over- and critically-damped systems

3.1 KIRCHHOFF'S LAWS

The basic laws which govern the interconnection of electrical elements in a circuit are known as *Kirchhoff's laws*. These laws arise from the conservation laws of energy and charge.

Consider what occurs at a junction point (more usually termed a *node*) when a charge (an electron, say) enters it. The charge may either be stored there or may leave it instantaneously. The former suggestion can be immediately dismissed because the node is infinitesimally small and the charge has a finite mass and size. So we may conclude that the charge must leave the node *instantaneously*.

In terms of currents — the total current entering the node is equal to the total current leaving it. To distinguish between these currents let us use an algebraic convention in which the current *leaving* the node is positive and the current *entering* it is negative. Then,

THE ALGEBRAIC SUM OF THE CURRENTS AT THE NODE IS ZERO.

Fig. 3.1 indicates that I_1, I_2, and I_3 are all positive and I_4 and I_5 are negative. Therefore

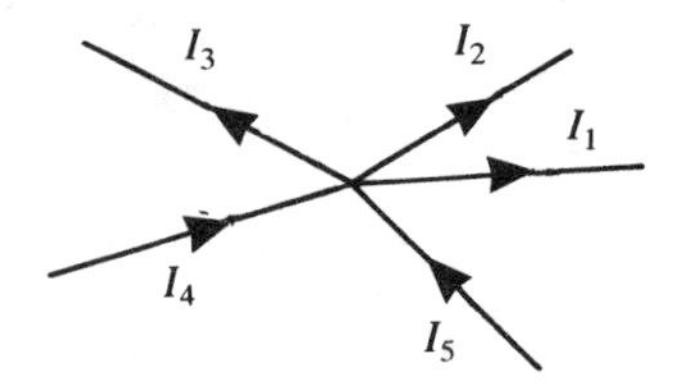

Fig. 3.1.

$$I_1 + I_2 + I_3 - I_4 - I_5 = 0$$

or

$$\sum_{s=1}^{3} I_s - \sum_{r=4}^{5} I_r = 0 \tag{3.1}$$

Next, consider two nodes A and B in a circuit characterized by potentials V_A and V_B. We can say unequivocally that the energy that 1 C of positive charge gains as it moves from A to B (if $V_A > V_B$) is the same along any path connecting A to B. If this were not so it would be possible to make a perpetual motion machine. How? First, by allowing the charge to go from A to B along a 'high-energy gain' path, and second by making it return along a 'low-energy loss' path. There would be a net gain in energy for the round trip. As nature does not give us *presents* of this kind we must conclude that

(a) THE ENERGY GAINED (LOST) BY 1 C OF CHARGE IN MOVING FROM A TO B IS INDEPENDENT OF THE PATH;
(b) THE CHANGE IN ENERGY FOR THE ROUND TRIP IS ZERO.

We may write, therefore, that

$$W_{AB} + W_{BA} = 0 \, .$$

However, since the charge involved in the process is 1 C, this reduces to

$$V_{AB} + V_{BA} = 0 \tag{3.2}$$

or, in general,

$$\sum V_{ij} = 0 \tag{3.3}$$

where the subscripts i and j stand for a neighbouring pair of nodes. Equation (3.2) says that:

> THE ALGEBRAIC SUM OF THE PD's AROUND A LOOP *AT ANY INSTANT* IS ZERO.

The phrase *at any instant* is important because it means that this rule applies to ac conditions also.

Fig. 3.2a depicts a circuital loop consisting of four resistors. Identify the various nodes, using letters, such as those used in the diagram.

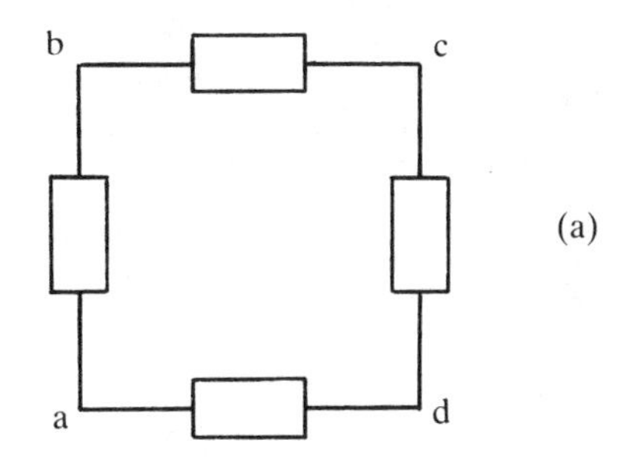

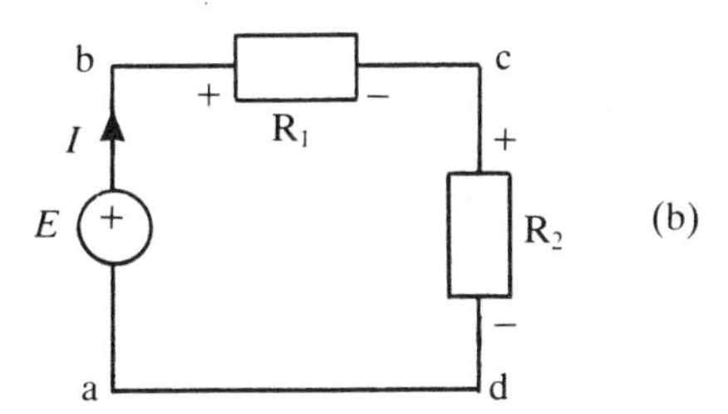

Fig. 3.2.

On applying (3.3) we have

$$V_{ba} + V_{ad} + V_{dc} + V_{cb} = 0 \tag{3.4}$$

Now introduce a voltage source, such as that shown in Fig. 3.2b. Note the high potential side of the voltage source; the positive sign on the voltage source indicates that node 'b' is at a higher potential than node 'a' and conventional current flows from b→c→d→a. It is sensible practice to include the appropriate signs on each side of the resistors as these will assist us when writing the complete form of (3.3) for the circuit. Hence substituting into (3.4) we obtain

$$E + 0 - V_{R2} - V_{R1} = 0$$

SUMMARY OF KIRCHHOFF'S LAWS

(i) The algebraic sum of the currents at a node is zero at any instant.

(ii) The algebraic sum of the emf's and potential differences around a loop is zero at any instant.

3.2 SOURCE TRANSFORMATIONS

Consider a 'real' voltage source of emf E_o and internal resistance R_o, as in Fig. 3.3a. This source is equivalent to an *ideal* voltage source of zero internal resistance and a resistance in series with it. Leads from the voltage source go to some external circuit. However, the nature of that circuit need not concern us at this moment.

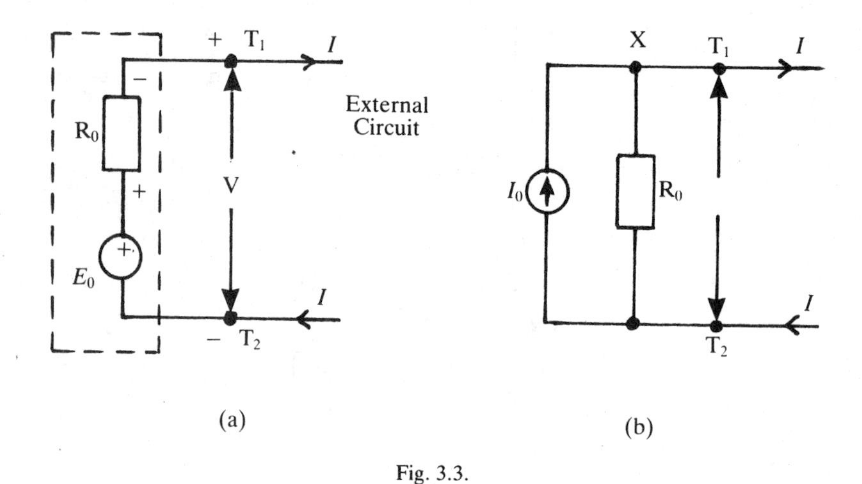

Fig. 3.3.

The current I produces a potential difference V between the terminals T_1 and T_2 of the source. Using Kirchhoff's voltage law, we may write

$$-E_o + IR_o + V = 0$$

or

$$V = E_o - I R_o \tag{3.5}$$

If one of the leads to the external circuit is disconnected then there is an open circuit and $I = 0$. This means that

$$V = E_o \, . \tag{3.6}$$

E_o can be measured by using a high-impedance voltmeter (through which no current flows) connected across the terminals. In actual fact a current flows through the voltmeter, but it is sufficiently small so that no significant pd is produced across the voltmeter itself.

Now we shall consider a circuit with a current source included, see Fig. 3.3b. First, applying Kirchhoff's current law at node X, we should note that the current flowing through the resistor R is $(I_o - I)$. Therefore, the pd between the terminals T_1 and T_2 is given by

$$V = (I_o - I)R_o \ . \tag{3.7}$$

In open circuit, $I = 0$, which means that if we stipulate that V has the same value in Figs 3.3a and b then

$$I_oR_o = E_o \ .$$

Now (3.7) can be rewritten

$$V = E_o - IR_o \tag{3.8}$$

We have ending up with an identical relation for V. In other words, a voltage source of emf E_o in series with a resistance R_o is equivalent to a current source $(I_o = E_o/R_o)$ in parallel with a resistance R.

Worked examples 3.1

Q1. Convert the voltage source to a current source in the circuit:

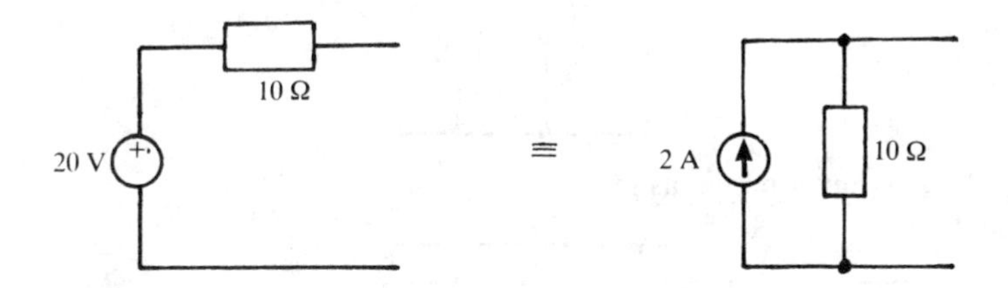

Q2. Convert the current source to a voltage source in the circuit:

Q3. By the principle of superposition:

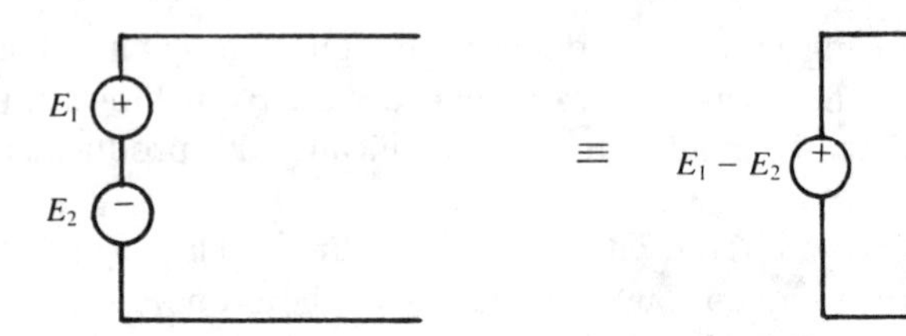

and

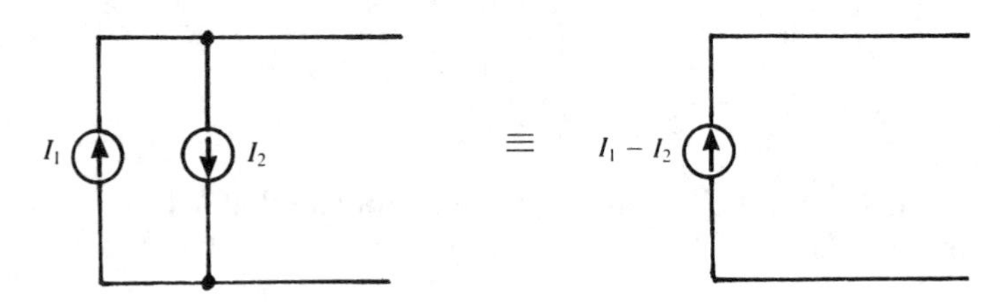

Q4. Reduce this series–parallel circuit to a simpler form.

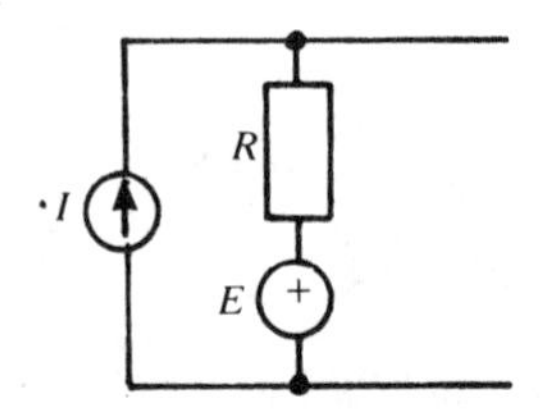

Replace the voltage source and series resistance R by a current source E/R and a parallel resistance R. That is

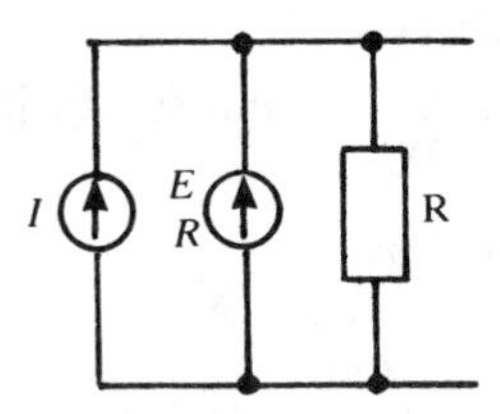

Combine the current sources, as in

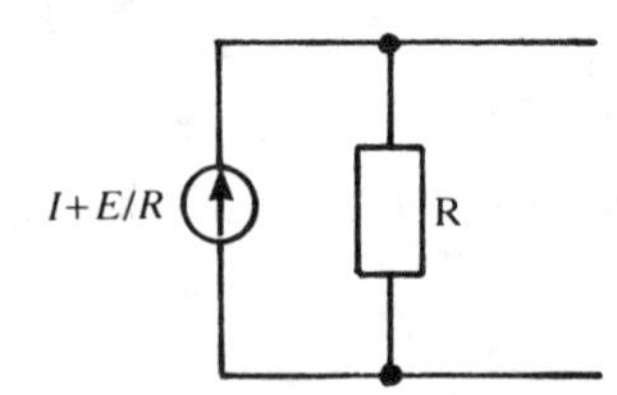

which is equivalent to

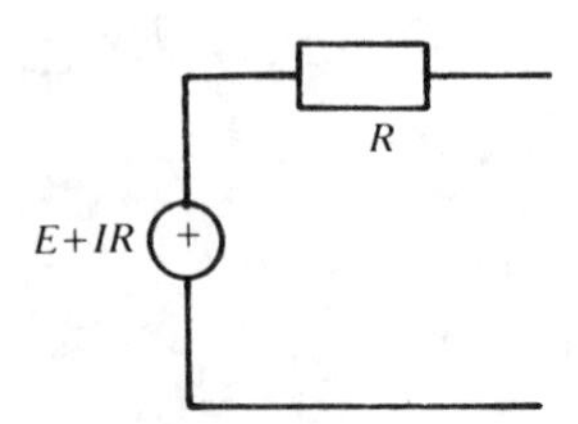

The alternative technique is to use the principle of superposition. With $E=0$, we have

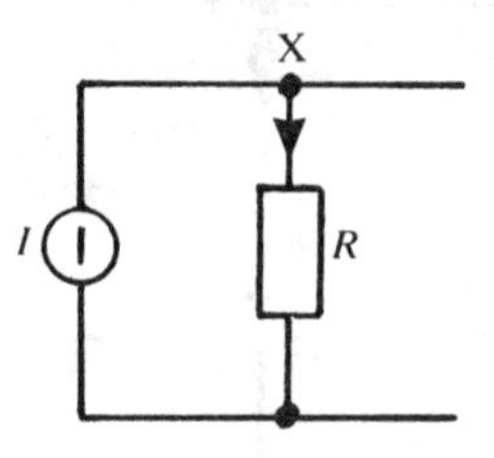

The potential at node X is $V_x = IR$. Note that the current can only flow downwards.
With $I=0$; we have

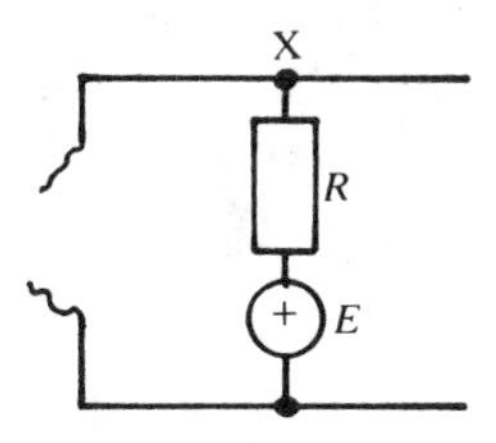

Now $V_x = E$. Therefore with both sources acting together,

$$V_x = E + IR$$

This is equivalent to the open circuit pd measured across the current source–voltage source combination.

It is appropriate at this stage to remind ourselves of the value of the emf and internal resistance of some typical voltage sources; see Table 3.1.

Table 3.1 — Emf's and internal resistance of some typical voltage sources

Source	E (volts)	R (Ω)
Weston-cadmium (standard)	1.01864	500
Lead acid accumulator	1.95	0.01
Ni-Fe alkaline	1.2	0.1
Dry cell	1.5	30

3.3 MAXIMUM POWER CONDITION

Fig. 3.4 shows a *practical* voltage source connected to a load resistor. In other words the resistor acts as the external circuit referred to in section 3.2. The signs are helpful because they indicate the high potential and low potential ends of the resistors. Hence applying Kirchhoff's voltage law in an anti-clockwise direction gives

$$E_o - V - I\,R_o = 0\,. \tag{3.9}$$

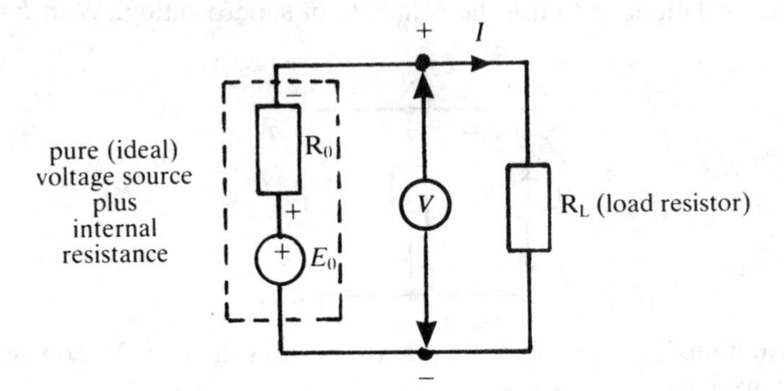

Fig. 3.4.

Also, as

$$V = I\,R_L \tag{3.10}$$

Substituting (3.10) into (3.9), we find that

$$E_o = I(R_o + R_L)\ .$$

Therefore, the rate of energy dissipation in R_L (the power P_L) is

$$\begin{aligned} P_L &= V\,I = I^2\,R_L \\ &= E_o^2\,R_L/(R_o + R_L)^2\ . \end{aligned} \tag{3.11}$$

The maximum power condition may be found by differentiating (3.11) with respect to R_L, and equating the result to zero. Doing this, we obtain

$$dP_L/dR_L = E_o^2\{(R_o^2 - R_L^2)/(R_o + R_L)^4\} = 0\ .$$

That is,

$$R_L = R_o\ . \tag{3.12}$$

Under this condition the pd across the load resistor is equal to $E_o/2$. Now, the maximum power can be calculated by substituting (3.12) into (3.11), when

$$P_L|_{max} = E_o^2/4R_o \tag{3.13}$$

Fig. 3.5 illustrates the way the power varies with the value of the load resistor. In general we can say that:

$$\frac{\text{rate of loss of energy in the load}}{\text{rate of loss of energy by the source}} = \frac{E_o^2 R_L/(R_o + R_L)^2}{E_o^2/(R_o + R_L)} = \tfrac{1}{2}$$

when the load is *matched* to the internal resistance.

3.4 LOOP (MESH) ANALYSIS: CONCEPT OF CIRCULATING CURRENTS

To illustrate the correct approach to this analytical technique we shall attempt to obtain the branch currents flowing in the network shown in Fig. 3.6. The currents flowing through the various branches of the network are I_1, I_2, and I_3. However, we *imagine* that the currents $\mathscr{I}_1$ and $\mathscr{I}_2$ flow around the loops ABEF and BCDE, respectively, in the directions shown. It is quite easy to relate the I's to the $\mathscr{I}$'s:

$$I_1 = \mathscr{I}_1$$
$$I_2 = \mathscr{I}_2$$
$$I_3 = \mathscr{I}_1 + \mathscr{I}_2 \ .$$

Using Kirchhoff's voltage law (KVL), we find for loop ABEF

$$E_1 - (\mathscr{I}_1 + \mathscr{I}_2)R_3 - \mathscr{I}_1 R_1 = 0 \tag{3.14}$$

and for loop BCDE

$$E_2 - (\mathscr{I}_1 + \mathscr{I}_2)R_3 - \mathscr{I}_2 R_2 = 0 \ . \tag{3.15}$$

Transposing (3.14) and (3.15) gives

$$\mathscr{I}_1(R_1 + R_3) + \mathscr{I}_2 R_3 = E_1$$
$$\mathscr{I}_1 R_3 + \mathscr{I}_2(R_2 + R_3) = E_2 \ .$$

These equations are easily solved to yield:

$$\mathscr{I}_1 = \frac{-E_2 R_3 + E_1(R_2 + R_3)}{(R_1 + R_3)(R_2 + R_3) - R_3^2} \tag{3.16}$$

and

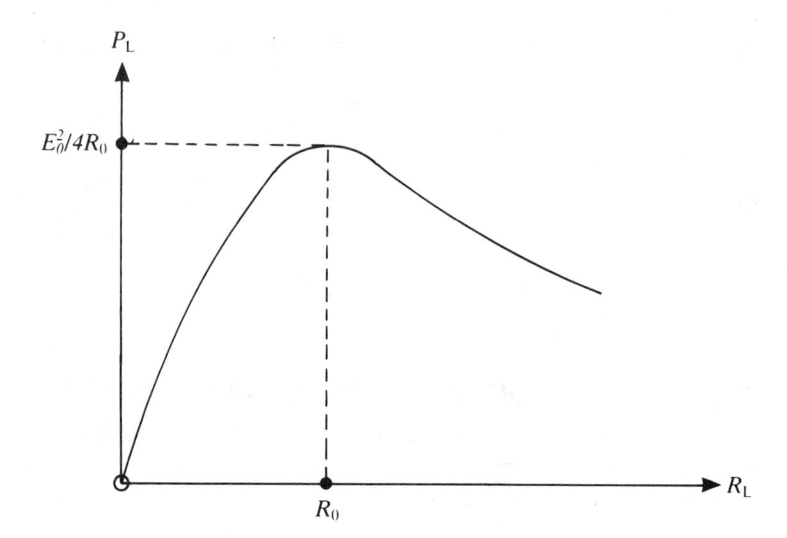

Fig. 3.5.

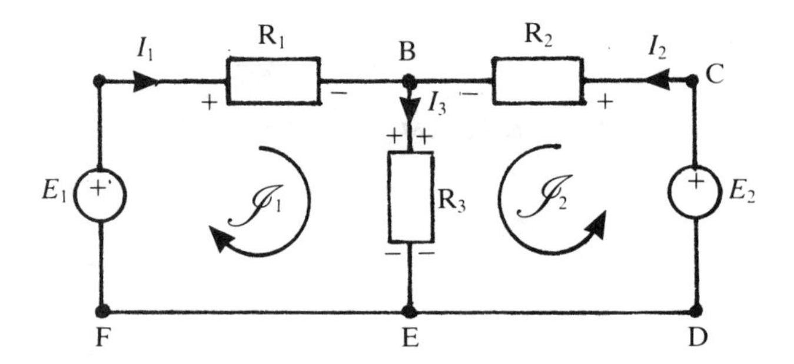

Fig. 3.6.

$$\mathcal{I}_2 = \frac{-(R_1 + R_3)E_2 + R_3 E_1}{(R_1 + R_3)(R_2 + R_3) - R_3^2}. \tag{3.17}$$

Now the branch currents can be evaluated.

If I_1 and I_2 are both found to be negative then this means that the loop currents actually flow in the opposite directions to those shown in the diagram. This analysis is fairly straightforward because there are only two loops to deal with. If the number of loops increases to six, say, then there will be six simultaneous equations to solve. Now the drawback to this approach becomes apparent. It should be applied only with relatively simple networks. In more complicated networks other techniques may have greater value, such as nodal analysis or Thévenin's theorem. Let us firstly discuss the same problem, using nodal analysis.

3.5 NODAL ANALYSIS

As we have just seen with *loop analysis*, the unknown currents are determined once a number of simultaneous equations have been written down, one for each loop in the network. A question which comes to mind is the following: can the currents be obtained from *one* equation, only, containing one unknown. The answer is yes — if we chose as the unknown, the potential at a specific node measured relative to a properly chosen reference potential. Let us see how this works in practice by relabelling Fig. 3.6.

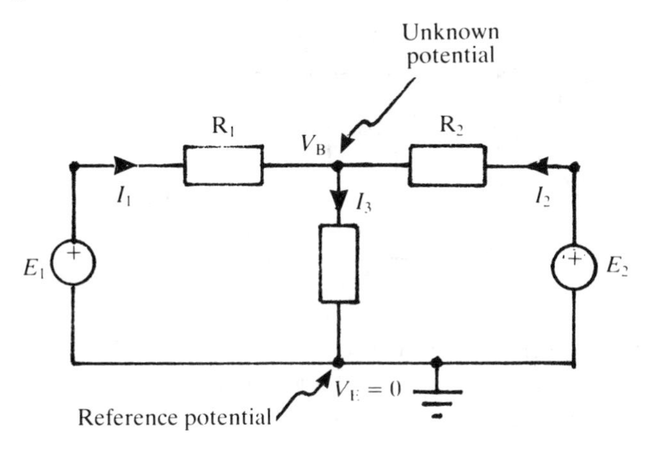

V_B is the unknown potential and V_E is at earth potential, which we will use as the reference potential. Kirchhoff's current law (KCL) can be applied at the node B to give:

$$I_3 - I_1 - I_2 = 0$$

or

$$V_B/R_3 - (V_A - V_B)/R_1 - (V_C - V_B)/R_2 = 0 \ . \tag{3.18}$$

On solving for V_B we obtain

$$V_B(1/R_1 + 1/R_2 + 1/R_3) = V_A/R_1 - V_C/R_2 \tag{3.19}$$

from which the branch currents are determined, using

$$I_1 = (E_1 - V_B)/R_1$$

$$I_2 = (E_2 - V_B)/R_2$$
$$I_3 = V_B/R_3 \ .$$

It should be noted that this technique cannot always be used. There is no simple rule which tells you when to use it, and with which networks it can best be used. Practice in, and experience of, the various techniques will allow you to unconsciously select the appropriate one to give a speedy solution.

Worked examples 3.2

Q1. Determine I, using Kirchhoff's voltage law in the following circuit:

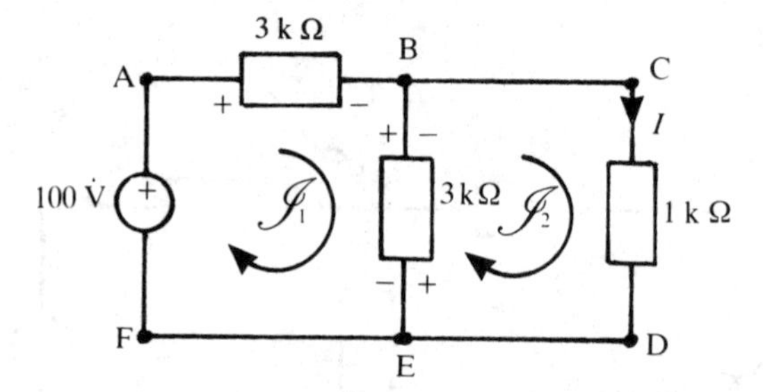

The loop currents are shown. I is identical with $\mathscr{I}_2$. We shall leave the value of the resistances in kΩ because the current will then be in mA.

For loop ABEF: $3\mathscr{I}_1 + 3(\mathscr{I}_1 - \mathscr{I}_2) = 100$ (1)

For loop BCDE: $\mathscr{I}_2 + 3(\mathscr{I}_2 - \mathscr{I}_1) = 0$ (2)

Simplifying (1) and (2) gives

$$6\mathscr{I}_1 - 3\mathscr{I}_2 = 100 \qquad (1a)$$

$$-3\mathscr{I}_1 + 4\mathscr{I}_2 = 0 \ . \qquad (2a)$$

From (2a) we obtain

$$\mathscr{I}_1 = 4\mathscr{I}_2/3$$

which, on substituting into (1a), gives

$$6 \times (4\mathscr{I}_2/3) - 3\mathscr{I}_2 = 100$$

or

$$5\mathcal{I}_2 = 100$$

and

$$\mathcal{I}_2 = 20 \text{ mA} .$$

Q2. Use Kirchhoff's laws to obtain I in the following circuit:

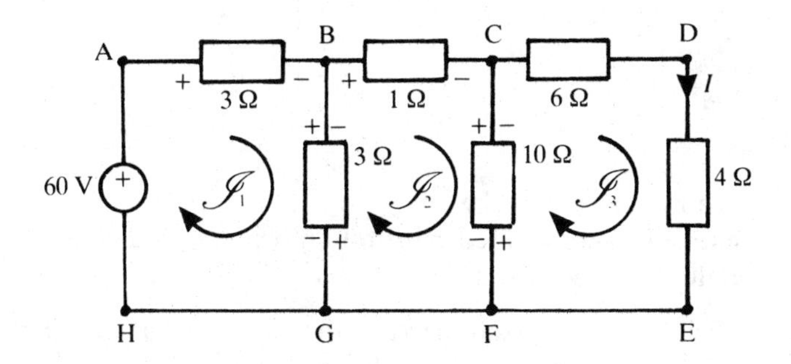

The loop currents are indicated. I is equivalent to $\mathcal{I}_3$.
As there are three loops, KVL is used three times:

Loop ABGH: $3\mathcal{I}_1 + 3(\mathcal{I}_1 - \mathcal{I}_2) = 60$ (1)
Loop BCFG: $1\mathcal{I}_2 + 3(\mathcal{I}_2 - \mathcal{I}_1) + 10(\mathcal{I}_2 - \mathcal{I}_3) = 0$ (2)
Loop CDEF: $10\mathcal{I}_3 + 10(\mathcal{I}_3 - \mathcal{I}_2) = 0$ (3)

The negative signs in the parentheses arise because the loop currents through the 3 Ω and 10 Ω resistors are in opposite directions. There are no sources of electrical energy in loops BCFG and CDEF, hence the right-hand sides are equal to zero.

Equations (1), (2), and (3) may be simplified to:

$$2\mathcal{I}_1 - \mathcal{I}_2 = 20 \quad (1a)$$
$$14\mathcal{I}_2 - 10\mathcal{I}_3 - 3\mathcal{I}_1 = 0 \quad (2a)$$
$$2\mathcal{I}_3 - \mathcal{I}_2 = 0 . \quad (3a)$$

From (3a) we obtain

$$\mathcal{I}_2 = 2\mathcal{I}_3 .$$

which may be substituted into (2a) to give

$$28\mathcal{I}_3 - 10\mathcal{I}_3 - 3\mathcal{I}_1 = 0 .$$

That is

$$6\mathcal{I}_3 = \mathcal{I}_1 .$$

Now we can express $\mathcal{I}_1$ and $\mathcal{I}_2$ in (1a) in terms of $\mathcal{I}_3$,

$$12\mathcal{I}_3 - 2\mathcal{I}_3 = 20$$

or

$$\mathcal{I}_3 = 2\text{ A} .$$

Thus I is 2 A because it is identical with $\mathcal{I}_3$.

As an alternative to the repeated substitution process used here, the method of determinants could have been used.

Q3. Use the principle of superposition to find the unknown currents in the following circuit. Check your answer by using any other method.

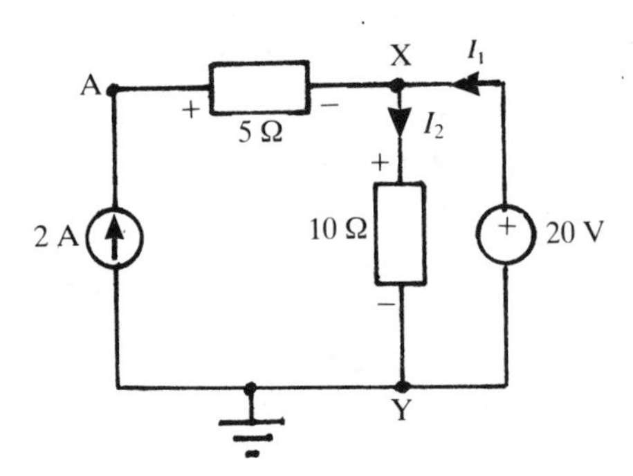

The effect of each source will be treated separately. Then the final answer will be obtained by combining the individual results. Thus with the current source absent, that is the branch AY is open circuited, we obtain

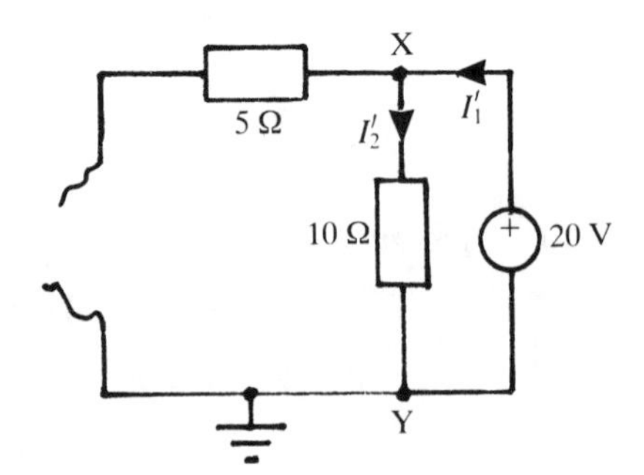

Immediately, we can see that $I_1' = I_2'$ is 2 A, because the current has only the one path to take.

With the voltage source absent, that is short-circuited, we have

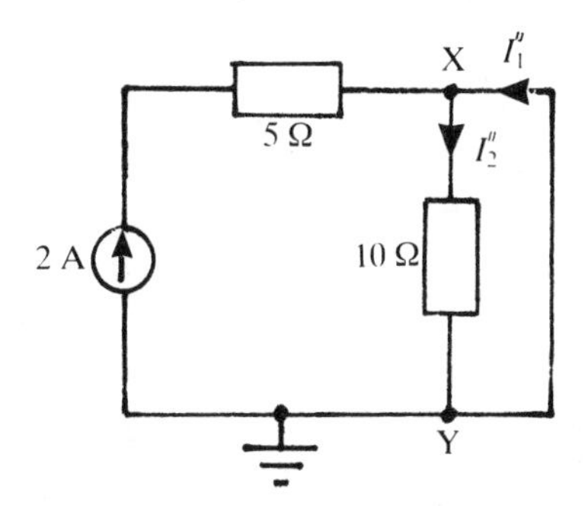

The 2 A current from the current source by-passes the 10 Ω resistor because it prefers the 'practically-zero' resistance path. (In problems of this sort the resistance of the connecting wire is always assumed to be zero unless otherwise stated.) Thus $I_1'' = -2$ A and $I_2'' = 0$. With both sources present, $I_1 = 2 - 2 = 0$ and I_2 is 2 A.

The alternative method which we can use as a check is, in fact, simpler than using the principle of superposition. The potential at node X in the original circuit diagram must be 20 V because it is connected directly to the positive terminal of the voltage source. Hence I_1 must be zero. Using Ohm's law, I_2 through the 10 Ω resistor must equal 20/10 or 2 A.

Q4. Use nodal analysis to determine I_1 in the circuit:

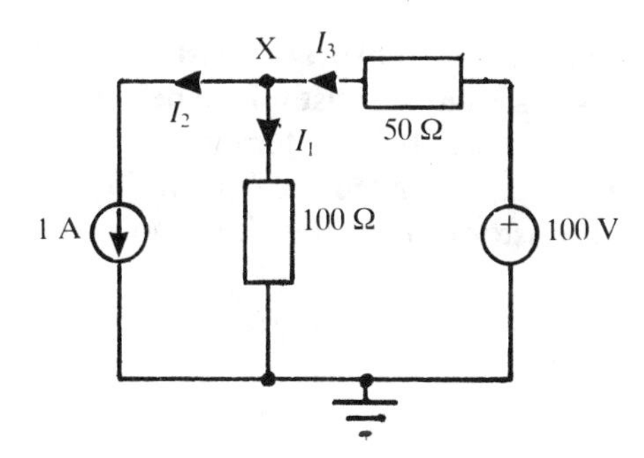

Earth potential is used as the reference potential. The directions of the currents are indicated in the diagram. The current out of the node X towards the current source can only be 1 A. Hence if the potential at node X is V, then applying KCL at X we obtain

$$1 + V_X/100 = (100 - V_X)/50$$

or

$$100 + V_X = 200 - 2\,V_X\ .$$

That is

$$V_X = 33\tfrac{1}{3}\text{ V}$$

Hence

$$I_1 = (33\tfrac{1}{3})/100$$
$$= \tfrac{1}{3}A\ .$$

3.6 THÉVENIN'S THEOREM

Kirchhoff's laws are relatively easy to solve when there are only a few loops in an electrical network. However, as the number of loops increases the arithmetical complexity also increases. Of course, one way out of the difficulty is to write a computer program. This is not really necessary, however, because Thévenin's theorem allows the networks to be analysed quite quickly once the mechanics of applying it are understood. It is an extremely powerful theorem, much valued by electrical engineers, for it applies with ac as well as dc sources. There is a corollary to Thévenin's theorem called Norton's theorem. We shall not consider it because a combination of Thévenin's theorem and a source transformation gives the same result.

Consider a 2-terminal linear network consisting of ideal voltage or current sources and a set of resistors. A load resistor is to be connected across the circuit at the points X and Y. Our aim may be to determine the current flowing through this resistor. For the moment, let us put the circuit inside a black box but with the connections to the load resistor broken at XY, as shown in Fig. 3.7a.

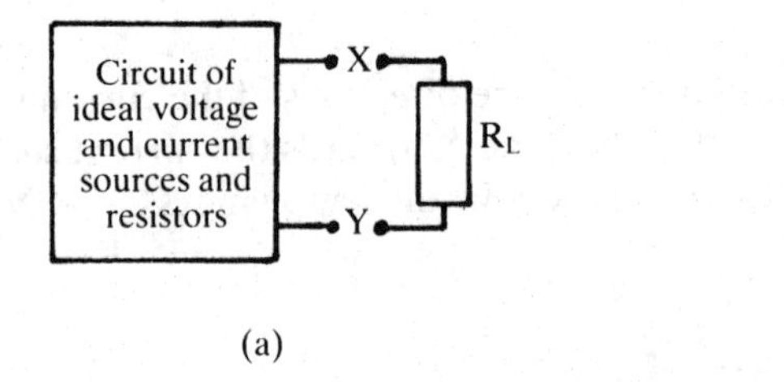

(a)

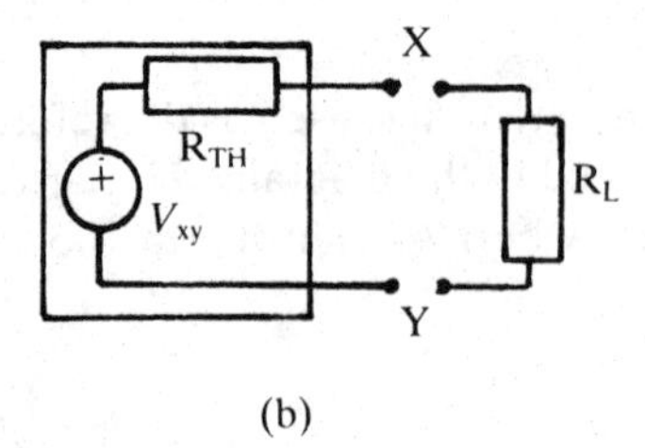

(b)

Fig. 3.7.

STATEMENT: Thvenin's theorem says that all linear 2-terminal networks with sources can be replaced by a single voltage source V_{XY} in series with an equivalent resistance R_{TH} — see Fig. 3.7b V_{XY} is the open-circuit potential difference between X and Y. It can be measured by using a high-resistance voltmeter, which passes very little current. R_{TH} is the net resistance between X and Y when all voltage and current sources are reduced to zero. It can be measured by using some kind of ohmmeter. We shall always assume that the sources are ideal, as stated above, but if this is not the case then the internal resistances of the sources would have to be taken account of.

We shall now do a number of examples to explain how the theorem works in practice.

Worked examples 3.3

Q1. To determine the current flowing through the 1 k Ω load resistor in Q1 of worked examples 3.2.

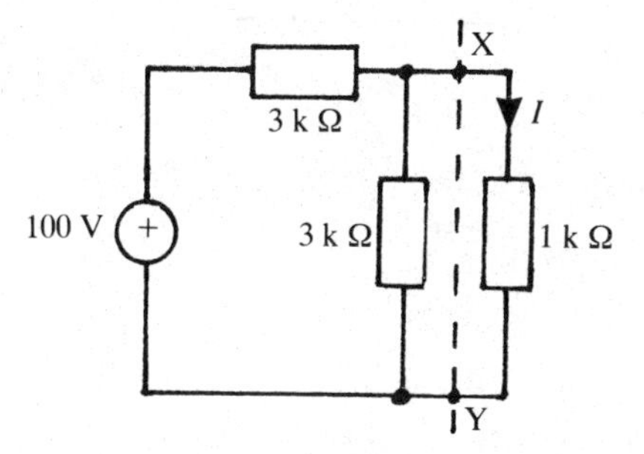

The load resistor is detached from the main part of the circuit (shown by the dashed line). The Thévenin open-circuit pd, V_{XY}, can be worked out fairly easily because one half of the source emf must be dropped across each 3 kΩ resistor. That is, V_{TH} is 50 V.

To obtain R_{TH}, short-circuit the voltage source to give

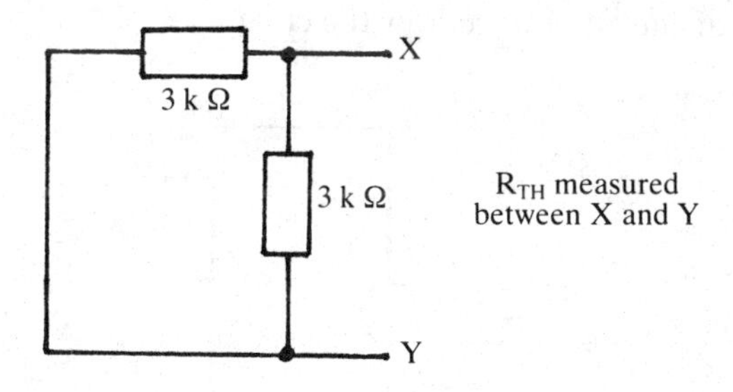

R_{TH} measured between X and Y

Then R_{TH} is calculated from relation (1.25). That is,

$$1/R_{TH} = \tfrac{1}{3}\,\text{k}\Omega + \tfrac{1}{3}\,\text{k}\Omega$$
$$= \tfrac{2}{3}\,\text{k}\Omega$$

from which

$$R_{TH} = 1.5\ \text{k}\Omega$$

Now the thévenised circuit can be constructed, with the load resistor connected:

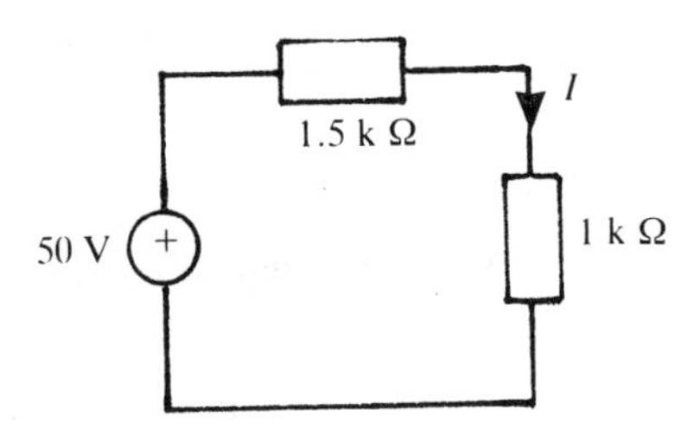

Using Ohm's laws we have

$$\begin{aligned} I &= 50/2\tfrac{1}{2}\ \text{k}\Omega \\ &= 20\ \text{mA} \end{aligned}$$

This is the same value of current as we obtained when using Kirchhoff's voltage law.

The reason for keeping the resistance in kΩ should now be clear. It permits the current to be expressed immediately in milliamperes.

☆ ☆

ASIDE: Before carrying on with some further examples, it is advisable to look at the so-called *potential divider* network. The 3 kΩ resistor-pair in this question formed a potential divider. Let us redraw the circuit.

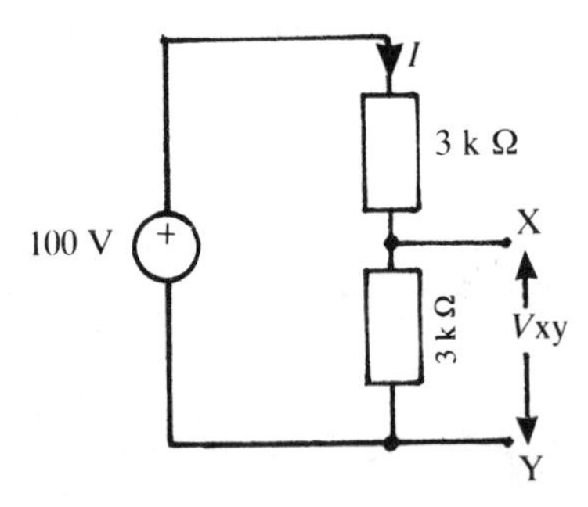

The object is to determine the pd between XY. A current I can flow only through both resistors, so by Ohm's law

$$100 = 6\ k\Omega \times I$$

and

$$V_{XY} = 3\ \mathrm{k}\Omega \times I\ .$$

Therefore, by division,

$$V_{XY}/100 = (3\ \mathrm{k}\Omega \times I)/(6\ \mathrm{k}\Omega \times I)$$
$$= \tfrac{1}{2}\ .$$

and

$$V_{XY} = 50\ \mathrm{V}$$

as we previously found.

Let us now generalize the circuit so that the resistors are R_1 and R_2, and the emf of the source is E — as in Fig. 3.8. Then

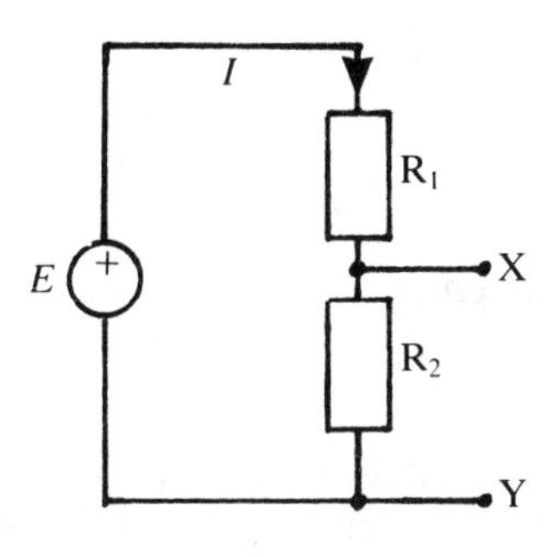

Fig. 3.8.

$$E = I.(R_1 + R_2)$$

and

$$V_{TH} = I.R_2$$

whence

$$V_{TH} = E.R_2/(R_1 + R_2) \tag{3.20}$$

This is the *potential-divider* rule. It will be used extensively when problem-solving with Thevenin's theorem.

☆ ☆

Q2. Determine the current I flowing through the 3 kΩ resistor in the following network.

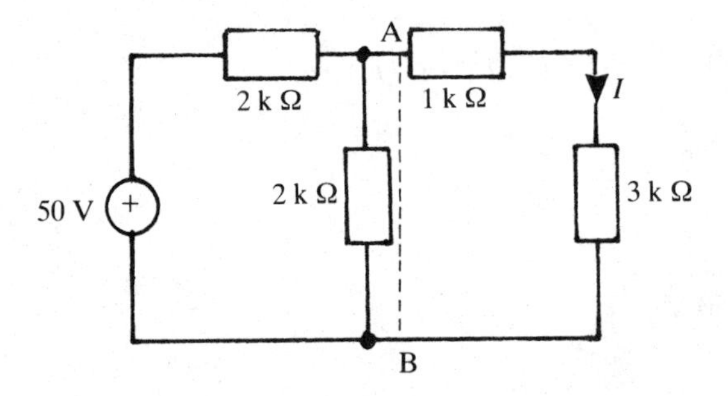

This network is identical with that in Q1; the 1 kΩ and 3 kΩ resistors add up to a single 4 kΩ load resistor. Thévenin's theorem must be applied to the network to the left of AB, treating the network to the right of AB as the load.

$$V_{TH} = V_{AB} = (\tfrac{2}{4}) \times 50 \text{ V} \qquad \text{(potential divider rule)}$$
$$= 25 \text{ V}$$

$$R_{TH} = 2 \text{ k}\Omega \| 2 \text{ km}$$
$$= 1 \text{ k}\Omega \qquad \text{(short-circuiting the voltage source)}$$

The symbol || will be used as shorthand for *in parallel with* and s/c for *short-circuit*.

So the network to the left of AB may be simplified, and reconnected to the remainder of the network to give

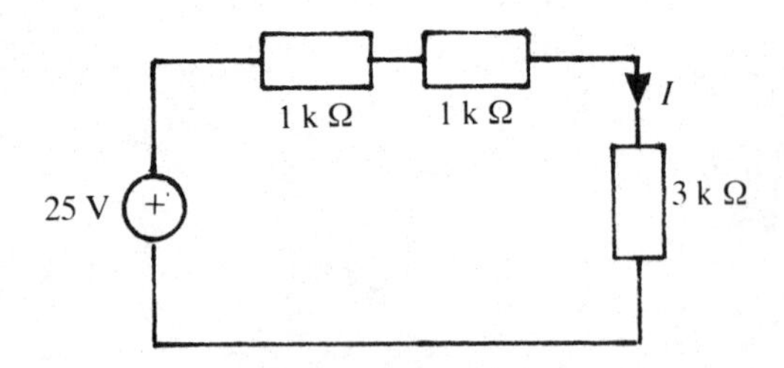

Hence

$$I = 25 \text{ V}/5 \text{ k}\Omega$$

$= 5\text{ mA}$.

Q3. Determine I in the following network:

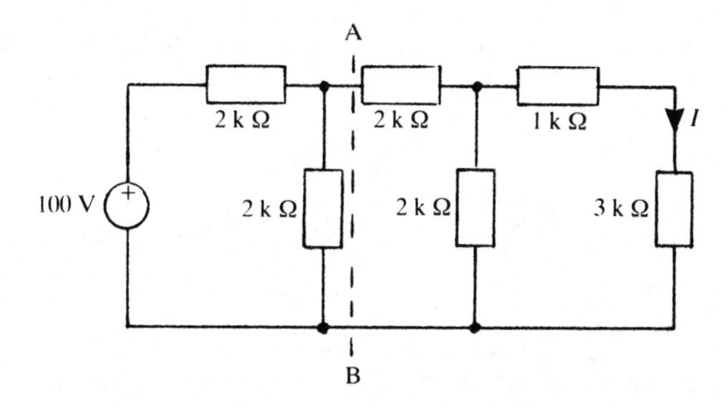

This network is slightly more complicated than Q.2 because of the additional loop. However, the initial procedure is similar. Thévenize the network to the left of AB — again treating the network to the right of AB as the load. Hence

$$V_{TH} = V_{AB} = (2/4) \times 100\text{ V}$$
$$= 50\text{ V}$$

$$R_{TH} = 2\text{ k}\|2\text{ k}\Omega$$
$$= 1\text{ k}\Omega\ .$$

Reconnect the load to give:

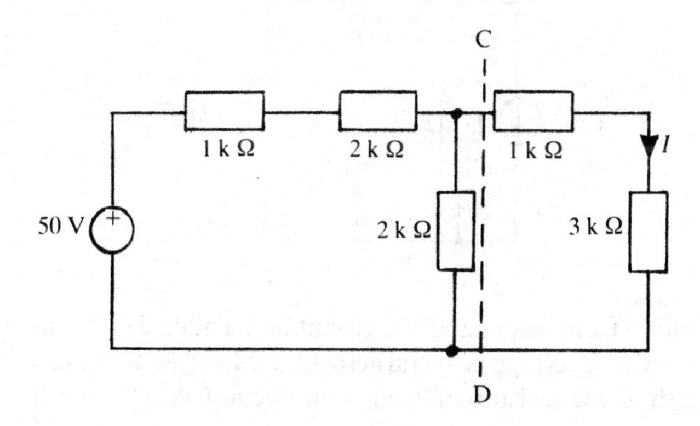

The $1\text{ k}\Omega$ and $2\text{ k}\Omega$ resistors have been separated in order to illustrate the reconnection process in detail. In future such resistors will be shown as a single combined resistor of value $3\text{ k}\Omega$. Now thévenize the network to the left of CD –

$$V_{TH} = V_{CD} = (\tfrac{2}{5}) \times 50 \text{ V}$$
$$= 20 \text{ V}$$
$$R_{TH} = 2 \text{ k}\Omega \| 3 \text{ k}\Omega \ .$$
$$= 1\tfrac{1}{5} \text{ k}\Omega \ .$$

Reconnect the remainder of the network to give the final simplified network:

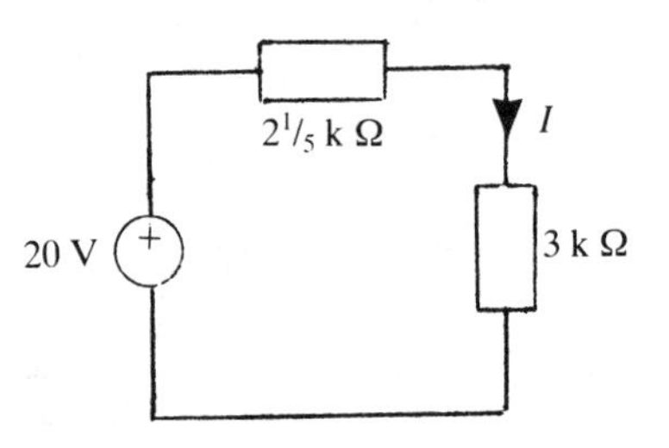

Now I is easily calculate from Ohm's law as

$$I = 20/5\tfrac{1}{5} \text{ k}\Omega$$
$$= 100/26 \text{ mA}$$
$$= 3.8 \text{ mA}$$

Q1 to Q3 indicate how Thévenin's theorem may be applied in stages to aid simplifications of a network. In a network consisting of eight loops there would be seven Thévenizing processes. This is not as horrendous a task as it appears because the calculations are quite simple, as the above examples illustrate.
Q4. Determine I in the following circuit which has a current source.

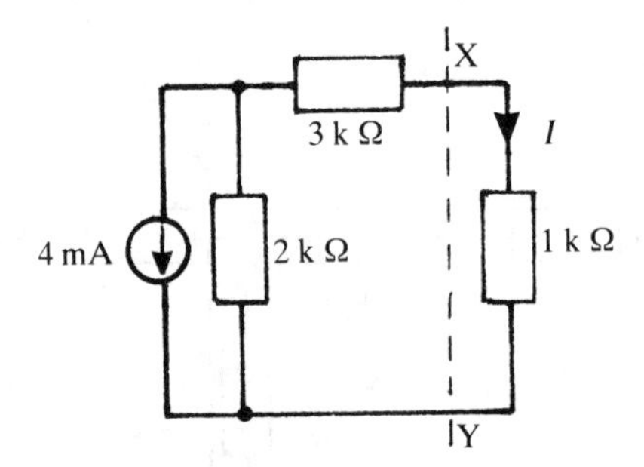

Detach the 1 kΩ load from the rest of the circuit and Thévenize the network to the left of XY. The Thévenin voltage is the pd across the 2 kΩ resistor because no current can flow through the 3 kΩ resistor (it is on open-circuit (o/c)).

$$V_{TH} = V_{XY} = -4 \text{ mA} \times 2 \text{ k}\Omega$$
$$= -8 \text{ V} \ .$$

The negative sign appears because the current flows upwards through the 2 kΩ resistor. This means that the potential at X is less than the potential at Y.

The Thévenin resistance is found by open-circuiting the current source. Hence

$$R_{TH} = 5\ \text{k}\Omega\ .$$

Reconnect the load to give the Thévenin equivalent circuit

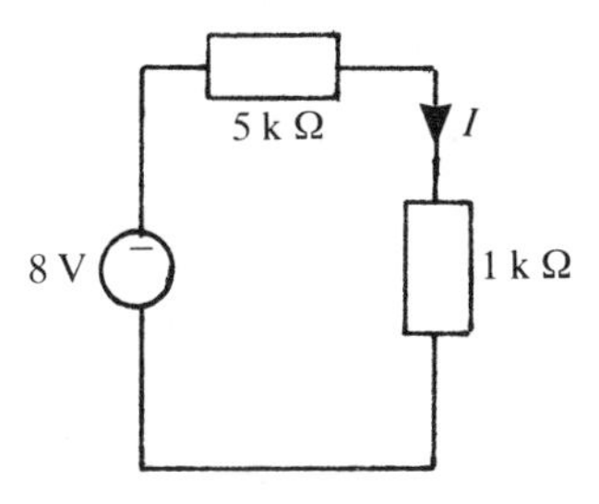

$$I = -8\ \text{V}/6\ \text{k}\Omega$$
$$= -1.33\ \text{mA}$$

The current flows in the opposite direction to that shown in the original diagram.

An alternative and speedier solution is to use a source transformation to convert the current source to a voltage source. The emf of the voltage source is calculated to be 4 mA × 2 kΩ, that is 8 V, with the 2 kΩ resistor appearing in series with it. The final circuit diagram is obtained immediately.

So we see that although Thévenin's theorem can greatly simplify complicated-looking circuits, it is, after all, only a tool, and like any tool needs to be used appropriately. For example, a chisel could possibly be used to drive a screw into a piece of wood, but it is not the most suitable tool to use! It is essential to attempt as many problems as possible. After a while, it will be perfectly obvious which technique gives the most suitable and speediest solution.

Q5. Determine the current I flowing through the 2 kΩ resistor in the following circuit:

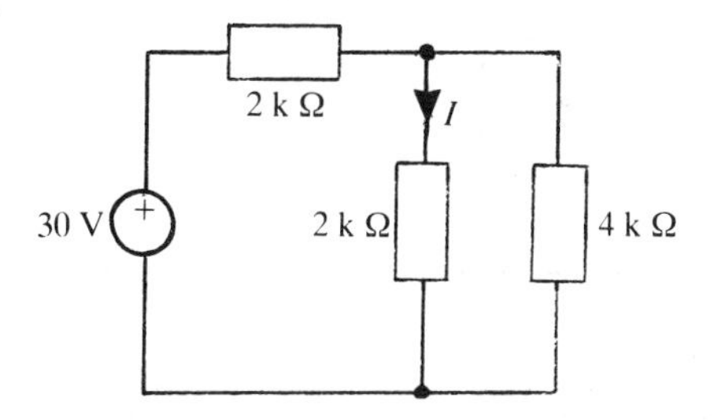

The 2 kΩ resistor cannot be disconnected to act as a load resistance because it is not at the end of the circuit. However, in this case the 2 kΩ and 4 kΩ resistors may be interchanged so that the 2 kΩ resistor becomes the load. It can be detached from the rest of the circuit in the usual way. Thus

$$V = (\tfrac{4}{6}) \times 30 \text{ V}$$
$$= 20 \text{ V}$$

and

$$R = 2 \text{ k}\Omega \| 4 \text{ k}\Omega$$
$$= 1\tfrac{1}{3} \text{ k}\Omega .$$

The circuit now becomes

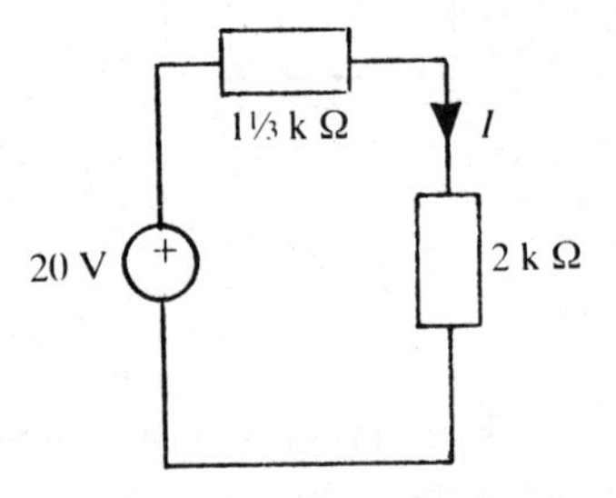

Hence

$$I = 20/3\tfrac{1}{3} \text{ mA}$$
$$= 6 \text{ mA} .$$

Q6. Calculate I in the following circuit:

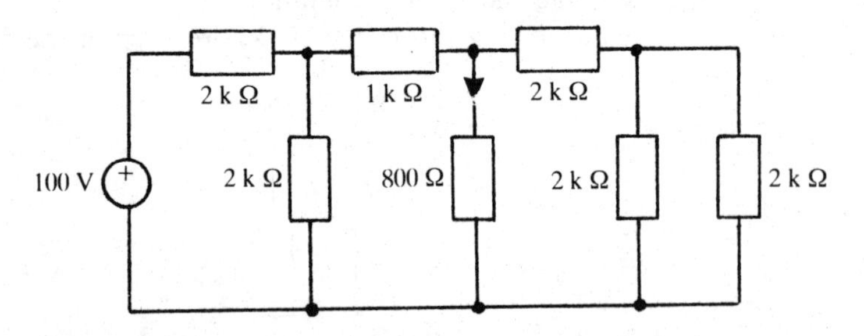

As with Q5, we notice that the resistor through which the current is flowing does not lie at the end of the circuit. However, through a simplification process the form of the circuit can be made equivalent to that of Q4. The two 2 kΩ end-resistors can be combined, using the parallel-resistor rule, to be 1 kΩ. Then we have

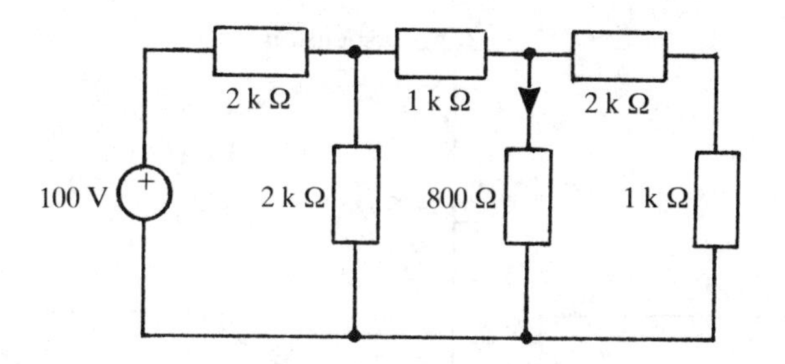

Now I can be calculated by using the procedure adopted in Q5: First, rearrange the positions of the 800 Ω and the 3 kΩ resistors, then Thévenize to, give

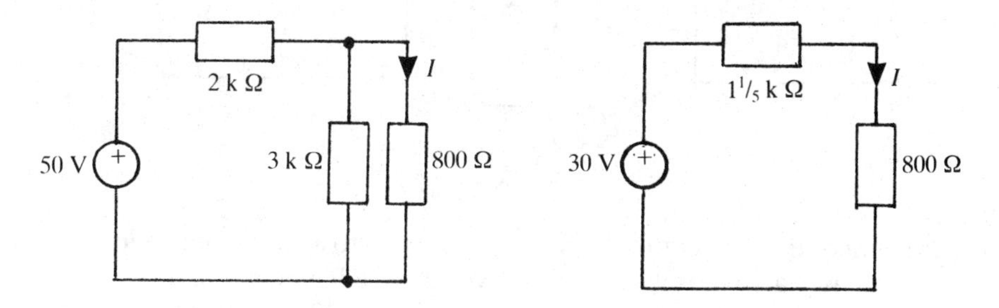

Thus

$$I = 30/2 \text{ mA}$$
$$= 15 \text{ mA} .$$

3.7 DC CURRENT AND VOLTAGE MEASUREMENT

There is no essential difference between an ammeter and a voltmeter, as the latter is a current-measuring device. This statement will become clearer as we proceed into this section.

The basic construction consists of a coil of wire, either single or multi-turns, suspended in a uniform magnetic field (~ 1.0 T), see Fig. 3.9. When a current flows through the coil, a force is produced on both vertical sides, giving rise to a couple. The coil rotates until the moment of the couple is balanced by a couple generated by the suspension itself; this is the elastic torque resulting from the twist in the suspension.

The sensitivity of the device is greatest if a small current ΔI in the coil produces a large deflection $\Delta\theta$. Mathematically, this is stated as:

$$\text{Sensitivity} = \Delta\theta/\Delta I \tag{3.21}$$

In the original d'Arsonval instrument a mirror, attached to the suspension, aided the

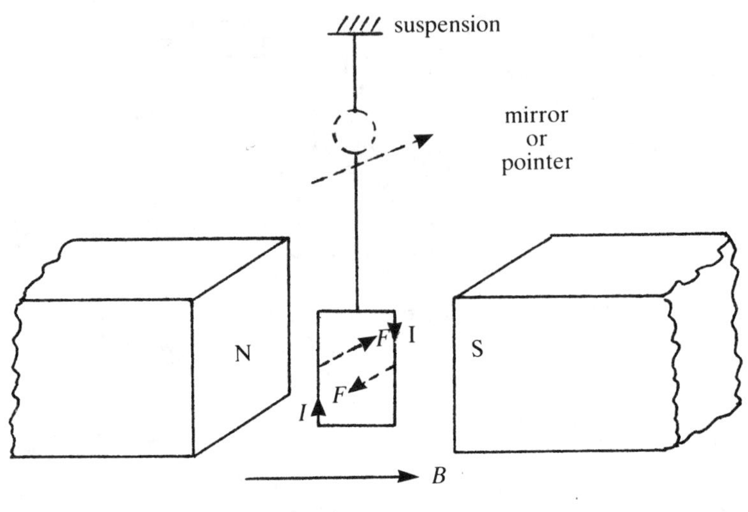

Fig. 3.9.

measurement of the deflection θ by reflecting light on to a graduated scale. Weston's modification uses a pointer which moves over the graduated scale.

Ammeters which measure currents up to about 20 mA pass all this current through the moving coil. Currents greater than this value may possibly burn out the coil. So to avoid this occurring, a *shunt* resistor of suitable value diverts a proportion of the current through itself.

The essential difference between an ammeter and a voltmeter is the resistance of the meter. For currents of 10 mA, or so, the meter resistance is ~1 Ω; for current ~100 μA the resistance is ~1 kΩ. The voltmeter, on the other hand, should have as high a resistance as possible in order to limit the current flowing through it.

The *ideal* ammeter has a coil of zero resistance and a scale graduated uniformly so that the current is directly proportional to deflection. The *real* ammeter, on the other hand, has a finite resistance coil, albeit as low as possible, and an almost linear scale.

3.7.1 Ammeter loading errors

This is an important error caused by the ammeter's coil having a finite resistance. It increases the overall resistance of a circuit with the result that the current is less than it should be. Let us look more closely at this statement. Fig. 3.10a depicts a black box containing a linear circuit. We wish to measure the current through this circuit, I_{WOM}, say, where WOM stands for *without meter*. Fig. 3.10b shows the *actual* set-up in which the ammeter has a resistance R_M. The current actually flowing is I_{WM}, where WM means *with meter*.

The next stage in the analysis is to Thévenize the circuits, see Fig. 3.11a and Fig. 3.11b. Now we see that

$$I_{WOM} = V_{TH}/R_{TH} \tag{3.22}$$

and

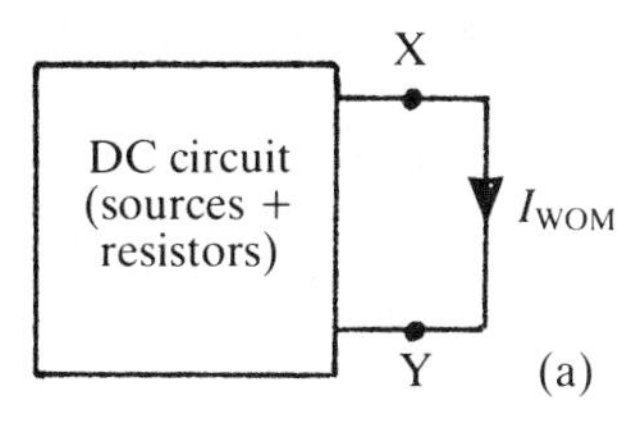

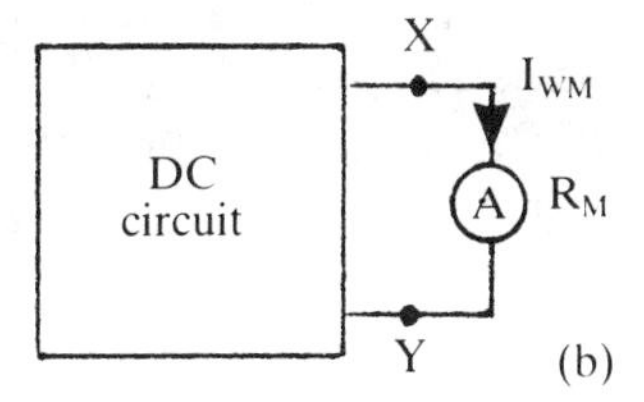

Fig. 3.10.

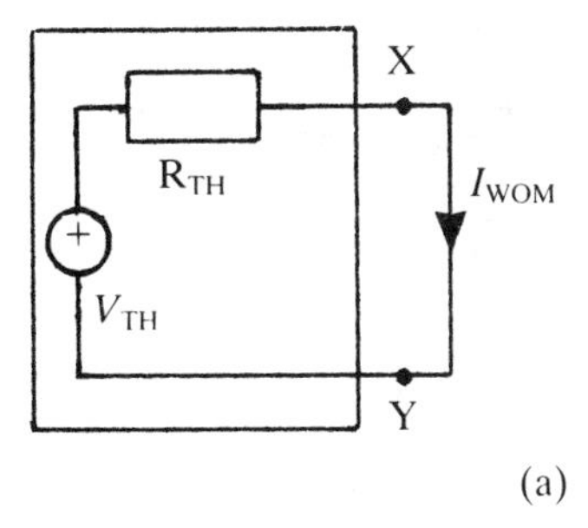

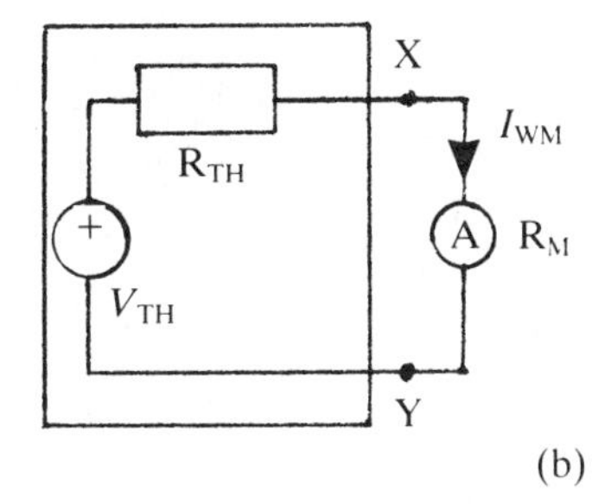

Fig. 3.11.

$$I_{WM} = V_{TH}/(R_{TH} + R_M) \tag{3.23}$$

The ratio of (3.22) and (3.23) is

$$\begin{aligned} I_{WM}/I_{WOM} &= R_{TH}/(R_{TH} + R_M) \\ &= 1/(1 + R_M/R_{TH}) \; . \end{aligned} \tag{3.24}$$

If R_{TH} is known, then I_{WOM} can be determined.

Equation (3.24) defines the ammeter accuracy. On the other hand, the loading error is

$$1 - \text{ammeter accuracy} \; . \tag{3.25}$$

It is not always necessary to calculate the actual value of the ammeter accuracy. There are some simple working rules which are just as valuable.

RULES:

(i) If R_{TH} is greater than a factor of 100 times R_M then the percentage accuracy will be better than 99%;
(ii) If R_{TH} is greater than a factor of 20 times R_M then a 95% accuracy will be achieved.

3.7.1.1 *Ammeter shunts*

An ammeter with a full-scale deflection (FSD) of I_{FS} can be used to measure currents I greater than this if an appropriate resistance R_S is placed in parallel with it — see Fig. 3.12. The current I divides at A, so that by Kirchhoff's current law, the current I_S

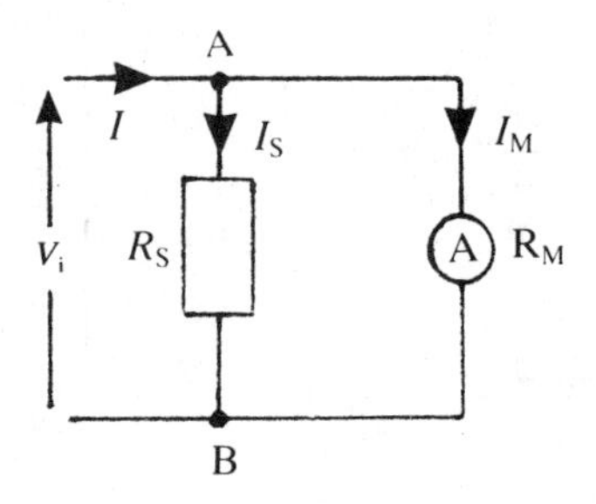

Fig. 3.12.

flowing through the shunt is $(I - I_M)$. The pd between AB must be the same across R_S and R_M. That is,

$$I_S.R_S = I_M.R_M$$

or

$$I_S = I_M.R_M/R_S \ .$$

Then

$$\begin{aligned} I &= I_M + I_S \\ &= I_M\{1 + R_M/R_S\} \ . \end{aligned} \tag{3.26}$$

Now the ammeter scale can be recalibrated in terms of I if I_M, R_S, and R_M are known. Of course, the maximum current that can be measured occurs for I_M equal to I_{FS}.

The input resistance of the shunted ammeter is defined by

$$V_i/I = R_S \| R_M = R_S R_M/(R_S + R_M) \ . \tag{3.27}$$

This is the resistance that any circuit to the left of the shunted ammeter will sense. It determines how much current will be drawn from that part of the circuit.

3.7.1.2 *The ammeter as a voltmeter*

The procedure is fairly straightforward. Place a resistor R of known value in series with the ammeter (see Fig. 3.13). Then calibrate the scale of the ammeter in terms of the potential difference V. This can be done if the current is accurately known. Hence

$$V = I(R + R_M) \tag{3.28}$$

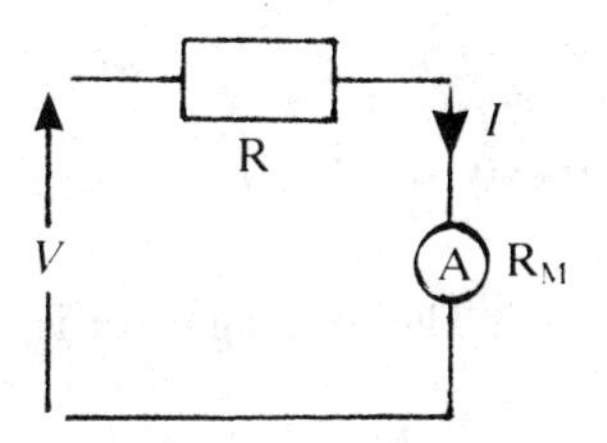

Fig. 3.13.

I can be varied from 0 to I_{FS}, at which value the pd will have its maximum value V_{FS}. The input resistance R_i of the voltmeter is V_{FS}/I_{FS} and the voltmeter sensitivity is $1/I_{FS}$. The latter has units of ohm volt^{-1}.

3.7.2 Voltmeter loading errors

We wish to determine the pd between X and Y in Fig. 3.14a. For convenience we shall call it V_{WOM}. It is also the pd across the resistor R. When a voltmeter is connected across R, see Fig. 3.14b, the pd becomes V_{WM} because the input resistance of the meter is now in parallel with R. On Thévenizing Fig. 3.14a and 3.14b

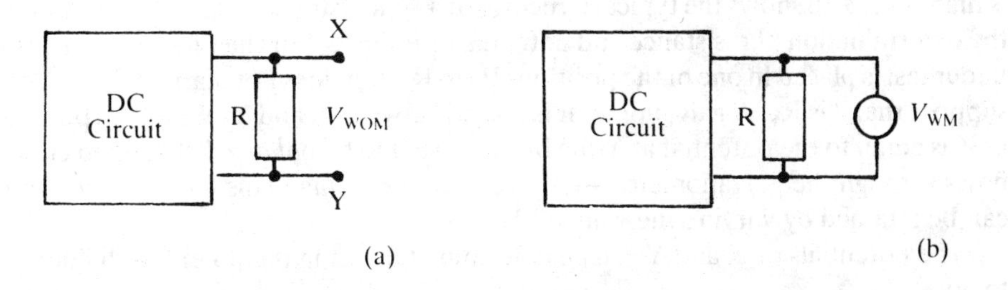

Fig. 3.14.

we obtain Fig. 3.15a and 3.15b. R_{TH}, is R in parallel with the resistive part of the dc circuit. Now we see that, because V_{TH} is equal to V_{WOM},

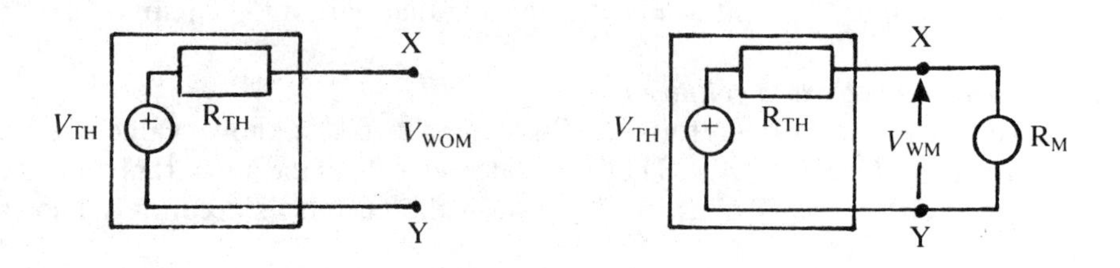

Fig. 3.15.

$$V_{WM}/V_{WOM} = \{R_M/(R_{TH} + R_M)\} \tag{3.29}$$

This is the voltmeter accuracy. The loading error is

(1 — voltmeter accuracy) .

Once again, there are some working rules which allow the degree of accuracy to be ascertained.

RULES:
(i) If R_M is greater than a factor of 100 times R_{TH} then the voltmeter accuracy is better than 99%;
(ii) If R_M is greater than a factor of 20 times R_{TH} then the accuracy is better than 95%.

3.8 DC WHEATSTONE BRIDGE

DC bridge techniques are extremely useful because they do not suffer from the loading errors of ammeters and voltmeters. Hence the accuracy attainable with them is high. Fig. 3.16 shows the typical structure of a Wheatstone bridge which is used for the determination of resistance and potential difference. For the former, the resistor under test is placed in one of the positions R_1 to R_4—it does not matter which. Let us suppose that it is R_1. A galvanometer is placed between X and Y. When the potential at X is equal to the potential at Y the bridge is said to be *balanced*. Then, no current flows through the galvanometer — it gives a *null* reading. This balanced condition can be attained by varying the value of R_2.

The potentials of X and Y can be determined by using the potential-divider rule to give

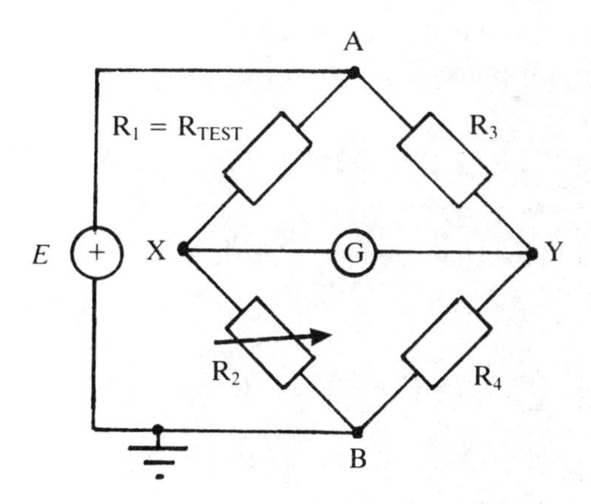

Fig. 3.16.

$$V_X = ER_2/(R_{TEST} + R_2) \tag{3.30}$$

and

$$V_Y = ER_4/(R_3 + R_4) \tag{3.31}$$

Equations (3.30) and (3.31) can be equated when the bridge is balanced to give

$$R_{TEST}/R_2 = R_3/R_4 \; . \tag{3.32}$$

Now the galvanometer can be removed or a load resistor inserted in its place without altering the balance condition.

High-value resistors, $> 1\ \text{M}\Omega$, are not measured with this bridge. Joule heating results in a rise in temperature, with a consequent change in resistance. Such resistors are generally measured by using an experimental arrangement entailing the rate of discharge of a capacitor.

If the Wheatstone bridge is unbalanced, that is $V_X \neq V_Y$, it may be Thévenized to give (see Fig. 3.17). Here, the Thévenin voltage V_{TH} is equal to the open-circuit pd

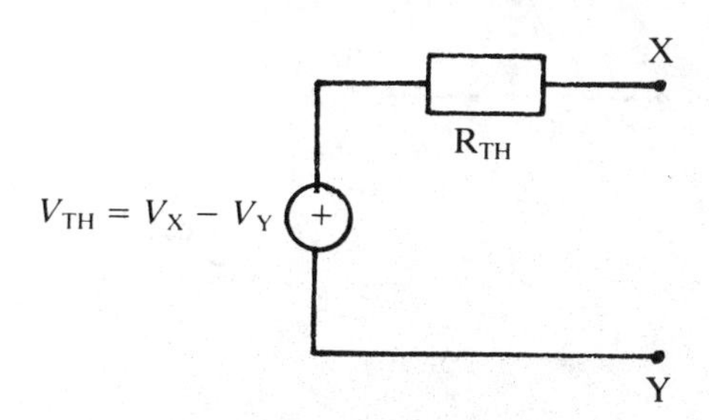

Fig. 3.17.

between X and Y, which is the absolute value of the difference between (3.31) and (3.32). The Thévenin resistance R_{TH} equals

$$R_1\|R_2 + R_3\|R_4 = R_1R_2/(R_1 + R_2) + R_3R_4/(R_3 + R_4)\ .$$

If a load resistor R_L is placed between X and Y the current flowing through it will be equal to

$$I_L = V_{TH}/(R_{TH} + R_L) = (V_X - V_Y)/\{R_1\|R_2 + R_3\|R_4 + R_L\}\ . \tag{3.33}$$

3.9 THE POTENTIOMETER

The traditional potentiometer with its linear 1 m board is no longer the familiar sight in scientific laboratories that it was. Instead, the wire is arranged in a helical coil and occupies far less space. Potentiometric applications are very important, for example, in comparing resistances and emf's, and for determining pH values and ionic concentrations.

The basic potentiometric arrangement consists of a potential-divider network. Fig. 3.18 depicts its use in the comparison of an unknown emf with a standard emf.

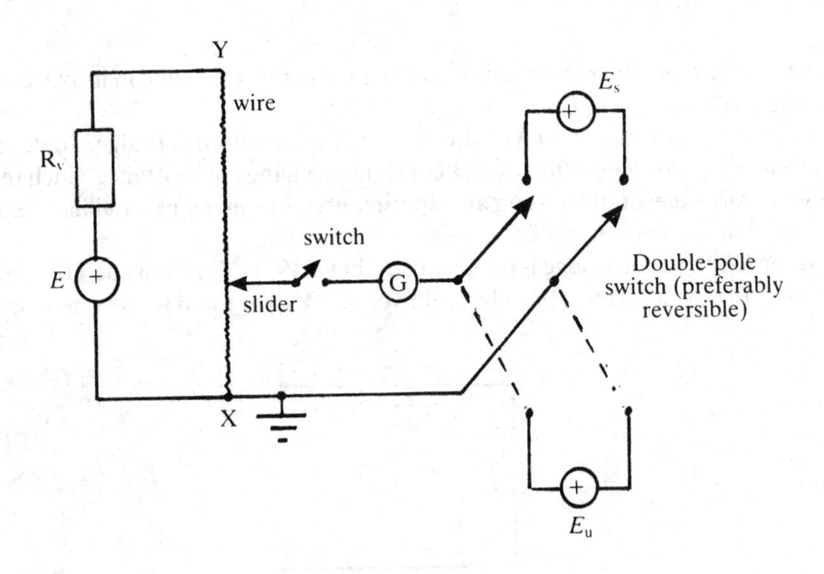

Fig. 3.18.

With the double-pole switch arranged so that the standard source of emf E_S is included in the circuit, the slider is moved along the wire until the galvanometer gives a null reading. When this occurs, the potential of the slider is equal to the potential of the positive electrode of the standard, E_S.

In section 1.3.1 we learnt that resistance is directly proportional to the length of a conductor. Hence, if balance occurs with the slider at A, the ratio XA/XY must equal R_{XA}/R_{XY}. Now by the potential-divider rule,

$$\begin{aligned} V_A = E_S &= V_W.R_{XA}/R_{XY} \\ &= V_W.(XA/XY) \end{aligned} \tag{3.34}$$

where $V_W(=E-IR_V)$ is the pd across the wire. Now switch-in the unknown emf. Determine the new balance point when the galvanometer again indicates a null reading. Suppose that this occurs with the slider at a point B. Then V_B equals the potential of the positive electrode of the unknown, E_U. By comparison with (3.34), the potential at B is given by

$$\begin{aligned} V_B = E_U &= V_W.R_{XB}/R_{XY} \\ &= V_W.(XB/XY)\ . \end{aligned} \tag{3.35}$$

Thus

$$E_S/E_U = XA/XB\ . \tag{3.36}$$

It is important that as much of the slide wire as possible is used whilst determining the balance positions. Errors in the exact determinations of points A and B can be very large, possibly $\sim \pm 25\%$ or so, when they lie close to end X. Another precaution which will help to eliminate any error due to the wire having a non-uniform cross-section is to reverse the direction of the current. Then the mean of the two balance lengths should be used in any subsequent calculations.

3.10 ANALYSING L–C–R CIRCUITS

3.10.1 Charging and discharging of an R–C circuit

Fig. 3.19 shows the basic circuit that we need to analyse. With the switch in position

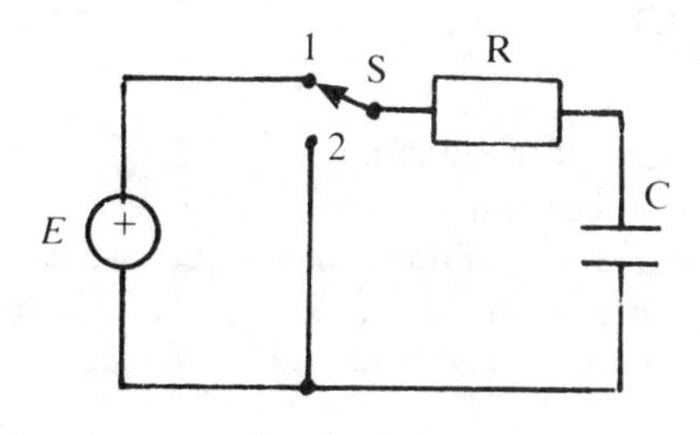

Fig. 3.19.

1, the resistor R and capacitor C are in series with the voltage source. A current flows in this part of the circuit and the capacitor, which we will assume to be uncharged to begin with, builds up charge on its plates. As a result, the pd between its plates increases until, in the limit, it equals the emf of the voltage source; the current gradually decreases to zero. Applying KVL, we have

$$E - v_C - v_R = 0 \tag{3.37}$$

v_R and v_C are the pd's across the resistor and the capacitor, respectively. Lower-case letters are used to denote these because the pd's vary with time. If at some time t, the current is i, then (3.37) becomes

$$E - iR - q/C = 0\ . \tag{3.38}$$

Differentiating (3.38) with respect to time t gives

$$R\ \mathrm{d}i/\mathrm{d}t + i/C = 0 \tag{3.39}$$

as E is a constant. Rearrange (3.39) and integrate the expression, i.e.

$$\int \mathrm{d}i/i = -\int \mathrm{d}t/RC$$

or

$$\ln\ i = -t/RC + \text{constant of integration } A'\ .$$

If A′ is replaced by ln A, where A is another constant, then on taking antilogs we obtain

$$i = A\ \exp(-t/RC)\ . \tag{3.40}$$

RC is called the *time constant* of the circuit.
UNITS: seconds (s) or minutes (min.)

What is A? As the capacitor is initially uncharged, we can say that *immediately* after the instant that the switch is thrown into position 1, there is no pd between the capacitor plates. We shall indicate this time as $t = +0$. So

$$E(t = +0) = iR$$

and

$$i(t = +0) = E/R \ . \tag{3.41}$$

We see from (3.40) that $i(t = +0) = A$. Hence,

$$A = E/R$$

and (3.40) is

$$i = (E/R)\exp(-t/RC) \ . \tag{3.42}$$

We now see that as $t \rightarrow \infty$, $i \rightarrow 0$, as required, since the capacitor will then be fully charged.

Rearranging (3.37) we have

$$\begin{aligned} v_C &= E - Ri \quad \text{(at any time } t\text{)} \\ &= E\{1 - \exp(-t/RC)\} \end{aligned} \tag{3.43}$$

The graph of v_C vs time is depicted in Fig. 3.20 as the *charging* curve. From a theoretical viewpoint, the capacitor requires an infinite time to become fully charged. In practice, however, it is almost fully charged after a few time-constants.

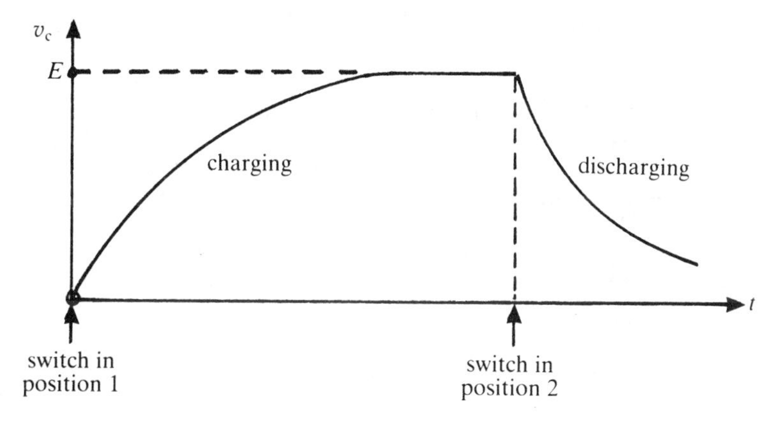

Fig. 3.20.

Now that the capacitor is fully charged, let us now throw the switch to position 2. By KVL, after a time t has elapsed,

$$v_R + v_C = 0 \ . \tag{3.44}$$

That is

$$Ri + q/C = 0 \ . \tag{3.45}$$

Differentiating with respect to time t gives

$$R \ \mathrm{d}i/\mathrm{d}t + i/C = 0$$

or

$$\mathrm{d}i/i = -\mathrm{d}t/RC \ .$$

Integration gives

$$i = B \exp(-t/RC) \ . \tag{3.46}$$

The constant of integration B can be determined in a similar way to A. *Immediately* after the instant when the switch is thrown to position 2, the pd across the capacitor is E. Once again, we shall call this instant $t = +0$. So, substituting E for v_C in (3.44), we can write

$$E + Ri(t = +0) = 0$$

where $i(t = +0)$ is the current flowing at this instant. Hence

$$i(t = +0) = -E/R \ . \tag{3.47}$$

But (3.46) informs us that at this time

$$i(t = +0) = B \ .$$

So at any time t after the switch is thrown to position 2 the current flowing in this part of the circuit is given by

$$i = -(E/R) \exp(-t/RC) \tag{3.48}$$

and

$$\begin{aligned} \upsilon_C &= -Ri \quad \text{(from (3.44))} \\ &= E\exp(-t/RC)\ . \end{aligned} \tag{3.49}$$

The variation of υ_C with time t is also shown in Fig. 3.20.

Worked examples 3.4

Q1. A circuit consists of a resistor of $10^{10}\,\Omega$ in series with a capacitor of $10\,\mu$F and a 1 kV voltage source. The capacitor is initially uncharged. After the capacitor has been fully charged, it is allowed to discharge itself through the resistor. Determine the pd across the capacitor after 24 hours.

The time constant RC is

$$10^{10} \times 10^{-5} = 10^5\ \text{s}\ .$$

Use (3.49) to determine υ_C, with $E = 1000$ V and $t = 24 \times 3600$ s. Then

$$\begin{aligned} \upsilon_C &= 1000\exp(-24 \times 3600/100000) \\ &= 1000\exp(-0.864) \\ &= 420\ \text{V}\ . \end{aligned}$$

So even after one day the capacitor still retains a substantial amount of charge on its plates. This example illustrates the care that must be taken with circuits which possess long time constants. Otherwise a serious, perhaps fatal, electrical shock may be incurred.

Q2. Determine the behaviour of the output pd in the following circuit. Let the pd across both R and C be $\upsilon_I(t)$ and the pd across the resistor alone be $\upsilon_o(t)$. It is assumed that the pd's are both time-dependent in order to obtain a general result. We shall suppose that $\upsilon_I(t)$ is applied to the circuit at time $t = 0$. Then we may write

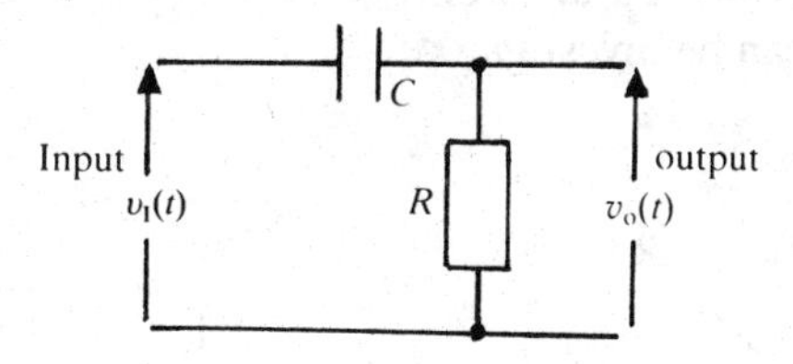

$$\begin{aligned} \upsilon_I(t) &= q/C + \upsilon_o(t) \\ &= (1/C)\int_{-\infty}^{t} i\,dt + Ri\ . \end{aligned}$$

The lower limit of integration is taken as $-\infty$ and not 0 in order to take account of the fact that the capacitor might be charged initially. For example,

$$(1/C)\int_{-\infty}^{t} i\,\mathrm{d}t = (1/C)\int_{-\infty}^{0} i\,\mathrm{d}t + (1/C)\int_{0}^{t} i\,\mathrm{d}t$$

$$= V' + (1/C)\int_{0}^{t} i\,\mathrm{d}t\,.$$

V' is, therefore, the constant pd between the plates of the capacitor before applying $v_I(t)$. We shall assume that this is not the case (that is, $V' = 0$) and, further, that $v_o \ll v_I$. Then

$$v_I = (1/C)\int_{0}^{t} i\,\mathrm{d}t$$

$$= (1/RC)\int_{0}^{t} v_o\,\mathrm{d}t$$

or

$$v_o = RC\,\mathrm{d}v_I/\mathrm{d}t\,.$$

The output voltage signal is a differentiated form of the input signal. Now some specific input signals can be investigated:

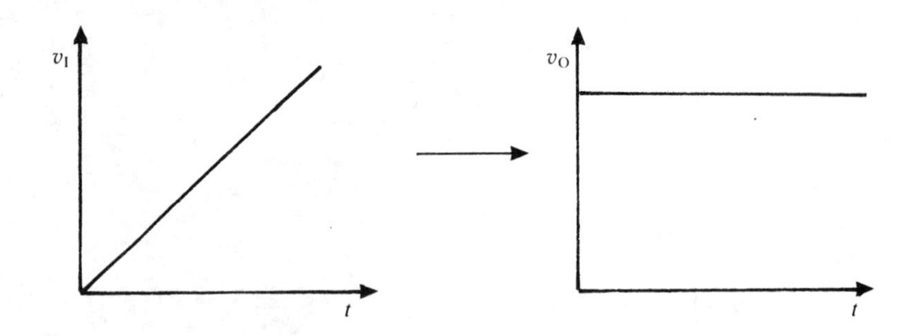

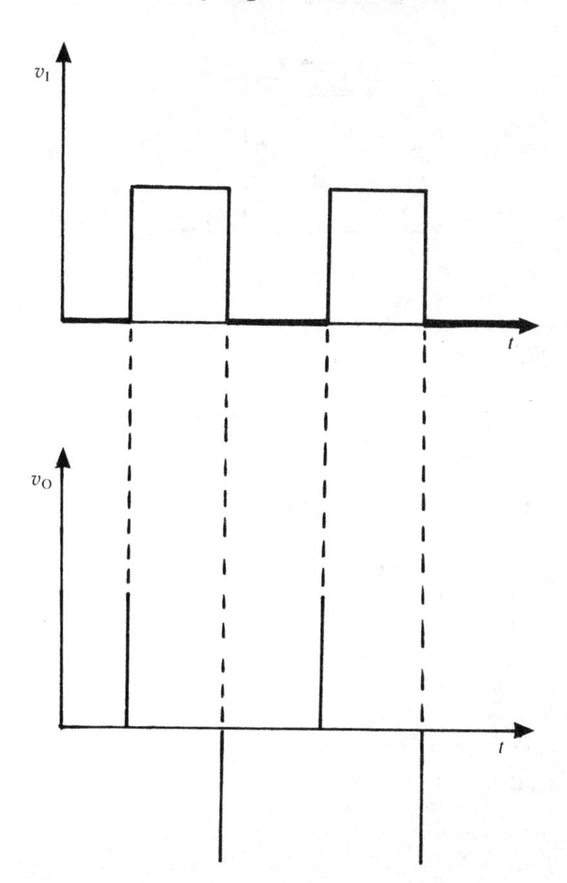

The spikes can act as timing pulses for counting purposes. In practice, the input signal is likely to be a distorted form of square wave with the result that the output signal looks like

No output will be obtained in the case of a dc input signal because the differential coefficient of a constant is zero. For obvious reasons the circuit is called a *differentiator*.

Q3. The output to the previous question is now taken across the capacitor. Investigate the form of the output as a function of time. Now we have

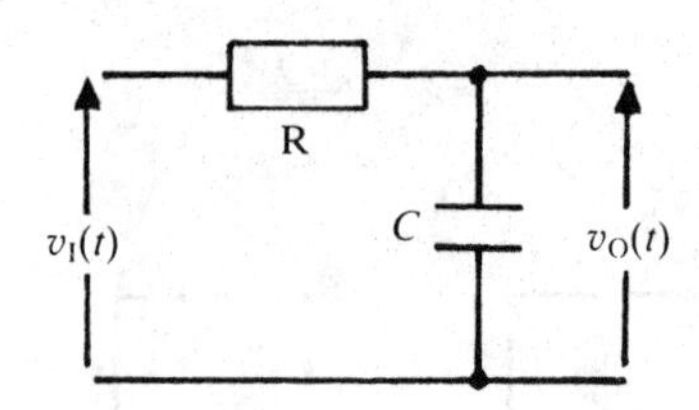

$$\begin{aligned} v_I &= Ri + v_o \\ &= Ri + q/C \\ &= Ri + (1/C)\int_{-\infty}^{t} i\mathrm{d}t \end{aligned}$$

Once again, assume that $v_o \ll v_I$, and that the capacitor is unchanged initially, then

$$v_I = Ri \text{ and } \int_{-\infty}^{t} i\mathrm{d}t = \int_{0}^{t} i\mathrm{d}t \ .$$

Therefore,

$$v_I = RC\mathrm{d}v_o/\mathrm{d}t$$

and

$$v_o = (1/RC)\int_{0}^{t} v_I\mathrm{d}t$$

The output voltage signal is now an integrated function of the input signal — hence the name of the circuit: the *integrator*. One important point to note about the integral — Fundamentally, it determines the area beneath the curve representing v_I and the time axis. With this information it should be now possible to determine the form of the output signal. Consider a dc input voltage:

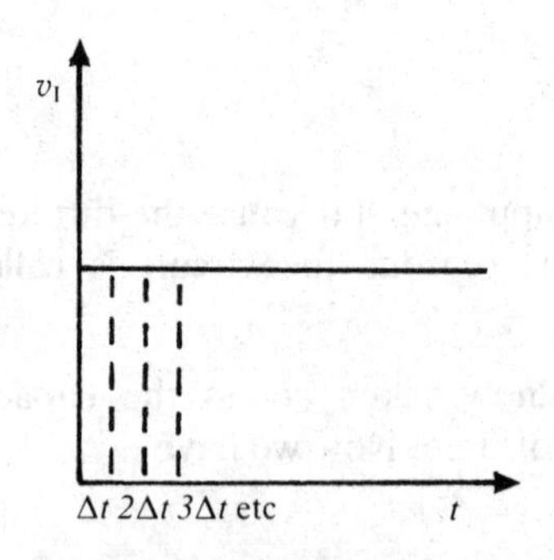

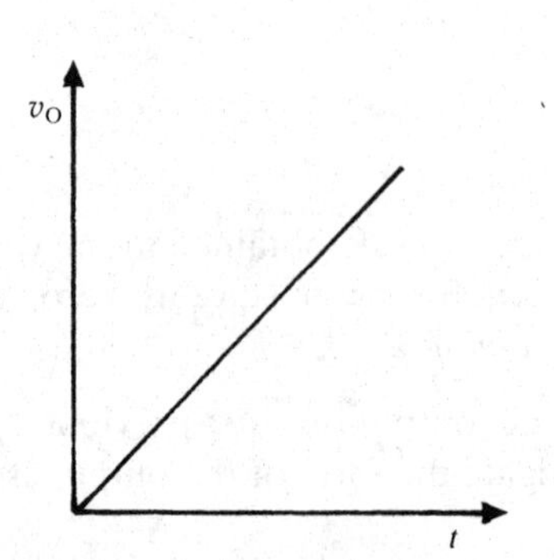

The output is called a RAMP function. To obtain it, divide the input dc voltage variation into a number of rectangles of infinitesimal width Δt. The area of the first rectangle is $v_I \Delta t$. This value divided by the scaling factor RC gives v_o. So plot this value on the second graph at $t = \Delta t$. The area of the first two rectangles on the input graph is $2v_I \Delta t$. Divide by RC, and plot the resulting value on the second graph at $t = 2\Delta t$. Repeat this procedure until all elementary rectangles under the input dc voltage signal have been considered.

In the case of a rectangular wave train, the output has a staircase appearance:

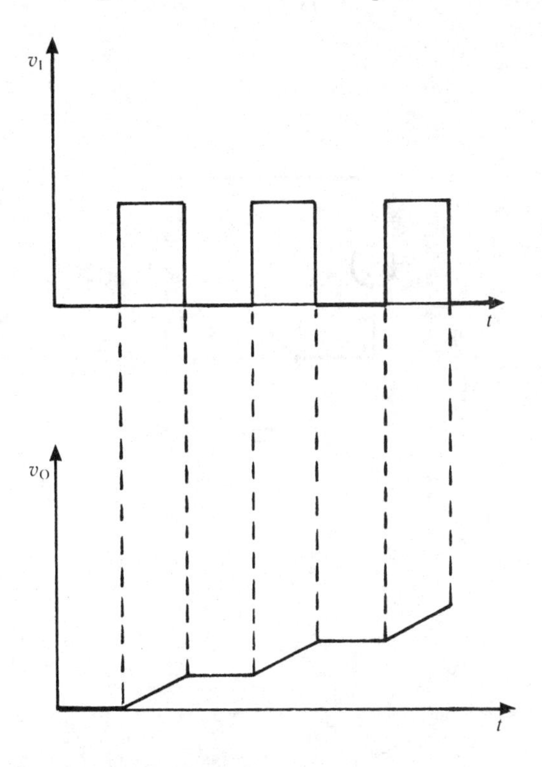

If the duration of the wave pulses are short, then the output steps will be correspondingly short and the overall appearance will be similar to a continuous RAMP function. This result is extremely interesting because we have found a way of converting a digital display to an analogue display.

3.10.2 The R–L circuit

Fig. 3.21 shows an inductor connected across a dc voltage source. The magnitude of the pd across the inductor is $L\, di/dt$ (section 1.4). So if the current is constant, di/dt is zero, and the inductor appears as a *short-circuit* to dc currents.

Now let us modify this circuit by introducing a switch S, as in Fig. 3.22. Immediately after S is closed, the whole of the emf E appears across the inductor. As in section 3.10.1 we shall indicate this instant as $t = +0$, then

$$E = L\, di/dt\big|_{t=+0} \qquad (3.50)$$

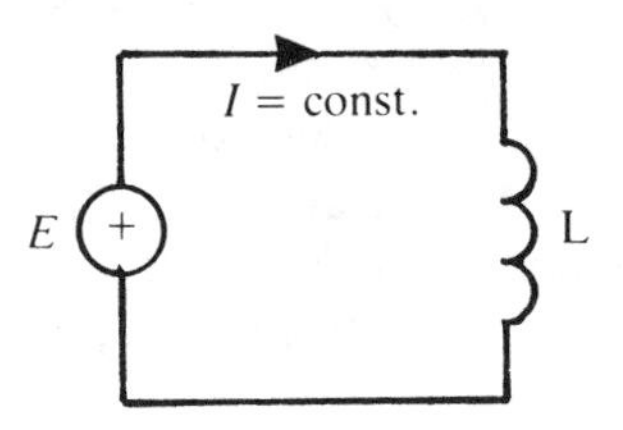

Fig. 3.21.

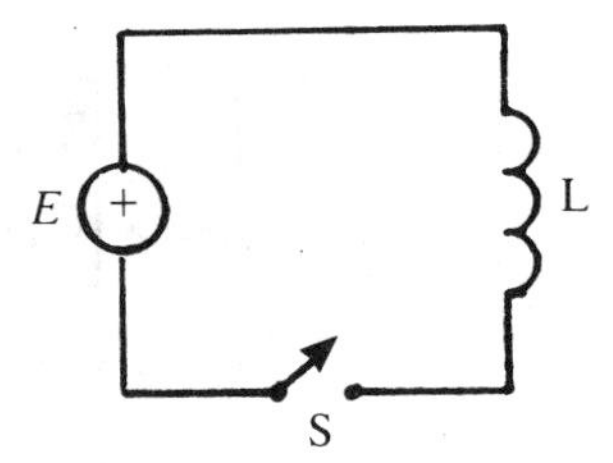

Fig. 3.22.

or

$$\mathrm{d}i/\mathrm{d}t|_{t=+0} = E/L \ .$$

Integrating this gives

$$i = Et/L + \text{constant} \ .$$

However, as no current flows at time $t = +0$, the constant of integration is zero, and

$$i = Et/L \ . \tag{3.51}$$

A graph of i vs t is linear. It passes through the origin and has the slope E/L.

According to (3.51) the current will increase indefinitely at the rate E/L. Why does this not happen in practice? The reason is that we have ignored the resistance of the coil. Therefore, Fig. 3.22 must be modified further to include a resistive component, see Fig. 3.23. Once again, immediately after S is closed ($t = +0$) the whole of the emf E appears across the inductor because the current has had no time to grow. By KVL,

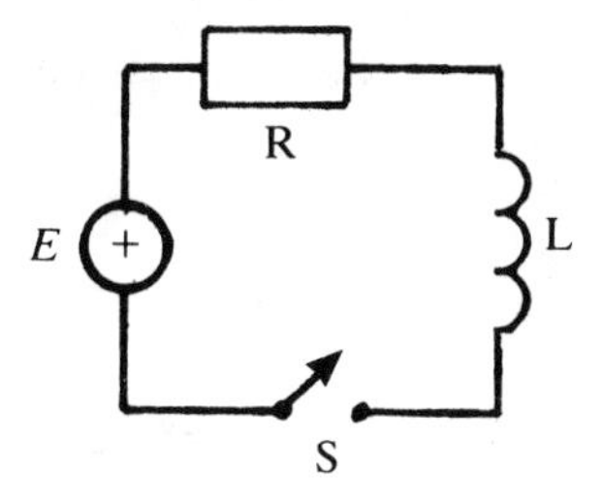

Fig. 3.23.

$$E - v_{\mathrm{L}} - v_{\mathrm{R}} = 0$$

and

$$E = v_{\mathrm{L}} + v_{\mathrm{R}}$$
$$= L\ \mathrm{d}i/\mathrm{d}t + \mathrm{R}i\ . \qquad (3.52)$$

On transposing terms in (3.52) we have

$$\mathrm{d}i/\{(E/R) - i\} = (R/L)\mathrm{d}t\ .$$

Integrating between the limits: $i = 0$ at $t = +0$ and $i = i'$ at $t = t'$ gives

$$-\ln\{(E/R - i'\} = (R/L)t' - \ln(E/R)$$

or

$$i' = (E/R)\ \{1 - \exp(-Rt'/L)\} \qquad (3.53)$$

There is no real need for the primes. These are used to indicate specific values of current and time only.

Fig. 3.24 depicts the variation of current with time. The slope of the curve at the origin is equal to E/L. Why? We originally stated that at time $t = +0$ the whole of the source emf E appears across the inductor. Therefore,

$$L\ \mathrm{d}i/\mathrm{d}t|_{t=+\mathrm{o}} = E$$

and

$$\mathrm{d}i/\mathrm{d}t|_{t=+\mathrm{o}} = E/L\ . \qquad (3.54)$$

As t increases, i tends to the value E/R. This result can be obtained by putting $t = \infty$ in (3.53).

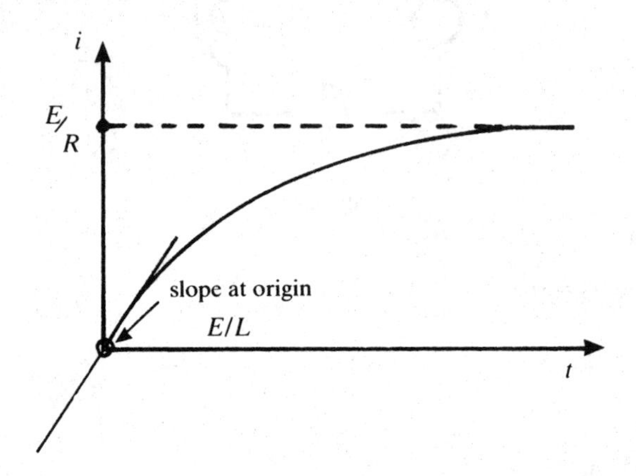

Fig. 3.24.

Next, let us consider current decay in an inductor using Fig. 3.25. Let the switch S

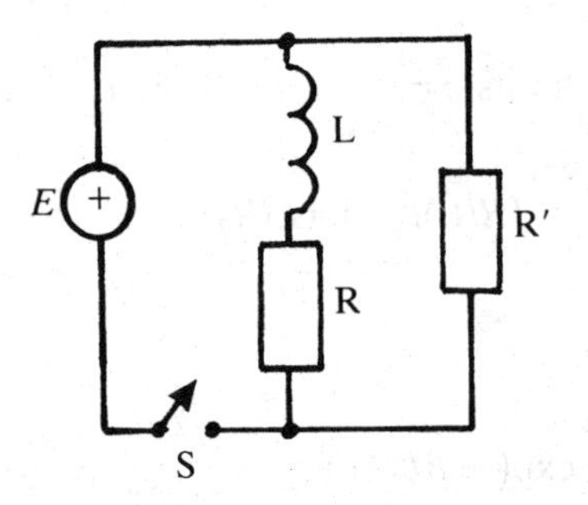

Fig. 3.25.

be closed for a sufficiently long time, so that a constant current I flows through the inductor. If the inductor possessed no resistance then this branch would short-circuit the resistor R'. R is included, therefore, to simulate the resistance of the coil.

Open S. At this instant, $t = +0$ and $i = I$. Applying KVL to the L–R–R′ loop gives

$$v_L + v_{R''} = 0$$

or

$$L\,\mathrm{d}i/\mathrm{d}t + R''i = 0 \tag{3.55}$$

where R'' is $(R + R')$. Transposing (3.55) we have

$$\mathrm{d}i/i = -(R''/L)\mathrm{d}t\ .$$

Integrate within the limits: $i = I$ at $t = +0$ and $i = I'$ at $t = t'$ to give

$$i' = I\exp[-(R''t'/L)] \tag{3.56}$$

(3.56) indicates that the current varies with time as shown in Fig. 3.26. L/R is called

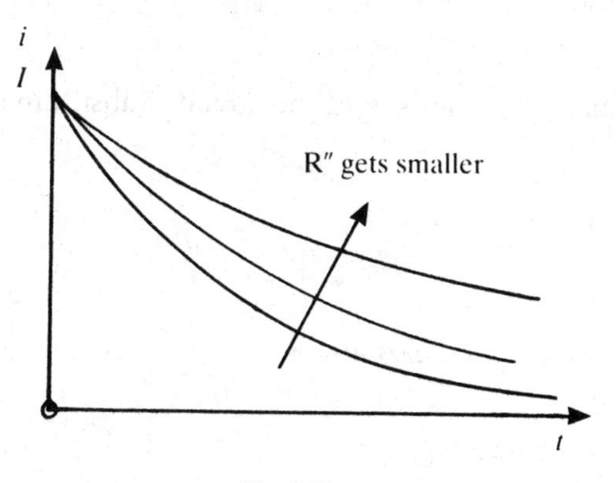

Fig. 3.26.

the time constant of the circuit. The current decays more rapidly as more resistance is included in the circuit. This is the opposite result to that stated for an RC circuit.

3.10.3 The L–C–R series circuit

It is the intention to discuss the series circuit of Fig. 3.27, only, in this section. The

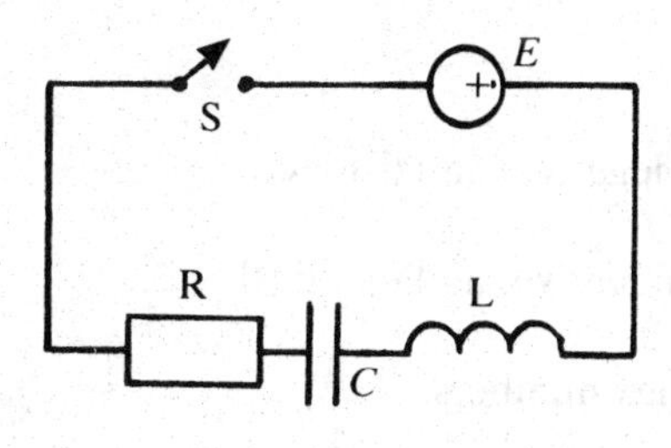

Fig. 3.27.

mathematics is more difficult and need be gone through in detail only by those students who are familiar with differential equations in calculus.

Apply KVL to the loop after the switch S is closed. This gives

$$E = v_R + v_C + v_L$$
$$= Ri + q/C + L\ di/dt \qquad (3.57)$$

On differentiating (3.57) with respect to time we obtain the 2nd order differential equation

$$L\ d^2 i/dt^2 + R\ di/dt + i/C = 0\ . \qquad (3.58)$$

Equation (3.58) can best be solved by assuming that the current i varies exponentially with time t. So let us put

$$i = A\ \exp(kt) \qquad (3.59)$$

where A is a constant characteristic of the circuit. Substituting for i in (3.58), we obtain

$$Lk^2 + Rk + 1/C = 0\ . \qquad (3.60)$$

This is a quadratic which has the solutions

$$k = \{-R \pm (R^2 - 4L/C)^{\frac{1}{2}}\}/2L\ . \qquad (3.61)$$

Let us call the solutions k_1 and k_2. So solutions of (3.58) are

$$i_1 = A_1\ \exp(k_1 t) \text{ and } i_2 = A_2\ \exp(k_2 t)\ ,$$

as is the sum of these — this you test by direct substitution into (3.58).

SPECIAL CASES

(i) $R^2 > 4L/C$;
k_1 and k_2 are real, negative, and distinct.

(ii) $R^2 = 4L/C$;
k_1 and k_2 are real, negative, and identical.

(iii) $R^2 < 4L/C$;
k_1 and k_2 are complex numbers.

(iv) $R = 0$;
k_1 and k_2 are purely imaginary.

The constants A and k can be determined from the initial conditions existing in the circuit. If the switch is closed at time $t = 0$ then the time just before it is closed is

$t = -0$ and the time just after it is closed is $t = +0$. Let us briefly summarize the initial conditions existing in resistors, inductors, and capacitors, because without a clear knowledge of these the L–C–R circuit cannot be analysed rigorously.

3.10.3.1 Initial conditions in circuit elements

For a resistor, the current and the pd are related through Ohm's law. If the pd across a resistor changes as a step function, such as might occur when a circuit switch is closed, then the current will vary in the same way, see Fig. 3.28. The current changes

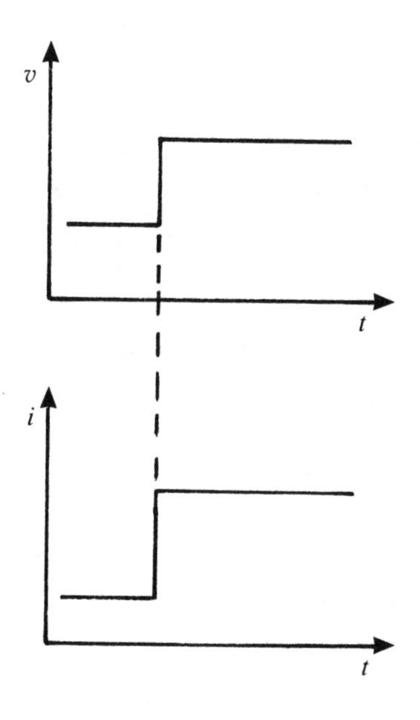

Fig. 3.28.

instantaneously if the pd changes instantaneously, and vice versa. The current through an inductor cannot change instantaneously because this implies that the pd across it is infinite (as can be seen from the relation $v_L = L\,di/dt$). Hence, closing a switch will not result in a current flowing at that instant. The inductor behaves as though it were an *open circuit* at time $t = +0$. If a current I_o flows in the inductor just as switching occurs then the inductor behaves like a current source I_o. This statement is equivalent to our earlier one, viz. the inductor acts like a short-circuit to dc currents.

The pd across a capacitor cannot change instantaneously, because this implies an infinite current (using $i = C\,dv_C/dt$). Hence closing a switch will not generate a pd across the capacitor at that instant. In other words the capacitor behaves like a *short-circuit*. If the capacitor is initially charged to a value Q ($= V_oC$) then it behaves like a voltage source of magnitude V_o.

Worked Examples 3.5

Q1. The switch in an R–C circuit is closed at time $t=0$. If the capacitor is initially uncharged, determine an expression for the current at some time $t>0$ by first setting up the relevant differential equation.

This problem gives us experience in (i) setting up a differential equation and (ii) using the equivalent circuit component description immediately after the switch is closed.

For all times $t>0$,

$$\begin{aligned} E &= v_R + v_C \\ &= Ri + q/C \;. \end{aligned} \tag{1}$$

Differentiating with respect to t gives

$$0 = R\,\mathrm{d}i/\mathrm{d}t + i/C \;. \tag{2}$$

The solution must be of the form

$$i = A\,\exp(kt) \;.$$

So using it in (2), we have

$$kR + 1/C = 0$$

from which

$$k = -1/RC \;.$$

Hence

$$i = A\,\exp(-t/RC) \;. \tag{3}$$

At time $t=+0$, the capacitor is short-circuited because there is no pd across it. Hence,

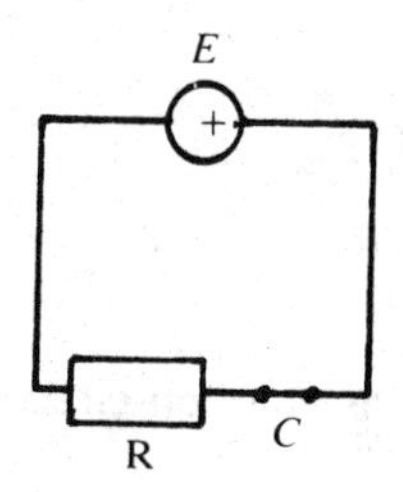

Thus

$$i(+0) = A = E/R$$

and (3) becomes

$$i = (E/R)\exp(-t/RC)$$

which is identical with (3.42).

Q2. To obtain the corresponding expression for a series R–L circuit. At time $t > 0$, the current i must satisfy the differential equation

$$\begin{aligned} E &= v_R + v_L \\ &= Ri + L\,\mathrm{d}i/\mathrm{d}t\,. \end{aligned} \tag{1}$$

Differentiating (1) with respect to t gives

$$0 = R\,\mathrm{d}i/\mathrm{d}t + L\,\mathrm{d}^2i/\mathrm{d}t^2\,. \tag{2}$$

Try

$$i = A\exp(kt)$$

as a solution of (2). Then

$$kR + k^2L = 0$$

from which

$$k_1 = 0 \text{ and } k_2 = -R/L\,.$$

Hence

$$i = A_1 + A_2 \exp(-Rt/L) \ . \tag{3}$$

At time $t = +0$ the inductor acts as an open-circuit

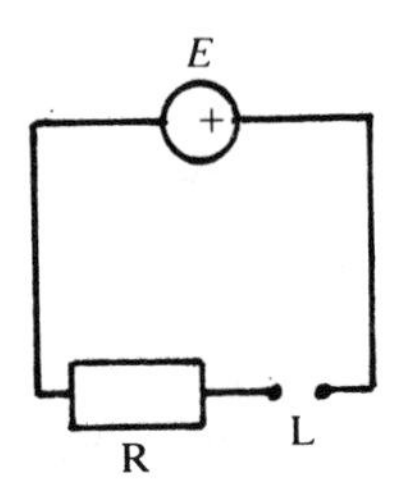

so that $i(+0) = 0$. This means that

$$A_1 + A_2 = 0 \ . \tag{4}$$

In addition all the source emf appears between the terminals of the open-circuit so that

$$v_L = L.di(+0)/dt = E \ . \tag{5}$$

That is, using (3),

$$L.(-RA_2/L) = E$$

and

$$A_2 = -E/R \ .$$

On substituting into (4) we obtain

$$A_1 = E/R \ .$$

Hence

$$i = (E/R)\,\{1 - \exp\,(-Rt/L)\} \ . \tag{6}$$

Q3. The series L–C–R circuit has components with values: $R = 1\ \Omega$, $L = \frac{3}{4}$ H, $C = 4$ F, and $E = 3$ V. Obtain a general expression for the current at any time t.
We can substitute the component values into the general solution of the 2nd order differential equation for the L–C–R circuit, that is (3.58). Then

$$
\begin{aligned}
(k_1, k_2) &= \{-1 \pm \sqrt{(1 - \tfrac{3}{4})}\}/(\tfrac{3}{2}) \\
&= 2(-1 \pm \tfrac{1}{2})/3\ .
\end{aligned}
$$

Therefore,

$$k_1 = -\tfrac{1}{3} \text{ and } k_2 = -1\ ,$$

and

$$i = A_1 \exp(-t/3) + A_2 \exp(-t)\ . \tag{1}$$

At $t = +0$, $i = 0$. Hence (1) reduces to

$$A_1 + A_2 = 0\ . \tag{2}$$

Also the equivalent circuit looks like

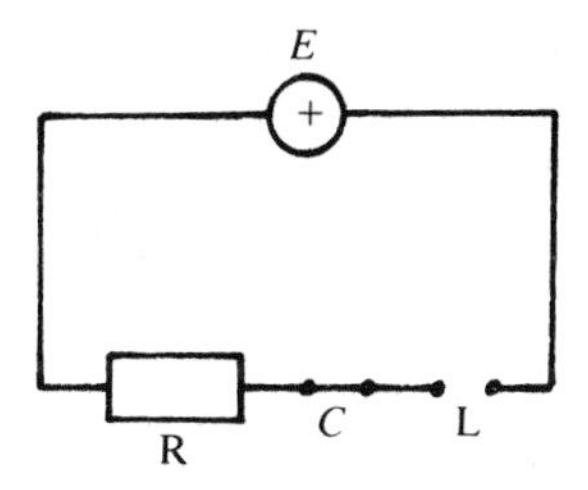

The whole of the emf appears across the inductor, so that we can write

$$v_L = L\ di(\text{or})/dt = E \tag{3}$$

which, after differentiating (1) with respect to t, becomes

$$A_1/3 + A_2 = -E/L = -4\ . \tag{4}$$

Equations (2) and (4) are simultaneous equations which can be solved to give

$$A_1 = 6 \text{ amps and } A_2 = -6 \text{ amps} .$$

Hence we arrive at the final solution

$$i = 6\,\{\exp(-t/3) - \exp(-t)\} . \tag{5}$$

The two component exponential terms are sketched below, together with the resultant.

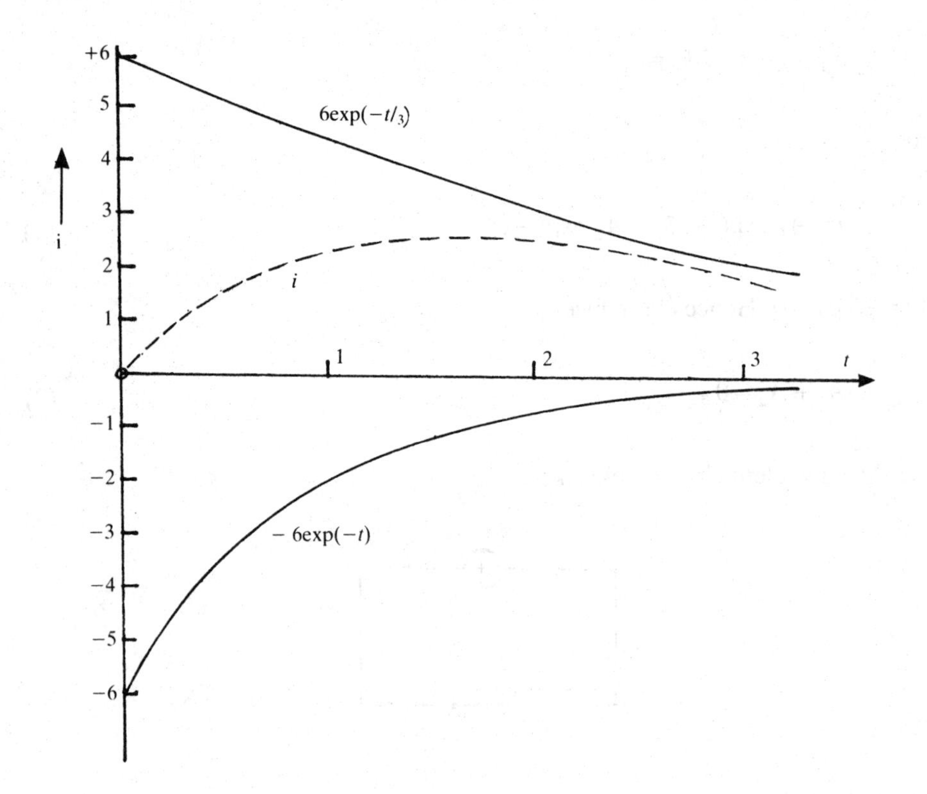

This circuit is described by an equation which has two real roots. It is an example of an *over-damped* system.

Q4. Re-analyse the series L–C–R circuit for components: $R = 2\,\Omega$; $L = 1$ H; $C = \frac{1}{5}F$, and $E = 3$ V.

On substituting the above values into (3.58), we obtain

$$\begin{aligned}(k_1,k_2) &= \{-2 \pm (4-20)^{\frac{1}{2}}\}/2\\ &= \{-2 \pm \sqrt{-16}\}/2\\ &= \{-2 \pm \text{j}4\}/2\\ &= -1 \pm \text{j}2 .\end{aligned}$$

So

$$k_1 = -1 + j2 \text{ and } k_2 = -1 - j2$$

Note that these are complex numbers of the form: $a + jb$ (see 4.6).

So the general expression for the current is

$$i = A_1 \exp\{-(1 - j2)t\} + A_2 \exp\{-(1 + j2)t\} \ . \tag{1}$$

It is convenient to expand the exponentials as

$$i = \exp(-t)\ [A_1 \exp(j2t) + A_2 \exp(-j2t)] \ . \tag{2}$$

With the help of deMoivre's theorem:

$$\exp(j\theta) = \cos\theta + j\sin\theta$$

we can rewrite (2) as

$$\begin{aligned} i &= \exp(-t)\ [A_1\ \{\cos 2t + j\sin 2t\} + A_2\{\cos 2t - j\sin 2t\}] \\ &= \exp(-t)\ [(A_1 + A_2)\cos 2t + j\ (A_1 - A_2)\sin 2t] \end{aligned} \tag{3}$$

As the current is a real quantity, all the individual terms must be real. In particular, $(A_1 + A_2)$ and $j(A_1 - A_2)$ are real.

For convenience, let us put

$$A_1 + A_2 = B_1$$

and

$$j(A_1 - A_2) = B_2 \ .$$

Then

$$i = \exp(-t)[B_1 \cos 2t + B_2 \sin 2t] \ . \tag{4}$$

Equation (4) can be expressed in an even more compact form as:

$$i = K \exp(-t) \sin(2t + \phi) , \tag{5}$$

To understand why, expand the sine term in (5) and equate with the non-exponential part of (4). Then

$$K \sin(2t + \phi) = K[\sin 2t \cos \phi + \cos 2t \sin \phi] = B_1 \cos 2t + B_2 \sin 2t$$

with

$$B_1 = K \sin \phi \text{ and } B_2 = K \cos \phi .$$

Therefore,

$$K = (B_1 + B_2)^{\frac{1}{2}}$$

and

$$\tan \phi = B_1/B_2 .$$

The value of ϕ can be determined by using the initial condition that $i = 0$ at $t = +0$. Then (5) becomes

$$K \sin \phi = 0$$

from which

$$\phi = 0$$

and, therefore,

$$i = K \exp(-t) \sin 2t . \tag{6}$$

The equivalent circuit at $t = +0$ is the same as that in Q3. Hence

$$v_L = L \,\mathrm{d}i(+0)/\mathrm{d}t = E . \tag{7}$$

Differentiating (6) with respect to t, we obtain

$$\mathrm{d}i/\mathrm{d}t = K \exp(-t)[2 \cos 2t - \sin 2t]$$

which at $t = +0$ is

$$\mathrm{d}i(+0)/\mathrm{d}t = 2\,K\ . \tag{8}$$

Substituting (8) into (7), we have

$$K = E/2L = \tfrac{3}{2}\ .$$

The current can finally be written as

$$i = \tfrac{3}{2} \exp(-t) \sin 2t\ . \tag{9}$$

If $\exp(-t)$ was equal to 1, then the current would vary sinusoidally with time. Its amplitude would be $\frac{3}{2}$. However, as the $\exp(-t)$ term also varies with time, the overall result is that the amplitude of the sinusoidal component decays exponentially. This is shown in the following diagram.

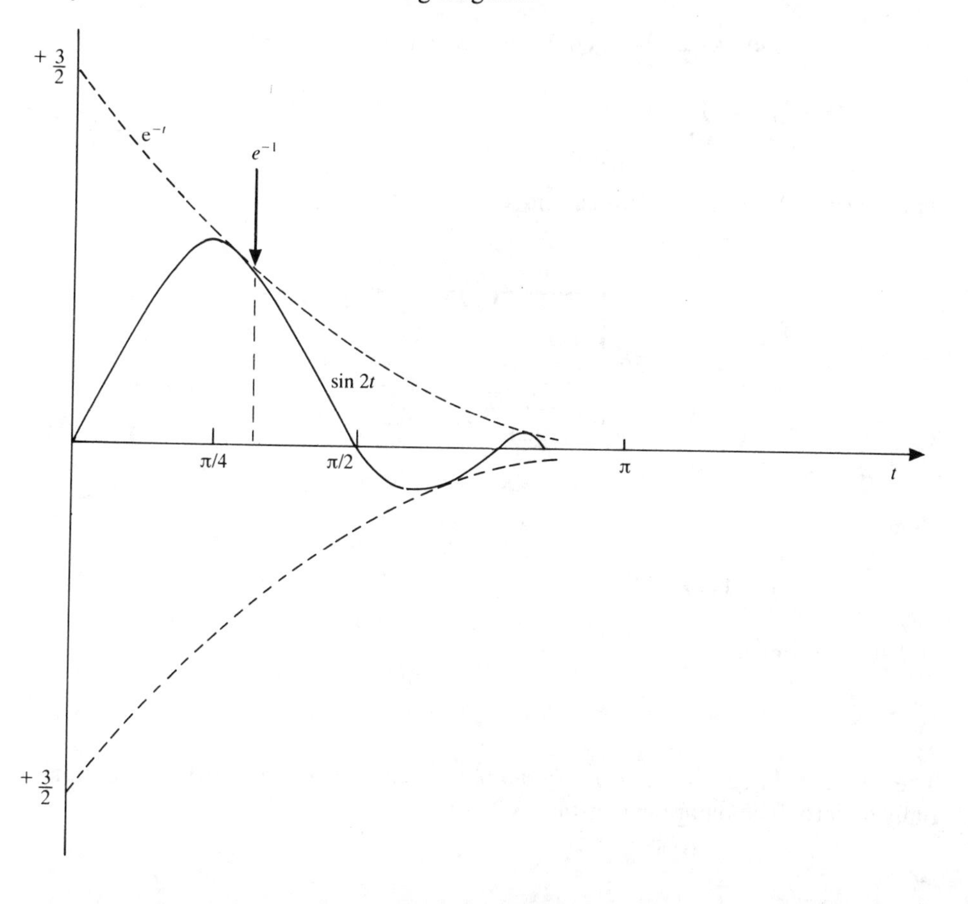

The current has the general form

$$i = K \exp(-\alpha t) \sin(\omega t + \phi) \qquad (10)$$

in which α is the *damping* coefficient and ω is the angular frequency of the sinusoidal term. (See Chapter 4 on ac theory for further information.) In $1/\alpha$ seconds the exponential envelope will have decreased to $\exp(-1)$ of its initial value K. Comparing (9) with (10) we see that α is $1\ \text{s}^{-1}$ (or 1 Hz) and ω is $2\ \text{rad.s}^{-1}$; the larger α is the sooner the sinusoidal term dies away. The $\exp(-1)$ point is indicated on the above diagram.

All circuits which are described by an equation with complex roots are said to be *oscillatory* or *under-damped*.

Q5. Q3 and Q4 refer to over-damped and under-damped systems, respectively. If, however, the component values are: $R = 4\ \Omega$, $L = 1\ \text{H}$, $C = \frac{1}{4}\text{F}$, and $E = 3\ \text{V}$, then we have an example of a *critically-damped* system. Suppose that a current of 1 A flows through the inductor at the instant when the switch is closed and that the capacitor is uncharged. Hence obtain an expression for $i(t)$.

Note that

$$R^2 - 4L/C = 16 - 4 \times 1/(\tfrac{1}{4}) = 0,$$

so that roots k_1 and k_2 are both equal to -2. Then

$$\begin{aligned} i &= (A_1 + A_2) \exp(-2t) \\ &= A \exp(-2t) . \end{aligned} \qquad (1)$$

At time $t = +0$, the equivalent circuit is

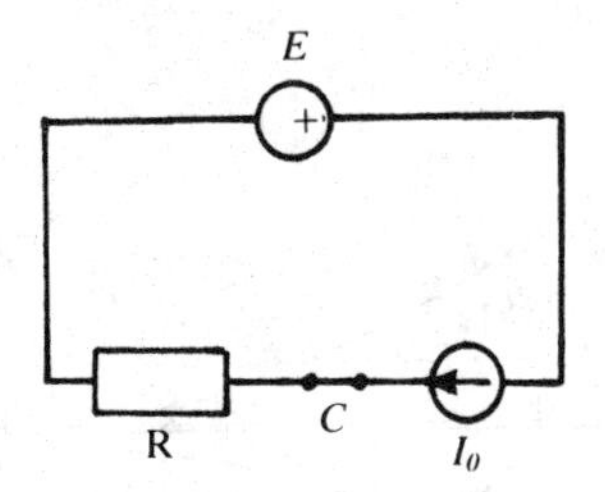

Hence

$$i(+0) = A = I_o = 1$$

and at other times t

$$i = 1.\exp(-2t) . \qquad (2)$$

The circuit behaves like a *critically-damped* system. The current decays exponentially to zero. The damping constant is $2\ \text{s}^{-1}$.

4

AC theory

Objectives

(i) To identify some basic properties of the ac current (and voltage) waveform
(ii) The meaning of root mean square (rms) current and voltage
(iii) How to convert between peak and rms values
(iv) To determine the phase relations between i and v for the resistor, inductor, and capacitor
(v) How to use a Bode plot to obtain the bandwidth of an electrical filter
(vi) To discuss complex numbers, the operator j, and the rotating vector (phasor)
(vii) The importance of the phasor diagram in ac analysis
(viii) To differentiate between impedance and resistance
(ix) To discuss electrical resonance in the series L–C–R circuit
(x) To define the quality factor as a frequency-discriminating parameter

4.1 THE SINUSOIDAL WAVEFORM

So far we have discussed three kinds of voltage waveforms. These are: the steady dc voltage; the step-voltage pulse (such as that encountered when a switch is closed in a circuit); the train of rectangular or square waves (for digital ON/OFF purposes). These are depicted in Fig. 4.1. Throughout this chapter, however, we will be concentrating on the sinusoidally-varying voltage (and current) waveform, similar to that of Fig. 4.2.a.

The term *sinusoidal* means that the waveform varies like a sinewave. For the moment, let us concentrate on the voltage waveform. Then Fig. 4.2 indicates that the instantaneous emf e of a voltage source varies with time according to the relation

$$e = E_p \sin \omega t \tag{4.1}$$

E_p is called the *amplitude* or the *peak* value of the waveform and ω is the angular frequency. Generally, e varies about 0 V, although it can vary about some other value by applying a suitable bias.

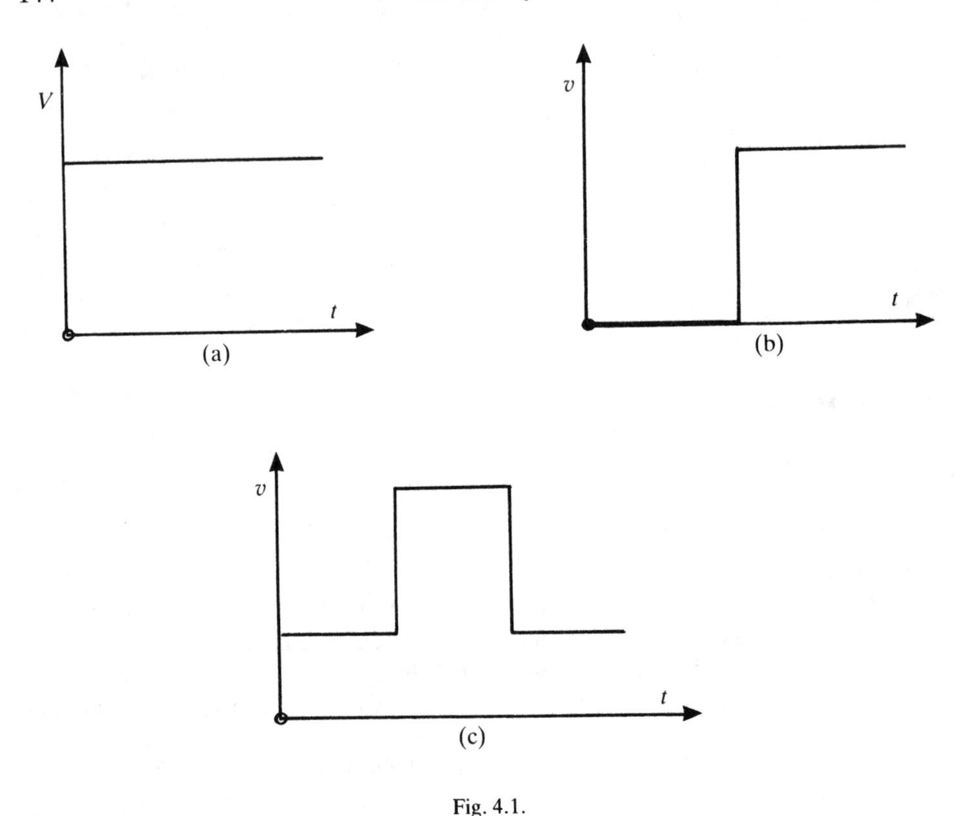

Fig. 4.1.

UNITS: e and E are in volts (V) and ω in radians s^{-1}.

Note that Fig. 4.2 is *periodic* or *cyclic*. This means that the waveform repeats itself after a time T — so that between 0 and T, T and $2T$, $2T$ and $3T$, etc. the waveform has an identical appearance. As a result, T is called the *period* of the voltage waveform. The waveform of Fig. 4.2 also repeats itself between t' and $t'+T$, $t'+T$ and $t'+2T$ . . . etc. In other words the period T is the important parameter and not the initial time t'.

The frequency f of the voltage waveform is the number of complete cycles that occur in unit time.

UNITS: Hz (short for Hertz) or cycles s^{-1}.
The UK mains frequency is 50 Hz or 50 cycles s^{-1}. The frequency f is related to the period T by

$$f=1/T \tag{4.2}$$

and ω, the angular frequency, is equal to $2\pi/T$ or $2\pi f$.

e may also be described by the more general sinusoidal expression

$$e=E_p \sin(\omega t+\phi) \tag{4.3}$$

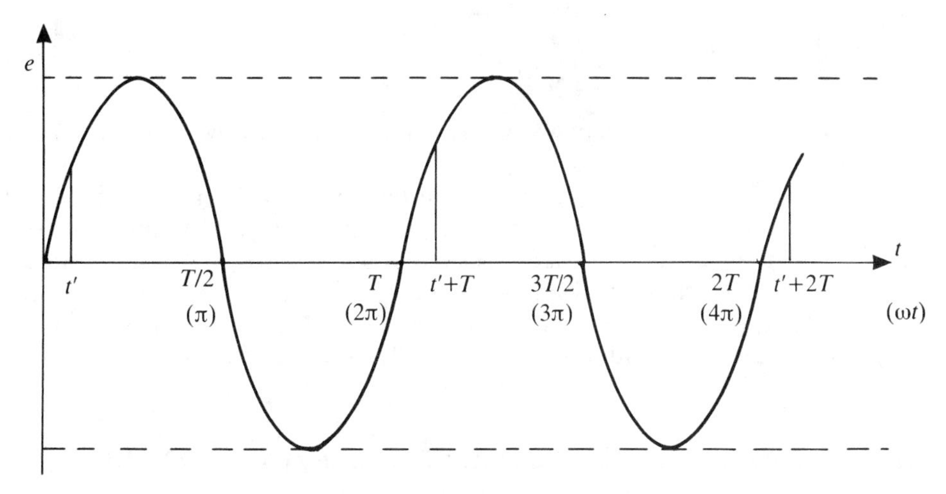

Fig. 4.2.

in which ϕ is the *initial phase* or the *epoch*. Now e does not equal 0 when t=0. For example, if $\phi=\pi/4$ then curve (a) in Fig. 4.3 is obtained. It is similar to Fig. 4.2 except

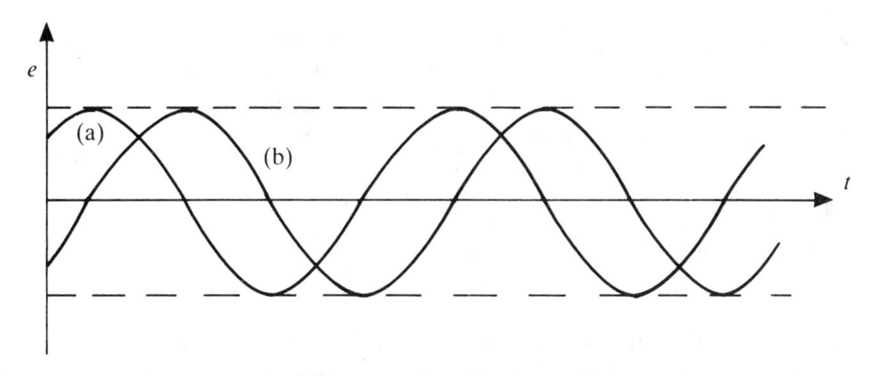

Fig. 4.3.

that the sinewave is shifted to the left through $\pi/4$ radians. That is, it *leads* Fig. 4.2 by $\pi/4$ radians. It is not necessary to plot the two sine waveforms, however, in order to determine the nature of the phase shift; the positive sign before ϕ in (4.3) gives this information. If ϕ was $-\pi/4$ then waveform (b) in Fig. 4.3 would be obtained; it *lags* Fig. 4.2 by $\pi/4$ radians.

Using an e vs t plot, the resultant of two waveforms, such as those represented by (4.1) and (4.3), can be determined by laboriously summing the instantaneous emf's at various values of t. Then to find the initial phase ϕ_{res} of the resultant, the waveform is shifted either to the right or to the left by an amount Δt until it looks like Fig. 4.2. Then

$$\phi_{\text{res}} = |\omega \Delta t| \tag{4.4}$$

Later, we shall use a more powerful technique which incorporates the *phasor* diagram.

The above discussion also applies to the temporal variation of the potential difference across a circuit component and the current through it. An instantaneous emf e has an instantaneous current i associated with it. When e reaches its maximum value E_p, i also reaches its maximum value I_p. However, the pd v across the component may or may not be in phase with the current i — as we saw in sections 1.4 and 1.5 with the inductor and capacitor. We shall return to this aspect in section 4.4.

4.2 ROOT MEAN SQUARE (RMS) VALUE OF CURRENT AND VOLTAGE

It is often necessary to compare a sinusoidally-varying current (voltage) with a dc current (voltage). This can be done by using the concept of the root mean square (rms) current (voltage). It is defined as:

THE dc CURRENT WHICH PRODUCES THE SAME HEATING EFFECT AS THE ac CURRENT.

In section 1.6 we briefly discussed Joule or resistive heating in terms of the rate at which one system does work on another. Relation (1.34) defines this as the power P. For a resistor R connected to a dc voltage source, electrons do work on the atoms composing the resistor, and P is given by the relation

$$\begin{aligned} P_{\text{dc}} &= I.V \\ &= I^2 R \end{aligned} \tag{4.5}$$

In the case of an ac voltage source, the current i through the resistor and the potential difference v across it change their values continuously. However, within a time interval Δt, i and v can be considered to be constant and the work done by the source on the electrons in the resistor in this time interval is equal to $i^2 R \Delta t$. The total work done by the electrons in the resistor over one complete cycle of the ac current is found by summing the work done over all the small time intervals N which make up the period T. That is

$$i^2 R N \Delta t \quad \text{or} \quad \sum i^2 R \Delta t.$$

Using the calculus, by making Δt sufficiently small, we may replace the summation by the definite integral. Then

$$P_{AC} = \int_0^T i^2 R\, dt. \tag{4.6}$$

If the average value of the power over one complete cycle is $\overline{P}_{ac}$ then the total work done by the electrons in heating the resistor is $\overline{P}_{ac}T$. This means that

$$\overline{P}_{ac}T = \int_0^T i^2 R\, dt$$

and

$$\overline{P}_{ac} = (1/T)\int_0^T i^2 R\, dt. \tag{4.7}$$

This is the average rate at which energy is transferred from the electrons to the lattice atoms of the resistor.

As e is sinusoidal, we naturally expect i to be. So let us put

$$i = I_p \sin \omega t.$$

Then substituting for i in (4.7) we have

$$\overline{P}_{ac} = (I_p^2 R/T)\int_0^T \sin^2 \omega t\, dt$$

$$= (I^2 R/2T)\int_0^T [1 - \cos 2\omega t]\, dt \tag{4.8}$$

(using the trigonometric relation:

$$\cos 2\,\omega t = 1 - 2\sin^2 \omega t).$$

On integrating (4.8) we obtain

$$\overline{P}_{ac} = I_p^2 \mathrm{R}/2\mathrm{T}[\mathrm{t} - (\sin 2\omega t)/2\omega]_o^T$$

$$= I_p^2 R/2. \tag{4.9}$$

For the heating effects due to dc and ac currents to be equal, we require (4.5) and 4.9) to be equal, that is

$$I_{\text{rms}} = I_p/\sqrt{2}. \tag{4.10}$$

Thus

THE rms CURRENT is equal to the peak current divided by $\sqrt{2}$.

It would have been perfectly legitimate to have worked throughout with the pd υ instead of current. For then,

$$P_{\text{dc}} = V^2/R \qquad \text{(a)}$$

and (4.11)

$$\overline{P}_{\text{ac}} = E_p^2/2R \qquad \text{(b)}$$

from which

$$V_{\text{rms}} = E_p/\sqrt{2}. \tag{4.12}$$

Voltmeters and ammeters which measure ac signals are calibrated in terms of the rms value for easy comparison with dc readings. We shall always indicate rms values with a prime, for example, I', E', or V'. In addition, in an ac circuit such as Fig. 4.4,

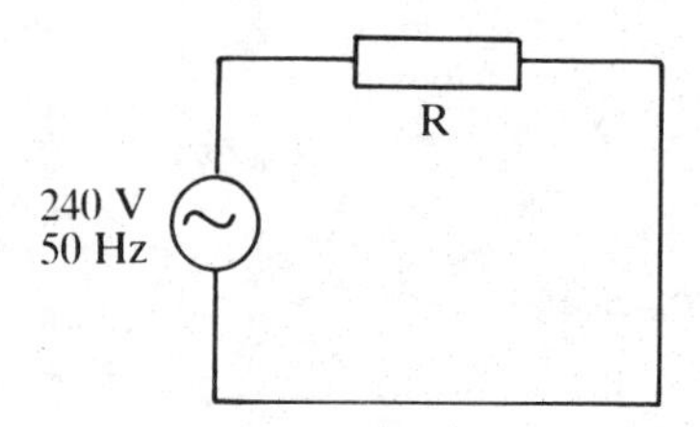

Fig. 4.4.

240 V is the rms value, that is E'. Unless otherwise stated, numbers quoted alongside the ac source symbol will always indicate the rms value. However, whilst reading other textbooks care should be taken to establish whether such numbers refer to the source's rms value or amplitude.

☆ ☆

AN ASIDE: The above proof is strictly correct only if the resistance R stays constant. It does not apply to either a heated filament, because its temperature varies with i, or a non-ohmic component.

☆ ☆

4.3 POWER FACTOR

Suppose that the instantaneous values of current and pd in some part of a circuit are given by:

$$i = I_{\text{p}} \sin \omega t \tag{4.13}$$

and

$$\upsilon = E_{\text{p}} \sin(\omega t + \phi) \tag{4.14}$$

at some time t. It is assumed that there is a phase difference between i and υ in order to obtain a general result for the power factor.

The instantaneous power p is given by

$$p = i\upsilon = I_{\text{p}}E_{\text{p}} \sin \omega t \sin(\omega t + \phi). \tag{4.15}$$

As time goes on, p varies because i and υ alter. For example, there will be times when p will be negative when that part of the circuit does more work on the rest of the circuit in a given time than the rest of the circuit does on it; electrical power is transferred to the rest of the circuit.

The average value of (4.15) over a complete cycle is

$$\begin{aligned}
\overline{P} &= (1/T)\int_{0}^{T} i\upsilon \, \mathrm{d}t \\
&= (I_{\text{p}}E_{\text{p}}/T)\int_{0}^{T} \sin \omega t . \sin(\omega t + \phi) \, \mathrm{d}t \\
&= (I_{\text{p}}E_{\text{p}}/T)\int_{0}^{T} \{\cos\phi . \sin\omega t + \sin\phi . \sin\omega t . \cos\omega t\} \, \mathrm{d}t \\
&= (I_{\text{p}}E_{\text{p}}/T)\int_{0}^{T} [\cos\phi . \sin\omega t \, \mathrm{d}t + (sin\,\phi . \sin 2\omega t/2) \, \mathrm{d}t] \\
&= (I_{\text{p}}E_{\text{p}}/2)\cos\phi \\
&= I'E'\cos\phi .
\end{aligned} \tag{4.16}$$

$\cos\phi$ is called the *power factor* of the circuit.

CASES:

(i) When $\phi = \pi/2$, $\overline{P} = 0$. Even when I' and E' are large $\overline{P}$ is still zero.

(ii) When $\phi = 0$, $\overline{P} = I'E'$ and the generation of electrical power is identical to that in a dc circuit.

Worked examples 4.1

Q.1 An inductor is connected to an ac voltage source of instantaneous emf e. Determine an expression for the instantaneous power developed in it.

With

$$i = I_p \sin \omega t \tag{1}$$

the pd across the inductor is

$$\begin{aligned} v_L &= L\, di/dt \\ &= \omega L\, I_p \cos \omega t, \end{aligned}$$

The instantaneous value of the power is

$$\begin{aligned} p_L &= i v_L \\ &= \omega L\, I_p^2 \sin \omega t \cos \omega t \\ &= 2\omega L\, I'^2 \sin \omega t \cos \omega t \\ &= \omega L\, I'^2 \sin 2\omega t, \end{aligned}$$

ωL is called the *inductive reactance* X. So

$$p_L = X_L I'^2 \sin 2\omega t, \tag{2}$$

You should note that p has the form

$$p_L = P_p \sin \omega_o t \tag{3}$$

in which P_p $(= X_L I'^2)$ is equivalent to the amplitude of a 'power' waveform of angular frequency ω_o, equal to twice the angular frequency ω of the current waveform.

Although the average power $\overline{P}$ is zero, the inductor will periodically increase its energy content when p_L is positive and decrease its energy when p_L is negative.

Q.2 Develop an expression for the instantaneous power in a capacitor.

Using

$$v_c = q/C$$

for the instantaneous pd across the capacitor, and

$$i = dq/dt$$

for the instantaneous current, we can obtain an expression for the instantaneous power, viz

$$\begin{aligned} p_c &= (q/C)\, dq/dt \\ &= (1/C)\, i \int i\, dt. \end{aligned} \qquad (1)$$

With the help of relation (1) in Q.1,

$$\begin{aligned} p_c &= (1/C)\, I_p \sin \omega t \int I_p \sin \omega t\, dt \\ &= -\,(I'^2/\omega C) \sin 2\omega t, \end{aligned}$$

$1/\omega C$ is called the *capacitive reactance* X_c. So

$$p_c = -\,X_c I'^2 \sin 2\omega t. \qquad (2)$$

Once again it has the form

$$p_c = P_p \sin \omega_o t.$$

The negative sign in (2) indicates that p_c is in antiphase with p_L; in an L-C circuit, at those times when the inductor has increased its energy the capacitor has its reduced, and vice versa.

4.4 THE PHASE RELATION BETWEEN CURRENT AND POTENTIAL DIFFERENCE FOR DISCRETE CIRCUIT COMPONENTS

4.4.1 The resistor

Consider Fig. 4.5 which consists of a single resistor connected to an ac voltage source. At every instant of time,

$$e = E_p \sin \omega t. \qquad (4.17)$$

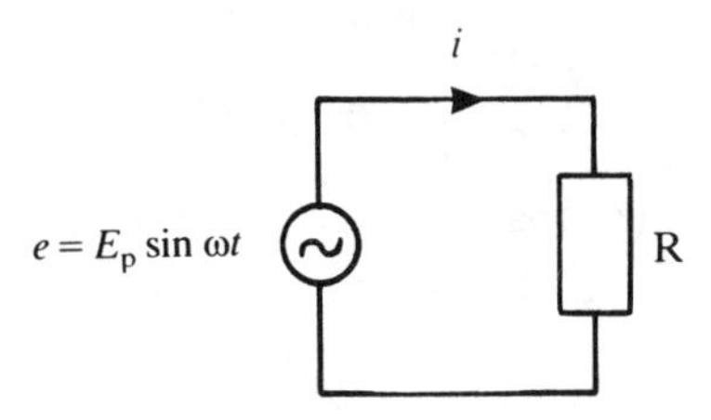

Fig. 4.5.

Treating R as a constant, we find that the pd across the resistor, by Ohm's law, is always given by

$$\begin{aligned} e &= v_R \\ &= iR \end{aligned} \tag{4.18}$$

Hence

$$i = (E_p/R)\sin\omega t. \tag{4.19}$$

This expression has the form

$$i = I_p \sin\omega t$$

in which the peak value of the current is E_p/R. The important point to note is that the instantaneous current i is always in phase with v_R because they both depend on t through the same sine function.

4.4.2 The capacitor

Now we require Fig. 4.6. At every instant of time

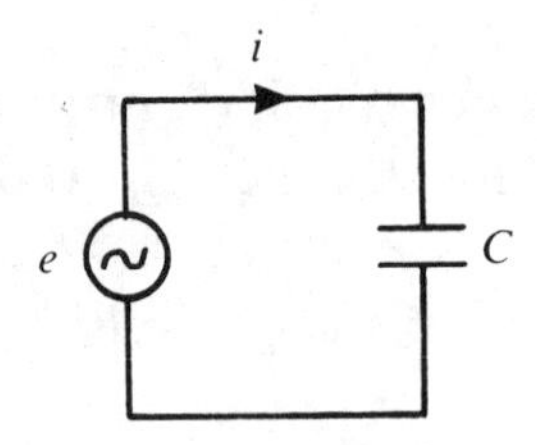

Fig. 4.6.

$$e = v_c$$
$$= q/C$$

or

$$q = E_p C \sin \omega t. \quad (4.20)$$

As

$$i = dq/dt$$

differentiating (4.20) with respect to t gives

$$i = \omega\ C\ E_p \cos \omega t$$
$$= \omega\ C\ E_p \sin(\omega t + \pi/2). \quad (4.21)$$

Comparing (4.17) with (4.21) we see that the current now leads the pd across C by $\pi/2$. The current amplitude I_p is given by

$$I_p = \omega\ C\ E_p. \quad (4.22)$$

We previously noted that $1/\omega C$ is the capacitive reactance X_c, so,

$$I_p = E_p/X_c. \quad (4.23)$$

Equation (4.23) greatly resembles an Ohm's law expression with X_c for the capacitor playing the same role as R for the resistor.

4.4.3 The inductor

Now

$$e = v_L$$
$$= L\ di/dt$$

therefore

$$di/dt = e/L$$
$$= E_p \sin \omega t/L$$

and

$$i = (E_p/L) \int \sin \omega t\ dt$$
$$= -(E_p/\omega L) \cos \omega t$$

$$= (E_p/\omega L) \sin (\omega t - \pi/2). \tag{4.24}$$

Now the current *lags* the instantaneous pd by the phase angle of $\pi/2$. The current amplitude I_p is $E_p/\omega L$. So once more we have an Ohm's law type expression with the inductive reactance ωL playing the same role as the pure resistance R.

4.5 COMBINATION OF CIRCUIT ELEMENTS

4.5.1 The R-C circuit

Our intention, here, is the same as in section 4.4, viz. to obtain an expression for the phase difference between the current i and the source emf e. Let us apply KVL to Fig. 4.7. (Remember that KVL is equally valid for ac purposes.) Then,

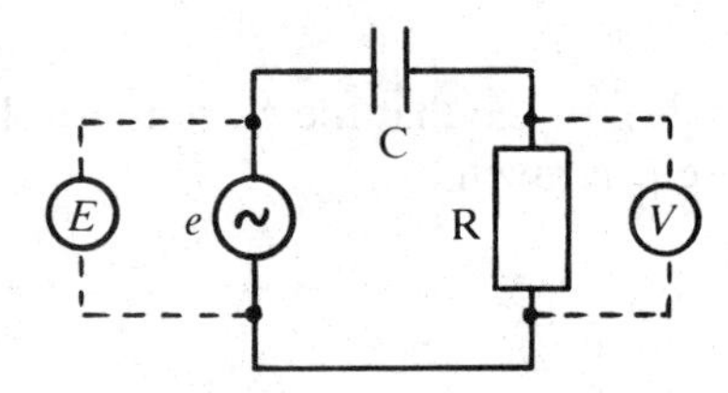

Fig. 4.7.

$$e - v_R - v_c = 0 \tag{4.25}$$

or

$$E_p \sin \omega t = Ri + q/C. \tag{4.26}$$

Differentiating (4.26) with respect to t, and transposing terms, we have

$$R\, di/dt + i/C = \omega\, E_p \cos \omega t. \tag{4.27}$$

This is a 1st order differential equation which can be solved relatively straightforwardly if you are familiar with this aspect of the calculus. However, we shall use a trick to help us solve it. As e has a sinusoidal variation with time, let us assume that i has also, and put

$$i = I_p \sin (\omega t + \phi) \tag{4.28}$$

where I and ϕ must be determined.

Substituting (4.28) into (4.27) gives

$$R\,\omega I_p \cos(\omega t + \phi) + (I_p/C)\sin(\omega t + \phi) = \omega E_p \cos \omega t. \tag{4.29}$$

Expand the terms in parentheses and transpose to give

$$\{R\cos\phi + (1/\omega C)\sin\phi - E_p/I_p\}\cos\omega t + \{(1/\omega C)\cos\phi - R\sin\phi\}\sin\omega t = 0. \tag{4.30}$$

By examining (4.30) at specific times *t* we will be able to determine ϕ. There are only two cases that we need to consider:

(i) At $t = 0$; *sin* $\omega t = 0$; $\cos\omega t = 1$

$$R\cos\phi + (1/\omega C)\sin\phi - E_p/I_p = 0 \tag{4.31}$$

(ii) At $t = T/4$; $\cos\omega t = 0$; $\sin\omega t = 1$

$$(1/\omega C)\cos\phi - R\sin\phi = 0. \tag{4.32}$$

Now, (4.30) will be *true* at all times *t* if (4.31) and (4.32) are both *true* because the latter are special cases of the former. (4.32) may be solved to give

$$\tan\phi = 1/\omega RC$$

or

$$\phi = \tan^{-1}(1/\omega RC). \tag{4.33}$$

Look at Fig. 4.8; the perpendicular and adjacent are obtained from (4.33). Hence

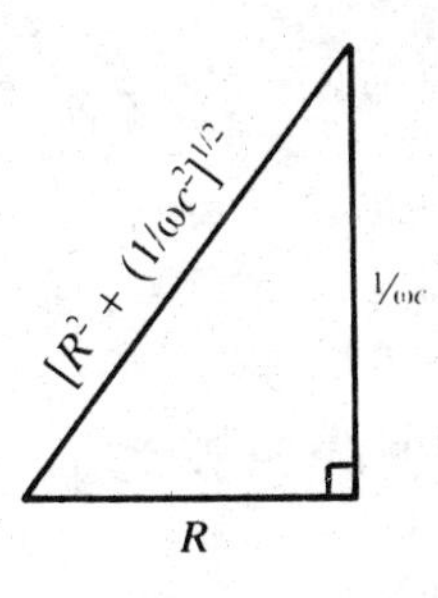

Fig. 4.8.

the hypotenuse can be found from Pythagoras' theorem. $\cos\phi$ and $\sin\phi$ may be read off immediately and substituted into (4.31) to obtain an expression for I_p.

$$R^2/[R^2 + (1/\omega C)^2]^{1/2} + (1/\omega C)^2.1/[R^2 + (1/\omega C)^2]^{1/2} - E_p/I_p = 0 \tag{4.34}$$

or

$$I_p = E_p/[R^2 + (1/\omega C)^2]^{1/2}.$$

Hence (4.28) may be written in full as

$$i = E_p/[\mathrm{R}^2 + (1/\omega C)^2]^{1/2}.\sin(\omega t + \phi)\ . \tag{4.35}$$

NOTES:

(i) ϕ is positive. This means that the current *always leads* the emf. Hence the circuit behaves predominantly as a capacitive circuit;

(ii) When $1/\omega RC$ becomes very small, that is, at high frequencies, ϕ tends to zero. Then i is in phase with e. The circuit behaves as a resistive circuit;

(iii) When $1/\omega RC$ becomes very large, at low frequencies, ϕ tends to $\pi/2$ and the circuit is *purely* capacitive.

4.5.1.1 The high-pass filter

Suppose that ac voltmeters record the rms pd across the resistor and the ac source, as shown in Fig. 4.7. Let us call these V' and E', respectively. The ac source is the input to the R–C circuit, the pd across R is the output, and the ratio V'/E' is the ac voltage gain. The instantaneous output pd is, by Ohm's law,

$$\begin{aligned} \upsilon_R &= iR \\ &= E_p R/[R^2 + (1/\omega C)^2]^{1/2}.\sin(\omega t + \phi) \end{aligned}$$

and the rms value of the pd is

$$V' = E_p R/[2\{R^2 + (1/\omega C)^2\}]^{1/2}. \tag{4.36}$$

In addition,

$$E' = E_p/\sqrt{2}. \tag{4.37}$$

Therefore the ac voltage gain A is given by

$$\begin{aligned} A &= R/[R^2 + (1/\omega C)^2]^{1/2} \\ &= 1/[1 + (1/\omega RC)]^{1/2}. \end{aligned} \tag{4.38}$$

A *Bode* plot can now be constructed which depicts how A varies as the source frequency ω (or f) is varied. Suppose $R = 100\,k\Omega$ and $C = 0.16\,\mu F$, the the Bode plot looks like Fig. 4.9. You can see that A has a maximum value of 1 above 10 Hz, when all frequencies in the input signal appear at the output. That is, they are not *attenuated*. Below 10 Hz, substantial attentuation occurs. Where does this figure of 10 Hz come from? It is referred to as the *cut-off*, *half-power*, *break*, or *corner* frequency at which the voltage gain equals $1/\sqrt{2}$. Now, the denominator of (4.38) will equal $\sqrt{2}$ when

$$1/\omega RC = 1$$

or

$$\omega_{co} = 1/RC \tag{4.39}$$

where the subscript 'co' is used to indicate the *cut-off* frequency. Of course, in terms of pure frequency,

$$f_{co} = 1/2\pi RC. \tag{4.40}$$

A equal to $1/\sqrt{2}$ is also known as the 3 dB point (pronounced '3-dee bee'). dB is the abbreviated form of *decibel* which is used as a measure of *noise*. In electronics, it is defined in terms of a power gain or a voltage gain. In the example that we have been considering, the voltage gain in dB is expressed by the relation

$$A_{dB} = 20 \log_{10} A \tag{4.41}$$

Then for an A of $1/\sqrt{2}$,

$$\begin{aligned} A_{dB} &= 20 \log_{10} (1/\sqrt{2}) \\ &= 20 \log_{10} 0.707 \\ &= -3. \end{aligned}$$

The minus sign occurs because A is fractional.

Why use the dB? The voltage gain in Fig. 4.9 is actually plotted on a logarithmic scale which may give a distorted appearance to the Bode plot. The dB linearizes the scale along the voltage gain axis. Also it is more convenient to talk in terms of a voltage gain of -3 dB than 0.707, for example.

Worked examples 4.2

Q.1 Using the R–C circuit of section 4.5.1, obtain an expression for the instantaneous pd across: (i) the resistor and (ii) the capacitor at low frequencies.

(i) *The resistor*:
The instantaneous pd is given by

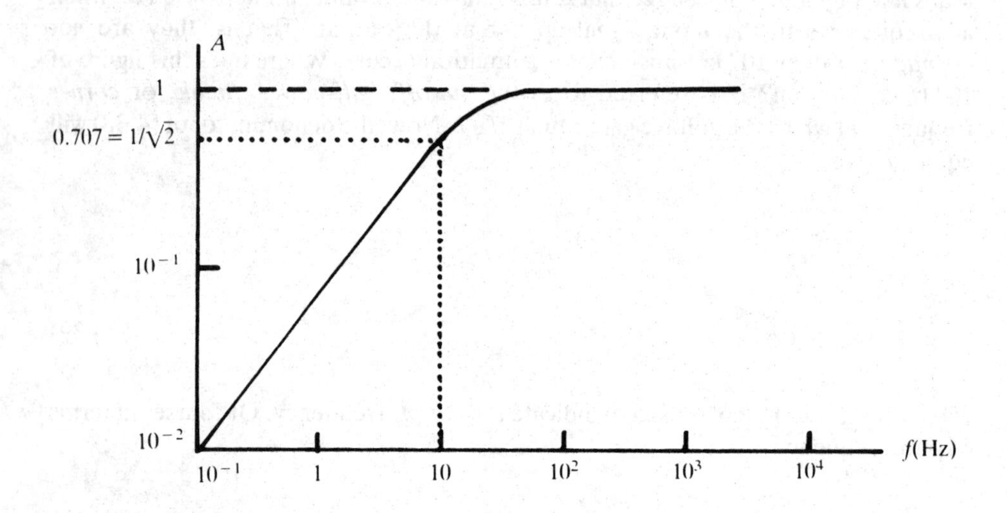

Fig. 4.9.

$$v_R = E_p R/[R^2 + (1/\omega C)^2]^{1/2} . \sin(\omega t + \phi) \quad (1)$$

with $\phi = \tan^{-1}(1/\omega RC)$.

We need to find the limiting value of v_R as $\omega \rightarrow 0$. Rewrite (1) as

$$v_R = E_p \omega RC/[1 + (\omega RC)^2]^{1/2} . \sin(\omega t + \phi)$$

Then, as $\omega \rightarrow 0$, the demoninator $\rightarrow 1$ and $\phi \rightarrow \pi/2$. So

$$v_R = E_p \omega RC \sin(\omega t + \pi/2)$$
$$= E_p \omega RC \cos \omega t. \quad (2)$$

As

$$e = E_p \sin \omega t$$
$$de/dt = E_p \omega \cos \omega t \quad (3)$$

which, on substituting into (1), gives

$$v_R = RC\, de/dt. \quad (4)$$

Thus the circuit acts like a differentiator, as occurs with a dc source.

(ii) *The capacitor*:
The instantaneous pd is

$$\begin{aligned} v_c &= i\,X_c \\ &= i/\omega C \\ &= (E_p/\omega C)/[R^2 + (1/\omega C)^2]^{1/2}.\sin(\omega t + \phi) \\ &= E_p/[1 + (\omega RC)^2]^{1/2}.\sin(\omega t + \phi). \end{aligned} \tag{5}$$

As $\omega \to 0$, $\phi \to 0$, as in (i) and the denominator $\to 1$. Therefore,

$$v_c = E_p \sin \omega t \tag{6}$$

This result implies that for frequencies close to zero, near the dc region, the capacitor can be approximated by an open circuit. In other words, the capacitor 'blocks off' dc current.

Q.2 Determine the voltage gain across the resistor in an R–L circuit.

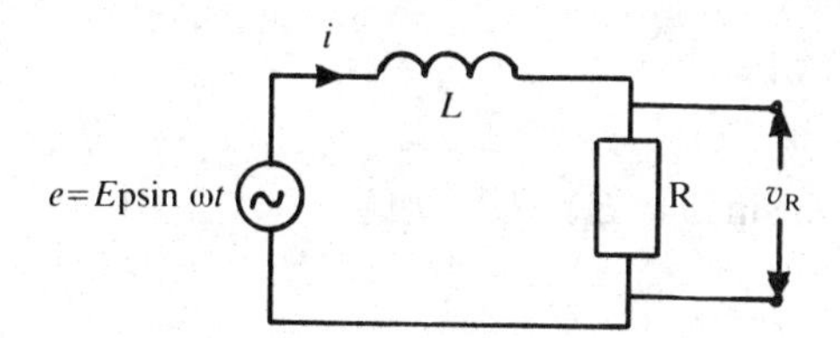

By KVL

$$e - v_R - v_c = 0$$

which leads to

$$di/dt + (R/L)i = (E_p/L).\sin \omega t. \tag{1}$$

Following the same procedure as with the R–C circuit, we will assume that

$$i = I_p \sin(\omega t + \phi)$$

and substitute it into (1) to give

$$\omega I_p \cos(\omega t + \phi) + (R/L)I_p \sin(\omega t + \phi) = (E_p/L)\sin \omega t. \tag{2}$$

Expand the trigonometric terms and then examine the resulting expression at:

(a) $t = 0$ ($\sin \omega t = 0$) when

$$\omega I_p \cos \phi + (RI_p/L) \sin \phi = 0 \tag{3}$$

and

$$\tan \phi = -\omega L/R$$

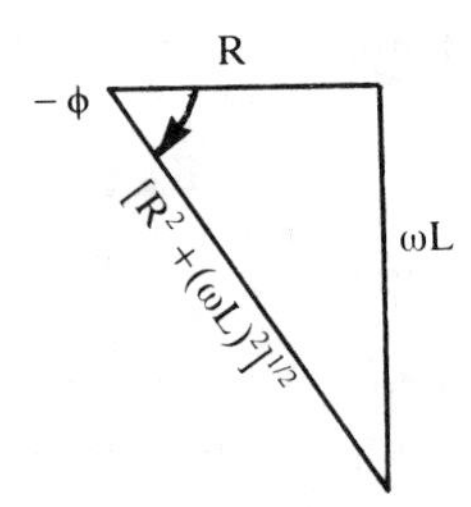

(The angle ϕ is measured in a clockwise direction below the horizontal.)

(b) $t = T/4$ ($\cos \omega t = 0$) when

$$(R/L) \cos \phi - \omega \sin \phi = E_p/LI_p \tag{4}$$

which becomes

$$I_p = E_p/[R^2 + (\omega L)^2]^{1/2}$$

using values of $\sin \phi$ and $\cos \phi$ obtained from the above triangle. Hence

$$\begin{aligned} v_R &= iR \\ &= E_p/[1 + (\omega L/R)^2]^{1/2} . \sin (\omega t + \phi) \end{aligned} \tag{5}$$

The cut-off frequency ω_{co} is R/L.

For $\omega \to 0$ (dc), $\phi \to 0$ and

$$v_R = E_p \sin \omega t, \tag{6}$$

The whole of the source emf appears across the resistor. That is, the voltage gain is 1.

This means that the inductor behaves like a short circuit — a result we obtain in section 3.10.2.

For $\omega \rightarrow \infty$ the denominator in (5) is infinite and $v_R \rightarrow 0$. Thus the R–L circuit acts like a low-pass filter.

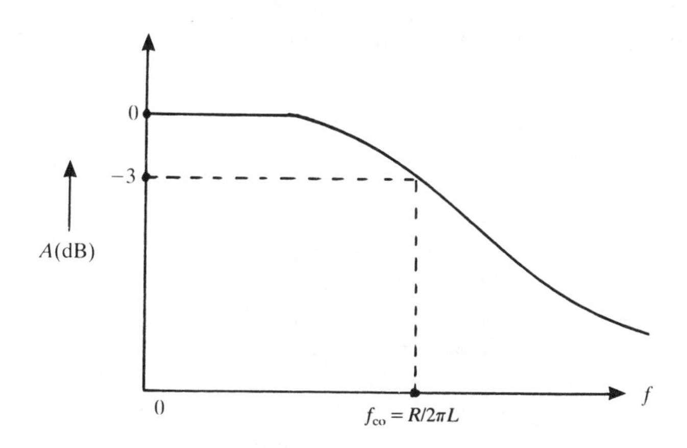

4.6 AC ANALYSIS WITH COMPLEX NUMBERS

The analysis present in the previous section for determining the phase difference between current and potential difference is long and tortuous. A shorter and more universal technique employs complex numbers. Before illustrating the general procedure, let us briefly review some properties of complex numbers

4.6.1 Summary of the properties of complex numbers

A complex number P can be written:

$$P = A + \mathrm{j}B \tag{4.42}$$

where A and B are real numbers and j is an *operator*. Look at Fig. 4.10. The

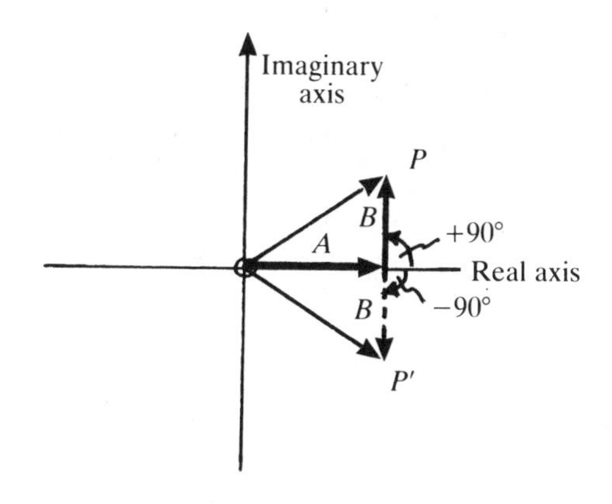

Fig. 4.10.

horizontal axis is called the *Real* axis and the vertical axis the *Imaginary* axis. Imagine a tiny arrow fixed at the origin of the coordinate system and pointing along the positive real axis. Move the arrow a distance A along this axis. Then, because the sign before j is positive, rotate the arrow anticlockwise through 90° and move it a distance B along this direction. We have arrived at point P in the *Argand diagram*.

To determine the position of P′, the mirror image of P in the *Real* axis, defined by

$$\mathrm{P}' = A - \mathrm{j}B,$$

move the arrow along the positive real axis a distance A, rotate it clockwise through 90°, and then move it a distance B to point P′ (also shown in Fig. 4.10).

The complex number is another way of describing a Cartesian vector **OP**; point P is the tip of the vector and A and B are the magnitudes of two component vectors (equivalent to **x** and **y**). The operator j indicates that vector **y** of magnitude B is at right angles to vector **x** of magnitude A; the sign, + or −, before the j gives the direction of the **y** vector. A translation from O to P is the vector sum of the translational vectors $A\hat{\mathbf{i}}$ and $B\hat{\mathbf{j}}$. (Do not confuse the unit vector $\hat{\mathbf{j}}$ with the complex operator j.)

What does j^2 mean? Now j^2 is the same as j(j1). Starting with the j outside the brackets, first rotate the tiny arrow at the origin through 90° in an anticlockwise direction and then, for the j inside the brackets, give it a second 90° anticlockwise turn. The arrow now points along the negative real axis. Move it 1 unit along this axis. The geometrical transformation represented by j^2 is, therefore, equivalent to −1. In a similar way, j^3 can be shown to be the same as −j and j^4 to +1.

Another way of representing the position of point P in the Argand diagram is to use polar coordinates. That is, let N be the displacement of P from the origin and let θ be the angle the radius vector from O to P makes with the positive real axis. Then,

$$A = N\cos\theta \quad \text{and} \quad B = N\sin\theta$$

and

$$P = N(\cos\theta + \mathrm{j}\sin\theta) \tag{4.43}$$

N is called the *modulus* of the *radius vector* **OP** and θ is called the *argument*. Thus

$$\left.\begin{aligned} N &= (A^2 + B^2)^{1/2} \\ \theta &= \tan^{-1}(B/A) \end{aligned}\right\} \tag{4.44}$$

(4.43) can also be written

$$P = N\exp(\mathrm{j}\theta) \tag{4.45}$$

(The equivalence of (4.43) and (4.45) is de Moivre's theorem (see Q1. in Worked examples 4.3)) or

$$P = N \exp \angle\theta \tag{4.46}$$

(4.45) and (4.46) are used interchangeably in many electronics textbooks. We shall use only (4.45).

As the ac currents and pds vary sinusoidally with time, let us put

$$\theta = \omega t.$$

Then, using (4.45) with $N = E_p$, we can write

$$\mathscr{E} = E_p \exp(j\omega t) = E_p(\cos \omega t + j \sin \omega t) \tag{4.47}$$

$\mathscr{E}$ is called the *complex emf* of the voltage source.

As time goes on, the angle θ increases. Point $\mathscr{E}$, lying at a radial distance of E_p units from the origin in the Argand diagram, therefore rotates in an anticlockwise direction with an angular velocity ω. We cannot represent the sequential movement of point $\mathscr{E}$ on a *single* diagram, but we can indicate the direction of rotation of $\mathscr{E}$, as shown in Fig. 4.11. This is quite acceptable. Remember that the axes remain fixed at all times.

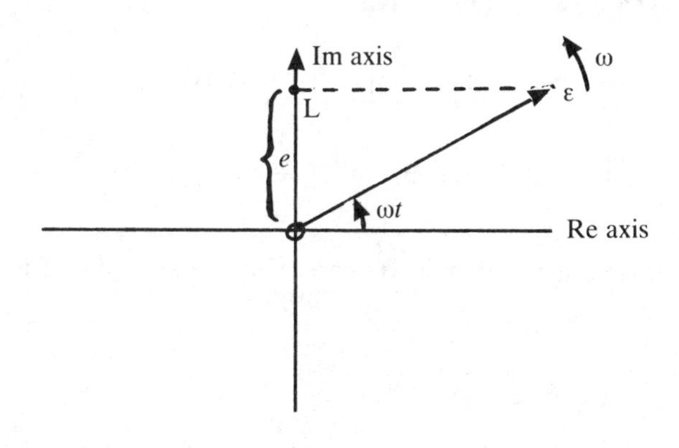

Fig. 4.11.

We have previously represented the instantaneous emf of an AC source by

$$e = E_p \sin \omega t.$$

This is the imaginary component of the complex quantity $\mathscr{E}(= E_p \exp(j\omega t))$.

So, on an Argand diagram, we can think of a point $\mathscr{E}$, a radial distance E_p from the origin, rotating anticlockwise. Then, at any instant of time, the displacement of the projection of $\mathscr{E}$ on the imaginary axis from O will give e (see Fig. 4.11). You should be able to see that as $\mathscr{E}$ rotates through 360°, the distance of L from O will vary periodically, taking all values between O and E_p, as we require. Thus we shall write in the following sections:

$$e = \text{Im}\,\mathscr{E} = \text{Im}\,E_p \exp(j\omega t). \tag{4.48}$$

The symbol Im stands for the *imaginary* part of $E_p \exp(j\omega t)$.

Fig. 4.11 is usually referred to as a *phasor* diagram. The rotating vector **OP** is called a *phasor*.

Worked examples 4.3

Q.1 In section 4.6.1 a complex number P has been expressed as (4.43) and (4.45), viz.

$$P = N(\cos\theta + j\sin\theta)$$

and as

$$P = N\exp(j\theta).$$

Prove that these expressions are equivalent.

First, we need to realize that $\exp x$ may be written as

$$\exp x = 1 + x + x^2/2! + x^3/3! + x^4/4! + x^5/5! + \tag{1}$$

So the way to tackle this problem is to express both $\cos\theta$ and $\sin\theta$ as a power series, viz.

$$\cos\theta = 1 - \theta^2/2! + \theta^4/4! - \theta^6/6! + \ldots \tag{2}$$

and

$$\sin\theta = \theta^3/3! + \theta^5/5! - \ldots \tag{3}$$

We can replace x by $j\theta$ in (1) so that

$$\begin{aligned} \exp(j\theta) &= 1 + j\theta + (j\theta)^2/2! + (j\theta)^3/3! + (j\theta)^4/4! + \ldots \\ &= 1 + j\theta - \theta^2/2! - j\theta^3/3! + \theta^4/4! + j\theta^5/5! - \ldots \end{aligned} \tag{4}$$

using the facts that $j^2 = -1$, $j^3 = -j$, $j^4 = +1$ etc.

Now multiply (3) by j to give

$$j \sin \theta = j\theta - j\theta^3/3! + j\theta^5/5! - \ldots \tag{5}$$

Adding (2) and (5) we find that

$$\cos \theta + j \sin \theta = 1 + j\theta - j\theta^2/2! - j\theta^3/3! + \theta^4/4! + j\theta^5/5! + \ldots \tag{6}$$

which is identical with (4).

$$\exp(j\theta) = \cos \theta + j \sin \theta$$

is called de Moivre's theorem.

Q.2 A complex number P is described by the relation

$$P = (1 - j2)/(3 + j2)$$

Rewrite it in the form $A + jB$ and $N \exp(j\theta)$. Plot P on an Argand diagram.

The first move in problems of this kind is to ensure that the denominator is a real quantity. We can do this here if we multiply the numerator and denominator by $(3 - j2)$. This is called the *complex conjugate* of $(3 + j2)$. All that is needed to determine the complex conjugate is to alter the sign before the j operator. So now we have

$$P = \frac{(1 - j2)(3 - j2)}{(3 + j2)(3 - j2)}.$$

The denominator is $9 + 4 = 13$. Therefore

$$P = \frac{3 - j8 + j^2 4}{13} = \frac{-1 - j8}{13}.$$

This is now of the form $A + jB$, in which

$$A = -1/13 \quad \text{and} \quad B = -8/13.$$

It is relatively straightforward to convert this to the polar form because the modulus N is given by

$$N = (A^2 + B^2)^{1/2}$$
$$= (1 + 64)^{1/2}/13 = 8.06/13 = 0.62$$

and the argument θ by

$$\theta = \tan^{-1}(B/A) = \tan^{-1}(-8/13)/(-1/13)$$
$$= 82.9° \text{ or } 262.9°.$$

The second angle is the required one because A and B are both negative in the third quadrant of the Argand diagramn. In radians, θ is 1.46π. Hence

$$P = 0.62 \exp[j(1.46\pi)].$$

P can be plotted in an Argand diagram if we graduate the axes in units of 2/13, for convenience. Then

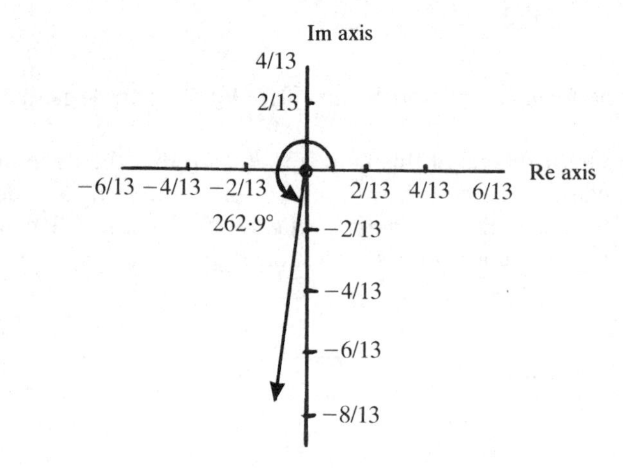

4.6.2 The resistor

Referring to Fig. 4.5, we can write, using KVL:

$$e = E_p \sin \omega t = v_R = iR. \tag{4.49}$$

Now in order to indicate complex notation, we shall use curly script from here onwards. Hence

$$\mathscr{E} = E_p \exp(j\omega t) = \mathscr{V}_R = \mathscr{I}R \tag{4.50}$$

and

$$\mathscr{I} = E_p \exp(j\omega t)/R \tag{4.51}$$

Because e has been represented by the imaginary part of $E_p \exp(j\omega t)$, the measured current is also the imaginary part of $\mathscr{I}$. So

$$\begin{aligned} i &= \text{Im}[E_p \exp(j\omega t)/R] \\ &= (E_p/R) \sin \omega t \end{aligned}$$

which is (4.19). Once again, we obtain the result that the current i is in phase with the emf e.

The phasor diagram looks like Fig. 4.12 at some arbitrary time t. The phasor for

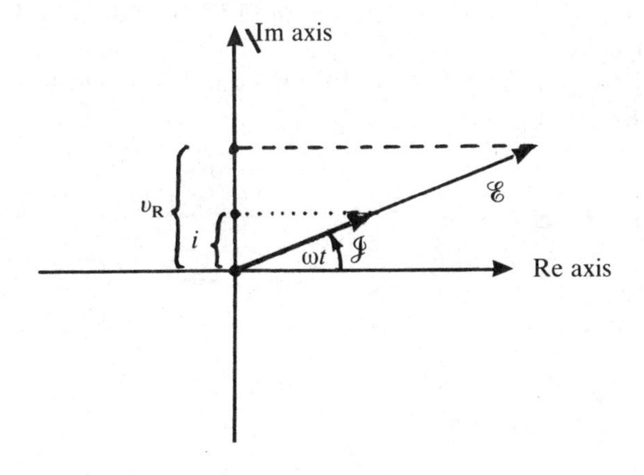

Fig. 4.12.

the complex source emf has a length of 4.0 cm whereas the phasor for the complex current has a length which depends on the value of the circuit resistance but is obviously less than 4.0 cm. The important point about them is that they are coincident — because they are in phase. The projection of the tip of the $\mathscr{E}_p$ vector onto the imaginary axis gives v_R whereas the projection of the $\mathscr{I}$ vector gives i. At $t = 0$, $\omega t = 0$ and both phasors lie along the positive real axis. Then $v_R = iR = 0$. At $t = T/4$, $\omega t = \pi/2$ and both phasors lie along the positive imaginary axis. Then $v_R = E_p$ and $i = E_p/R$.

4.6.3 The capacitor

Referring to Fig. 4.6 we can write, by KVL:

$$e = E_p \sin \omega t = v_c = q/C. \tag{4.52}$$

In complex notation this is

$$\mathscr{E} = E_p \exp(j\omega t) = \mathscr{V}_c = (1/C)\int_{-\infty}^{t} \mathscr{I}\, dt \tag{4.53}$$

which gives

$$\begin{aligned}\mathscr{I} &= C E_p d\{\exp(j\omega t)\}/dt \\ &= j\omega C E_p \exp(j\omega t). \end{aligned} \tag{4.54}$$

(The reason for the limits of integration in (4.53) was discussed in Q.2 of *Worked examples* 3.4.)

We need to convert j into an exponential form. This we can do by using de Moivre's theorem:

$$\exp(j\theta) = \cos\theta + j\sin\theta.$$

Then for $\theta = \pi/2$,

$$\exp(j\pi/2) = 0 + j.1. \tag{4.55}$$

So replace j in (4.54) by exp $(j\pi/2)$ to give

$$\mathscr{I} = \omega C E_p \exp\{j(\omega t + \pi/2)\}. \tag{4.56}$$

Thus the instantaneous current i is the imaginary component of (4.56), that is,

$$i = \omega C E_p \sin(\omega t + \pi/2)$$

which is (4.21).

The phasor diagram is given in Fig. 4.13. Note that the $\mathscr{E}$-phasor lies at an angle ωt to the positive real axis and the $\mathscr{I}$-phasor, of magnitude $\omega C E_p$, lies at an angle of $\pi/2$ ahead of the $\mathscr{E}$-phasor. Both phasors, of course, are rotating anticlockwise with angular velocity ω about the origin.

The complex form of the capacitive reactance is given by

$$\mathscr{X}_c = 1/\mathrm{j}\omega C \tag{4.57}$$

Therefore,

$$\mathscr{V}_c = \mathscr{X}_c \mathscr{I} \tag{4.58}$$

4.6.4 The inductor

The relevant equations are:

$$\mathscr{E} = E_p \exp(\mathrm{j}\omega t) = \mathscr{V}_L = L\,\mathrm{d}\mathscr{I}/\mathrm{d}t$$

$$\mathscr{I} = (E_p/L) \exp\int_{-\infty}^{t} (\mathrm{j}\omega t)\,\mathrm{d}t$$

$$= (E_p/\mathrm{j}\omega L) \exp(\mathrm{j}\omega t). \tag{4.59}$$

$1/\mathrm{j}\omega L$ is the complex inductive reactance $\mathscr{X}_L$.

Now we need to put 1/j in (4.59) into exponential form. $\mathrm{j}^2 = -1$, $-\mathrm{j}^2 = +1$ and $1/\mathrm{j} = -\mathrm{j}^2/\mathrm{j} = -\mathrm{j}$. Using de Moivre's theorem, $-\mathrm{j}$ is identical with $\exp(-\mathrm{j}\pi/2)$. Hence (4.59) can be rewritten

$$\mathscr{I} = (E_p/\omega L) \exp\{\mathrm{j}(\omega t - \pi/2)\} \tag{4.60}$$

Then i is obtained from

$$i = \mathrm{Im}\ (E_p/\omega L) \exp\{\mathrm{j}(\omega t - \pi/2)\} = (E_p/\omega L) \sin(\omega t - \pi/2)$$

which is (4.24).

In general,

$$\mathscr{V}_L = \mathscr{X}_L \mathscr{I} \tag{4.61}$$

Fig. 4.14 gives the phasor diagram.

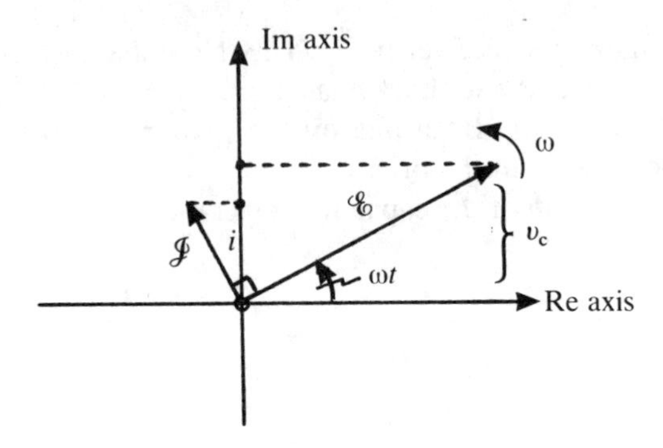

Fig. 4.13.

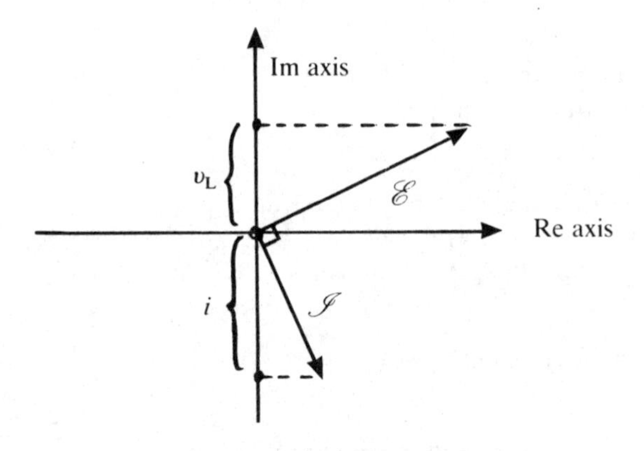

Fig. 4.14.

4.6.5 The L–C–R series circuit

For each component in the series circuit of Fig. 4.15 we know the following

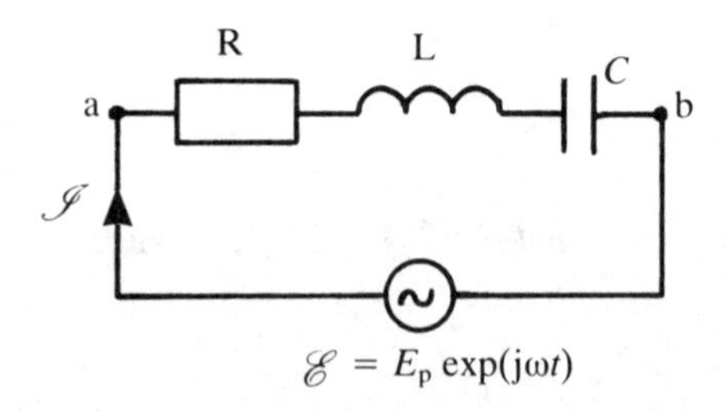

Fig. 4.15.

information:

$$\mathcal{V}_R = \mathcal{I}R$$
$$\mathcal{V}_L = \mathcal{I}\mathcal{X}_L \qquad (4.62)$$

$$\mathscr{V}_c = \mathscr{I}\mathscr{X}_c$$

$\mathscr{I}$ is in phase with $\mathscr{V}_R$, leads $\mathscr{V}_c$ and lags $\mathscr{V}_L$. Hence the pd between nodes (a) and (b) cannot be found by summing the measured pd across each component separately. It is essential to work in terms of complex quantities, for then the phase differences are automatically taken care of. Thus the complex pd between nodes (a) and (b) is

$$\begin{aligned}\mathscr{E} &= \mathscr{I}(R + j\omega L + 1/j\omega C) \\ &= \mathscr{I}[R + j(\omega L - 1/\omega C)] \\ &= \mathscr{I}\mathscr{Z} \\ &= \mathscr{I}Z \exp(j\phi)\end{aligned} \tag{4.63}$$

where $\mathscr{Z}$ is called the *complex ac impedance*, Z is the modulus, equal to $[R^2 + (\omega L - 1/\omega C)^2]^{1/2}$, and ϕ is the argument, equal to $\tan^{-1}\{(\omega L - 1/\omega C)/R\}$.

The phasor diagram for each component is known, and the current is the same everywhere in the circuit. So to obtain the phasor diagram for the series combination let us rotate the individual phasor diagrams until the current $\mathscr{I}$ lies along the same direction for each one. As $\mathscr{I}$ is in phase with $\mathscr{V}_R$, we arrived at Fig. 4.16. The

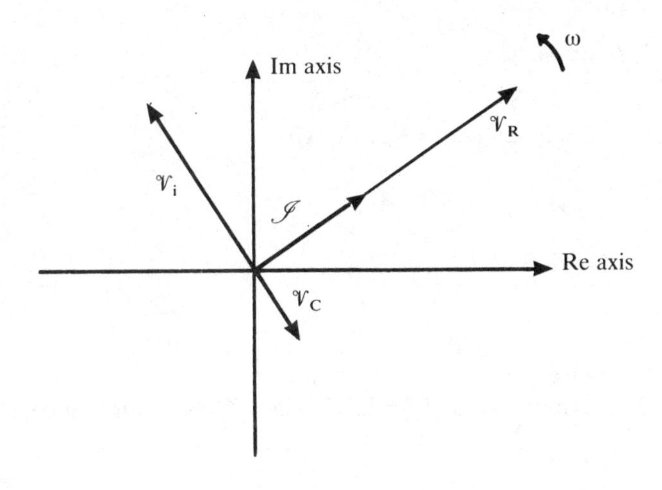

Fig. 4.16.

resultant pd $\mathscr{E}$ between (a) and (b) can now be found by vector summation. This is not as difficult as it may sound. It has been assumed in Fig. 4.16 that $\mathscr{V}_L$ is longer than $\mathscr{V}_c$, which means that there is a resultant $(\mathscr{V}_L - \mathscr{V}_c)$, along this direction, as shown in Fig. 4.17. Now $\mathscr{E}$ can be found by using the *parallelogram rule*. It can be seen that $\mathscr{E}$ leads the current $\mathscr{I}$ by the phase angle ϕ. If, on the other hand, $\mathscr{V}_c$ is greater than $\mathscr{V}_L$ then $\mathscr{E}$ lags the current.

Going back to (4.63), as $\mathscr{E} = E\exp(j\omega t)$, we have

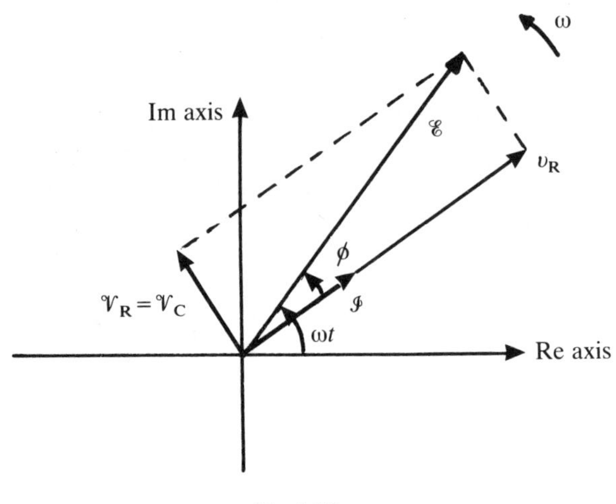

Fig. 4.17.

$$\begin{aligned}\mathscr{I} &= E_p \exp(j\omega t)/Z \exp(j\phi) \\ &= E_p \exp[j(\omega t - \phi)]/[R^2 + (\omega L - 1/\omega C)^2]^{1/2}. \end{aligned} \tag{4.64}$$

Since

$$\begin{aligned} i &= \text{Im}\mathscr{I} \\ &= I_p \sin(\omega t - \phi) \end{aligned}$$

the peak current I_p is related to E_p through

$$\begin{aligned} I_p &= E_p/Z \\ &= E_p/[R^2 + (\omega L - 1/\omega C)^2]^{1/2}. \end{aligned} \tag{4.65}$$

4.6.5.1 Resonance

I_p will take its maximum value $I_o (= E_p/R)$ when Z has its minimum value, equal to R. This occurs for

$$\omega_o L - 1/\omega_o C = 0 \tag{4.66}$$

which gives

$$\omega_o = 1/\surd(LC). \tag{4.67}$$

The subscript o in the above indicates that this is the resonance frequency. At resonance, an ac ammeter will record an rms current I_o' given by

$$I'_o = I_o/\sqrt{2} = E_p/R\sqrt{2} = E'_o/R. \tag{4.68}$$

The argument ϕ is zero at resonance, by (4.66). Hence the phasor diagram at resonance consists of $\mathcal{V}_R$ alone because the reactance terms are equal and opposite (see Fig. 4.18a). The latter are shown dashed. The corresponding phasor diagrams for $\omega < \omega_o$ and $\omega > \omega_o$ are depicted in Fig. 4.18b and Fig. 4.18c.

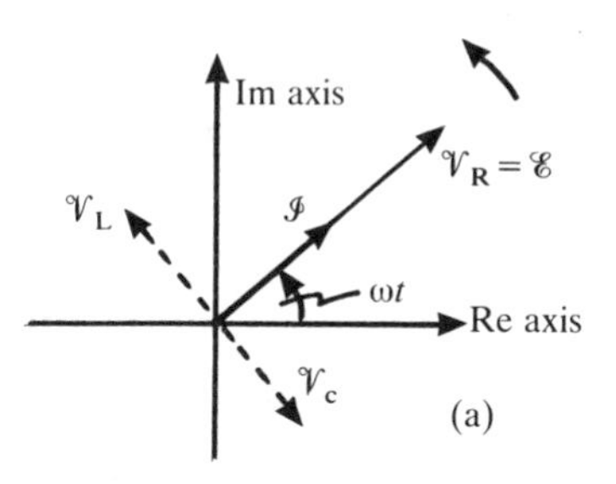

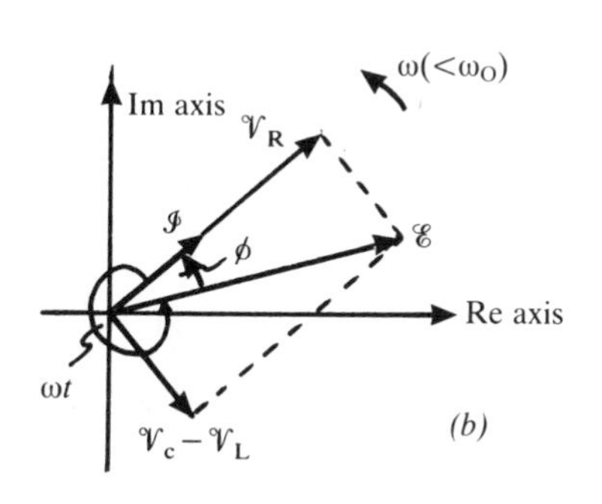

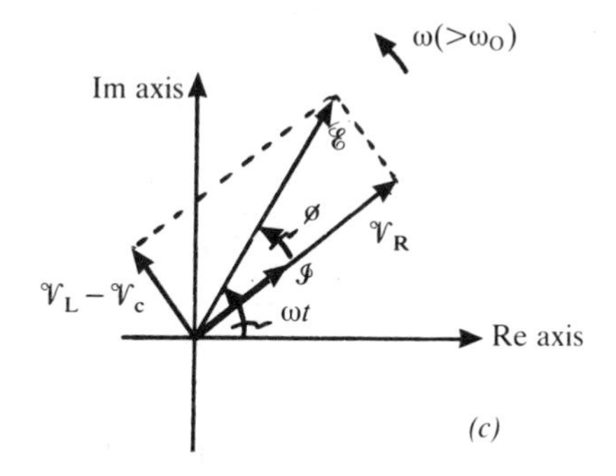

Fig. 4.18.

The rms current I' varies with ω according to Fig. 4.19. Note that at frequencies below ω_o, the circuit is largely capacitive because $1/\omega C$ is larger than ωL, but above ω_o the circuit is inductance-dominated. The curve is *not* symmetrical about ω_o.

The bandwith of the resonance curve is the range of frequencies between the 3 db points on either side of the maximum, as shown.

4.6.5.2 Quality factor

We shall make use of the series resonance circuit of Fig. 4.15. Now the rms pd across the inductor at resonance is

$$(V'_o)_L = X_L I'_o$$

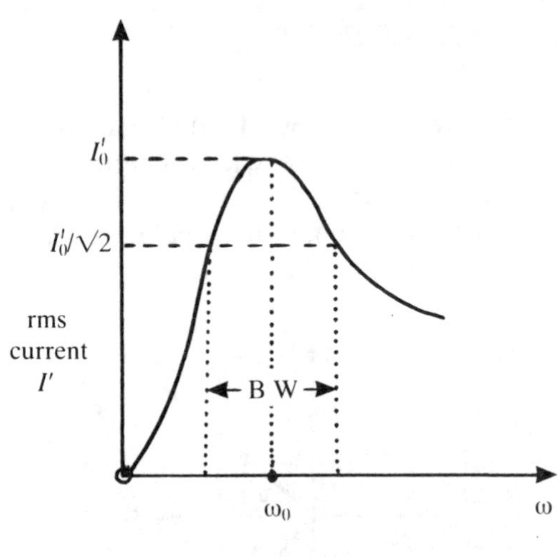

Fig. 4.19.

$$= \omega_o L I'_o = \omega_o L E'_o/R \tag{4.69}$$

and the rms pd across the capacitor at resonance is

$$(V'_o)_c = X_c I'_o$$
$$= I'_o/\omega_o C = E'_o/\omega_o RC. \tag{4.70}$$

The right-hand sides of (4.69) and (4.70) are equal at resonance (although the complex pds are in antiphase). So

$$\omega_o L/R = 1/\omega_o RC. \tag{4.71}$$

They are alternative ways of defining the *quality factor Q* of a resonance circuit. Note that both of them are dimensionless quantities.

DEFINITION:
The value of Q tells us by how much the electromagnetic energy of the inductor (or capacitor) has increased in one cycle of the ac source compared with the decrease in energy of the resistor.

Suppose that the value of R is altered whilst keeping L and C constant. If R is small then I' will be high; Q will be large because R appears in the denominator of (4.71). The resonance curve is sharply peaked, as you can see from Fig. 4.20. However, as R is increased I' gets smaller as does the value of Q. Now the resonance curve is much broader; the bandwidth has increased. This figure leads to an

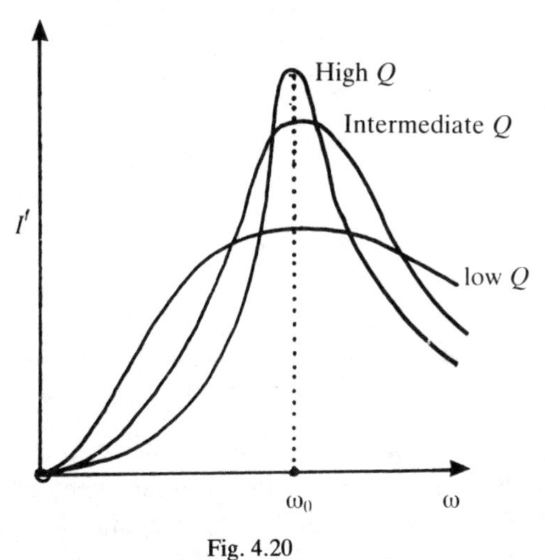

Fig. 4.20

alternative way of describing the quality factor. The higher the value of Q the easier it is to 'home-in' on a particular frequency. It allows us to discriminate between radio stations operating on frequencies which lie close together — a problem that we have probably all experienced from time to time.

Worked examples 4.4

Q.1 Two phasors are given by:

$$z_1 = 3\sin(\omega t + \pi/4) \quad \text{and} \quad z_2 = 4\sin(\omega t + \pi/6)$$

Find the resultant phasor $z = z_1 + z_2$.

Let us put z_1 and z_2 into complex form, as

$$\mathfrak{X}_1 = 3\exp[\mathrm{j}(\omega t + \pi/4)] = 3\exp(\mathrm{j}\omega t)\exp(\mathrm{j}\pi/4) \tag{1}$$

$$\mathfrak{X}_2 = 4\exp[\mathrm{j}(\omega t + \pi/6)] = 4\exp(\mathrm{j}\omega t)\exp(\mathrm{j}\pi/6). \tag{2}$$

We shall put $t = 0$ for then $\exp\mathrm{j}(\omega t) = 1$. The remainder of each expression now allows us to locate $\mathfrak{X}_1$ and $\mathfrak{X}_2$ in the Argand diagram at the time $t = 0$. The resultant vector can be either found graphically by using the parallelogram rule, or calculated. Once this has been done, the vectors can be allowed to rotate with angular velocity ω in order to obtain their time-dependence; the moduli of the phasors always stay constant but their directions vary with time.

The Argand diagram looks like:

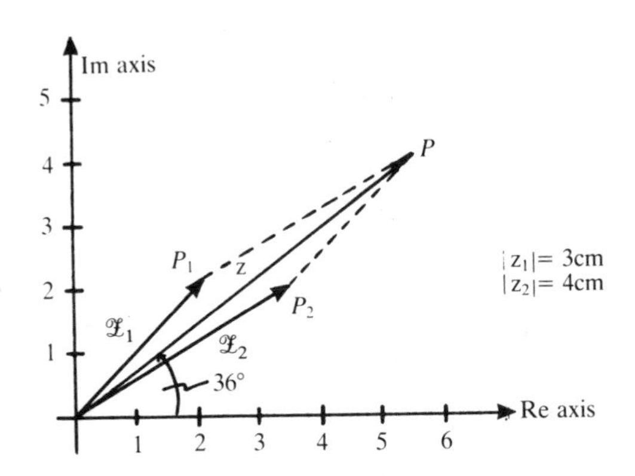

$\mathfrak{Z}_1$ makes an angle of $+45°$ with the real axis whilst $\mathfrak{Z}_2$ makes an angle of 30°. The parallelogram may be completed and the resultant vector $\mathfrak{Z}$ measured to give a modulus of 6.9 and an argument of 36° (= 0.2π radians).

The alternative approach is to make use of de Moivre's theorem, when

$$
\begin{aligned}
3\exp(j\pi/4) &= 3(\cos\pi/4 + j\sin\pi/4) \\
&= 3/\sqrt{2} + j\,3/\sqrt{2}
\end{aligned} \tag{3}
$$

(as $\cos\pi/4 = \sin\pi/4 = 1/\sqrt{2}$)
and

$$
\begin{aligned}
4\exp(j\pi/6) &= 4[\cos\pi/6 + j\sin\pi/6] \\
&= 2\sqrt{3} + j2
\end{aligned} \tag{4}
$$

(as $\cos\pi/6 = \sqrt{3}/2$ and $\sin\pi/6 = 1.2$).

The resultant $\mathfrak{Z}$ is found by adding the real terms and the imaginary terms to give

$$
\begin{aligned}
\mathfrak{Z} &= (3/\sqrt{2} + 2\sqrt{3}) + j\,(3/\sqrt{2} + 2) \\
&= 5.58 + j\,4.12.
\end{aligned} \tag{5}
$$

Hence the modulus z is $(5.58^2 + 4.12^2)^{1/2} = 6.94$ and the argument θ is $\tan^{-1}(4.12/5.58) = 36.4° = 0.2\pi$ radians. Hence

$$\mathfrak{Z} = 6.94\exp(j\,0.2\pi)$$

These values agree with the Argand diagram.

Q.2 Calculate the magnitude and phase angle of the impedance of an RC circuit operating at 1 kHz with: $R = 100\,\Omega$ and $C = 0.1\,\mu$F.

The complex impedance $\mathscr{Z}$ is

$$\mathscr{Z} = R + 1/j\omega C$$
$$= R - j/\omega C \tag{1}$$

Therefore

$$Z = [R^2 + (1/\omega C)^2]^{1/2} \quad (\text{with } \omega = 2\pi f = 2000\pi \text{ rad s}^{-1})$$
$$= [1 \times 10^4 + 1/(2 \times 10^3\pi \times 10^{-7})^2]^{1/2}$$
$$= 1595\,\Omega \tag{2}$$

The phase angle ϕ is

$$\phi = \tan^{-1}(-1/\omega RC)$$
$$= \tan^{-1}(-1/(2000\pi \times 100 \times 10^{-7}))$$
$$= \tan^{-1}[-100/2\pi]$$
$$= \tan^{-1}[-15.9]$$
$$= 360° - 86.4°$$
$$= 273.6°. \tag{3}$$

The Argand diagram looks like

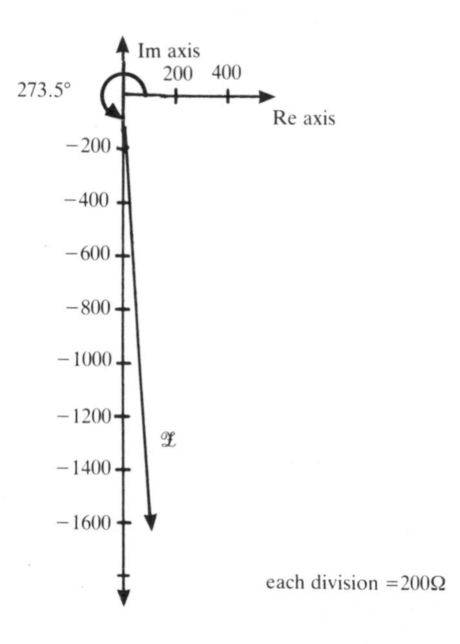

Q.3 In the L–C–R series resonant circuit:

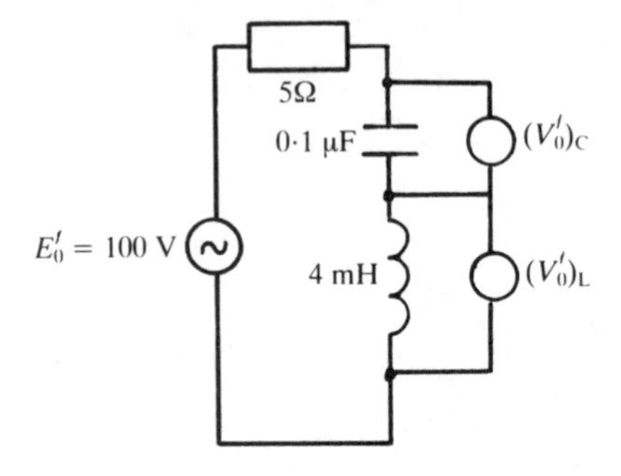

Determine the resonant frequency, the current flowing at resonance, and the rms pd's across L, R, and L–R.

The resonance frequency is given by

$$\begin{aligned}\omega_o &= 1/\sqrt{(LC)}\\ &= 1/\sqrt{4 \times 10^{-3} \times 10^{-7}}\\ &= 5 \times 10^4 \text{ rad s}^{-1}. \qquad (1)\end{aligned}$$

The current flowing at resonance is the rms value because the source pd of 100 V is an rms value (see Fig. 4.4 in section 4.2). Hence

$$\begin{aligned}I_o' &= E_o'/R\\ &= 100/5 \text{ A}\\ &= 20 \text{ A}. \qquad (2)\end{aligned}$$

The rms pd across the inductor at resonance is given by (4.69) as

$$(V_o')_L = \omega_o L I_o'.$$

Therefore

$$\begin{aligned}\omega_o L I_o' &= 5 \times 10^4 \times 4 \times 10^{-3} \times 20\\ &= 4000 \text{ V} \qquad (3)\end{aligned}$$

Similarly, the rms pd across the capacitor is given by (4.70) as

$$(V_o')_c = I_o'/\omega_o C.$$

Therefore

$$(V'_o)_c = 20/5 \times 10^4 \times 10^{-7}$$
$$= 4000 \text{ V}. \tag{4}$$

The rms pd across the combination is zero because $\mathcal{V}_L$ and $\mathcal{V}_c$ are always in antiphase.

You should note that $(V'_o)_L$ and $(V'_o)_c$ are very high, and much larger than E'_o. Each can be measured with an appropriate voltmeter. This calculation shows the need to use capacitors which are able to support 4 kV in such circuits without breakdown.

Q.4 Determine the complex impedance of the following circuit at 100 Hz.

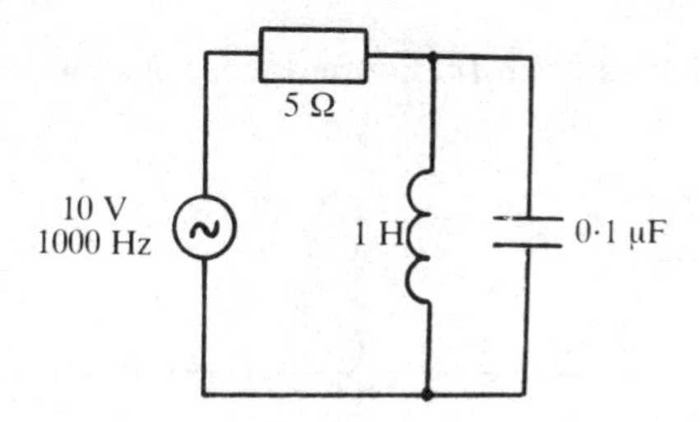

The rules for determining the complex impedance of a combination of circuit elements are the same as those for pure resistances. In this example, the circuit consists of a parallel L–C branch in series with a pure resistance. Hence

$$\mathcal{Z}_{tot} = R + \mathcal{Z}_{parallel}. \tag{1}$$

Now, $\mathcal{Z}_{parallel}$ is obtained as

$$1/\mathcal{Z}_{parallel} = 1/j\omega L + j\omega C$$
$$= (1 - \omega^2 LC)/j\omega L$$

so that

$$\mathcal{Z}_{parallel} = j\omega L/(1 - \omega^2 LC), \tag{2}$$

Hence

$$\mathcal{Z}_{tot} = R + j\omega L/(1 - \omega^2 LC). \tag{3}$$

The modulus Z is

$$[R^2 + \{\omega L(1 - \omega^2 LC)\}^2]^{1/2}. \tag{5}$$

On substituting the circuit values we find that

$$Z = 654.6\,\Omega \tag{5}$$

and the phase angle ϕ, defined by,

$$\tan^{-1}[\omega L/R(1-\omega^2 LC)],$$

is

$$\begin{aligned}\phi &= \tan^{-1}131\\ &= 89.6°\\ &\simeq 90°.\end{aligned} \tag{6}$$

The phasor diagram shows that the impedance phasor lies along the positive imaginary axis.

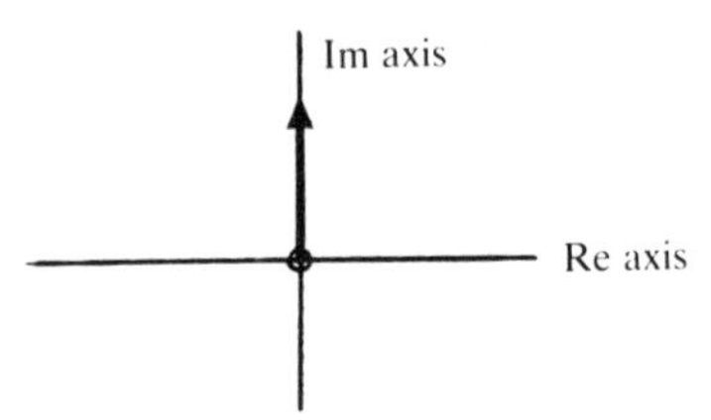

In addition,

$$\mathscr{E} = \mathscr{I}\mathscr{Z}_{tot}$$

and

$$\begin{aligned}\mathscr{I} &= \mathscr{E}/\mathscr{Z}_{tot}\\ &= (E_p/Z)\exp\{j(\omega t-\phi)\}.\end{aligned}$$

The rms current at 100 Hz is

$$\begin{aligned}E'/Z &= 10/654.6\text{ A}\\ &= 15\text{ mA}.\end{aligned} \tag{7}$$

It is interesting that Z is infinite when

$$1-\omega^2 LC = 0 \tag{8}$$

or

$$\omega_o = 1/\surd(LC)$$

$$= 1/\sqrt{10^{-7}}\ \text{rad s}^{-1}$$
$$= 316\ \text{rad s}^{-1}. \tag{9}$$

This means that the current is zero *irrespective of the amount of series resistance* in the circuit. The whole of the rms source emf appears across the L–C combination when this occurs. In other words, the ac voltage gain A is 1 at this value of frequency. At other frequencies A is <1. Hence ω_o is the resonance frequency for this circuit. The variation of A with ω can be worked out by noting that:

(i) At $\omega = 0$, $I' = 10/5 = 2\,\text{A}$. The capacitor blocks-off the dc current, and the inductor acts like a short-circuit. There is no pd across the L–C branch, and $A = 0$;
(ii) At $\omega = \infty$, all the current flows through the capacitor because its impedance is zero. It acts like a short–circuit. Once again, $A = 0$. Hence the variation of A with ω looks like:

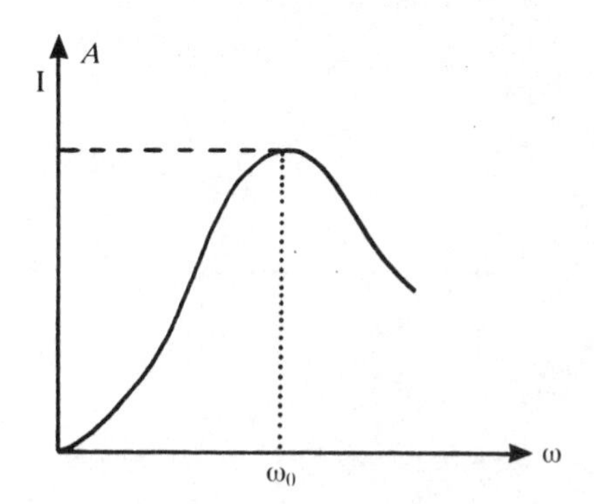

Q.5 Sketch the A vs ω curve for the L–C branch in the following circuit:

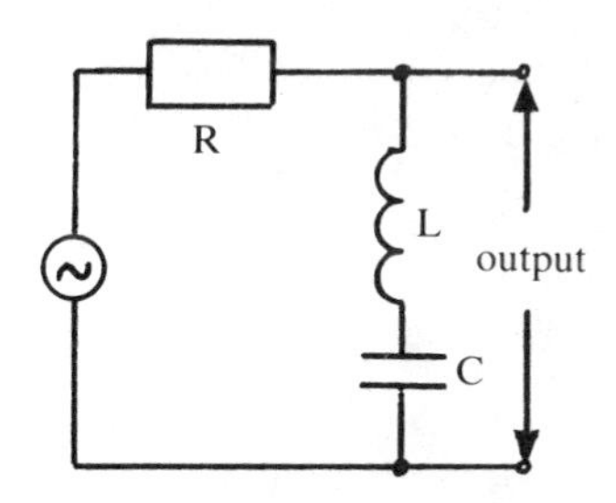

As

$$\mathscr{Z} = R + \text{j}(\omega L - 1/\omega C), \tag{1}$$

the resonance condition is

$$\omega_o L - 1/\omega_o C = 0$$

so that

$$\omega_o = 1/\sqrt{(LC)}$$

Therefore the impedance across L–C is zero at resonance. The rms output pd must be zero. On each side of ω_o the impedance increases and the rms pd must increase. So it is expected that the A vs ω curve looks like

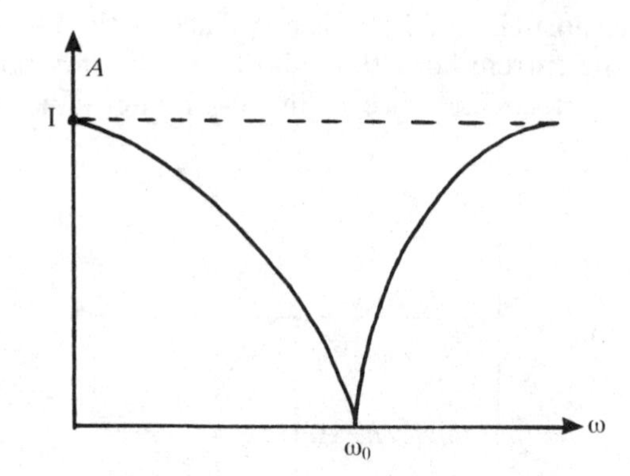

5

The semiconductor *P–N* junction diode and its circuits

Objectives

(i) To distinguish between intrinsic and extrinsic semiconductors, using an atomic description; doping
(ii) To discuss the (elementary) physics of the diode; the depletion layer
(iii) How to obtain the *I–V* characteristics of a diode
(iv) To meet the equation which governs the *I–V* characteristic
(v) To define the dc and dynamic resistance
(vi) To note the effect of temperature on the *I–V* characteristic
(vii) To analyse the behaviour of a number of basic diode circuits
(viii) To use the Zener diode as a voltage-regulating device
(ix) To have a brief introduction to the photo- and light-emitting diode.

5.1 SOME BACKGROUND INFORMATION

5.1.1 The atomic picture

To describe the atomic processes which occur in the semiconducting diode, it will be helpful to refer to one material in particular. Silicon is chosen here because of its total domination of the semiconducting commercial market over a number of years.

Silicon lies in Group IV of the periodic table of the elements, the same group as carbon. It has four outer electrons. The complete electronic structure is $1s^2 2s^2 2p^6 3s^2 3p^2$. A crystal of silicon can be thought of as a repeated pattern of *diamond lattice unit cells;* the unit cell is so named because it is also the repeating pattern in diamond. The unit cell is cubic with sides 0.543 nm in length, possessing atoms at each of its corners and at the centre of each face — just like a face-centred cubic crystal of gold. There is a significant difference, however, between the unit cells in silicon and gold — the silicon unit cell has four additional atoms. Two of the atoms are positioned one quarter way from the top face along one pair of opposite body diagonals, whilst the other two atoms are found one quarter way up the other pair of body diagonals. Each atom in the silicon lattice is surrounded by four nearest neighbours. In all, there are eight atoms per unit cell and 5×10^{22} atoms in each cubic centimetre of material.

There are two facts to consider now: (i) each silicon atom is surrounded by four nearest neighbours; (ii) each silicon atoms has four outer electrons. Putting these together we arrive at an atomic picture for silicon in which an electron from each atom forms what is termed a *covalent* bond. This means that each silicon atom *shares* electrons with each of its nearest neighbours.

This atomic picture is an ideal one because it really applies only at the absolute zero of temperature (0 K) in the absence of any defects and impurity atoms. At any temperature above 0 K *bond-breakage* can occur because some of the electrons have sufficient thermal energy to break free of their nucleus. Fig. 5.1 depicts the

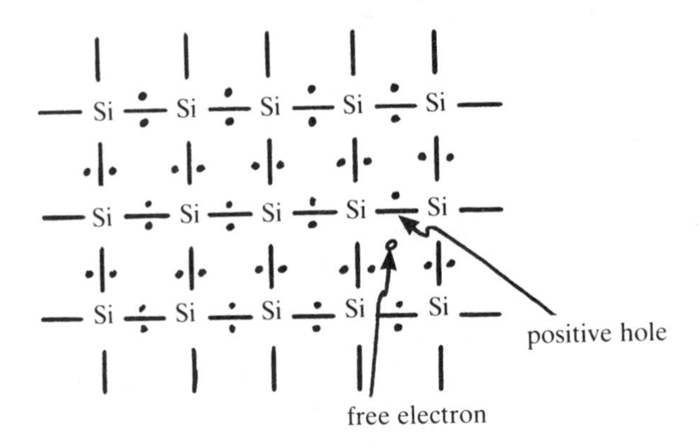

Fig. 5.1.

generation of a conduction electron. Two silicon atoms now have seven electrons associated with them instead of the normal complement of eight. Therefore, in this region of the crystal there is a net positive charge. A *positive hole* has been created. The presence of free electrons, created by bond-breakage, converts silicon into a conductor of electricity — albeit a poor one because there are only a small number of free electrons available.

From the discussion so far it is important to note that the number of positive holes generated is always equal to the number of free electrons. That is, in each unit volume of the crystal

$$n = p = n_{\text{i}} \tag{5.1}$$

where n_{i} is called the *intrinsic* concentration of charge carriers. Silicon is referred to as an *intrinsic* semiconductor.

A positive hole will disappear from place A in the crystal if a valence electron from a neighbouring bond fills it. But this electron leaves a positive hole at its previous location, B. So the positive hole has appeared to move from A to B. By carrying on with this process we can imagine the movement of a hole through the crystal.

5.1.2 Some relevant equations

The current density j in the crystal consists of a component due to electron movement and a component due to holes, such that

$$j = [\sigma_e + \sigma_h]F \tag{5.2}$$

which is equivalent to

$$j = q[n\ \bar{\upsilon}_e + p\ \bar{\upsilon}_h] \tag{5.3}$$

where q is the magnitude of the electronic charge, σ is the electrical conductivity, F is the electric field strength, and $\bar{\upsilon}_e$ and $\bar{\upsilon}_h$ are the mean speeds of the electrons and positive holes, respectively. The speeds are different because although the conduction electron is essentially free, the positive hole is not — the latter has to wait for a valence electron to enter it before any movement occurs.

The carrier mobility is defined as

$$\mu = \bar{\upsilon}/F \tag{5.4}$$

which allows (5.3) to be rewritten as

$$j = q[n\mu_e + p\mu_h]F$$

and

$$\sigma = (n\mu_e + p\mu_h). \tag{5.5}$$

In pure silicon, we can replace n and p by n_i so that (5.5) becomes

$$\sigma = n_i(\mu_e + \mu_h)q. \tag{5.6}$$

Substituting the values: $n_i = 1.5 \times 10^{16}\ \text{m}^{-3}$, $\mu_e = 0.135\ \text{m}^2\text{V}^{-1}\text{s}^{-1}$, $\mu_h = 0.048\ \text{m}^2\text{V}^{-1}\text{s}^{-1}$, we find that the electrical conductivity of silicon is $4.39 \times 10^{-4}\ \text{Sm}^{-1}$. The corresponding value for germanium is $2.17\ \text{Sm}^{-1}$.

As we have just observed, a positive hole moves through the crystal when a valence electron from a neighbouring lattice atom enters the region of positive charge. The hole will disappear entirely, however, if a conduction electron *recombines* with it. The rate of recombination depends on the concentrations of conduction electrons and positive holes.

5.1.3 Extrinsic conduction: doping

Pure silicon is unsuitable as a semiconductor because there are only a small number of conduction electrons present, even at room temperature. However, its electrical conductivity can be improved greatly by introducing a small amount of another atomic species into the crystal lattice. These atoms are termed *impurity* atoms, and the general process is known as *doping*.

5.1.3.1 n-type silicon

The addition of a Group V element, such as phosphorus or indium, to silicon produces *n-type* silicon. Phosphorus has five outer electrons, so that when a phosphorus atom *replaces* a silicon atom in the crystal lattice, four of these electrons are able to form covalent bonds but the fifth is *free* to wander through the lattice (see Fig. 5.2). The phosphorus atom is essentially fixed within the lattice so that it is

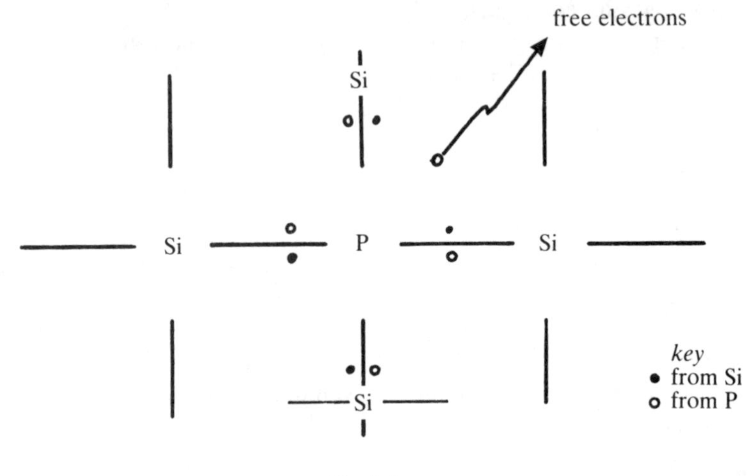

Fig. 5.2.

immobile, but as it has led to the generation of an electron it is called a *donor* atom with a net positive charge of + 1. However, as there is nothing to prevent an electron from entering the region around the phosphorus atom, this region is, on average, electrically neutral. Thus we see that one impurity atom generates one conduction electron. (The *n* in *n*-type is the first letter of the word negative).

Although the number of impurity atoms added to silicon is quite small, the effect on the electrical characteristics is substantial. For example, at room temperature, as we have stated, intrinsic silicon has an electrical conductivity of 4.4×10^{-4} Sm^{-1}, but by adding one indium atom for every 10 million silicon atoms this increases to about 40 Sm^{-1}.

5.1.3.2 p-type silicon

The addition of a group III element, such as boron, to silicon produces *p*-type silicon. As may be guessed, the *p* refers to the first letter of the word positive. Boron has only three outer electron, so only three covalent bonds can form with neighbouring silicon atoms, as Fig. 5.3 depicts. The only way that the fourth bond can form is if an electron from somewhere in the crystal completes the bonding. A positive hole is created in this doping process. As the boron atom will happily accept an electron to satisfy the bonding requirements, it is said to be an *immobile acceptor* atom.

The conduction electrons in *n*-type silicon and the positive holes in *p*-type silicon are known as *majority* carriers. However, there will still be positive holes in *n*-Si and electrons in *p*-Si produced through bond-breakage between the silicon lattice atoms. These are known as *minority* carriers because they exist in smaller numbers than the majority carriers. It is also worth remembering that neither *n*-Si nor *p*-Si possesses an overall electrical charge.

Now we are in a better position to understand the operation of the *pn* junction diode.

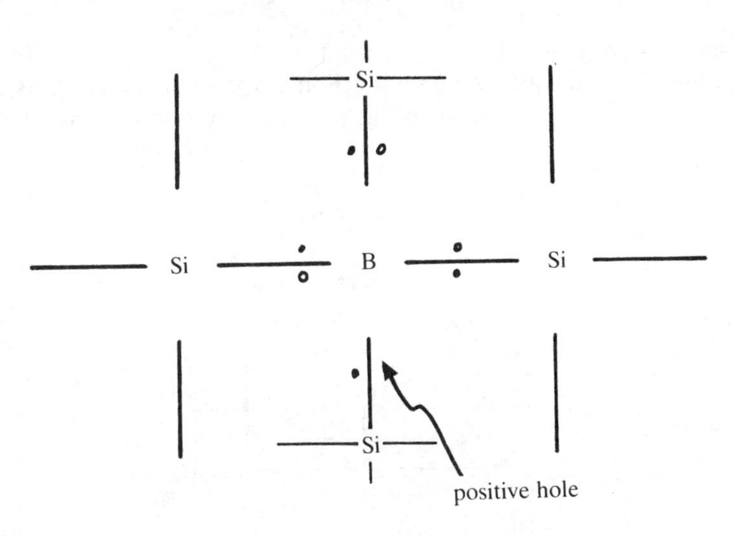

Fig. 5.3.

5.2 THE *pn* JUNCTION DIODE

The *pn* junction device is made from one piece of silicon. One half is doped appropriately to produce *n*-Si whereas the other half is of *p*-Si. Electrons flow across the *junction* from *n*-Si to *p*-Si. They leave behind them, in the region close to the junction, a number of positively-charged donor ions whilst creating on the other side of the junction a region of negatively-charged acceptor ions — as Fig. 5.4 illustrates.

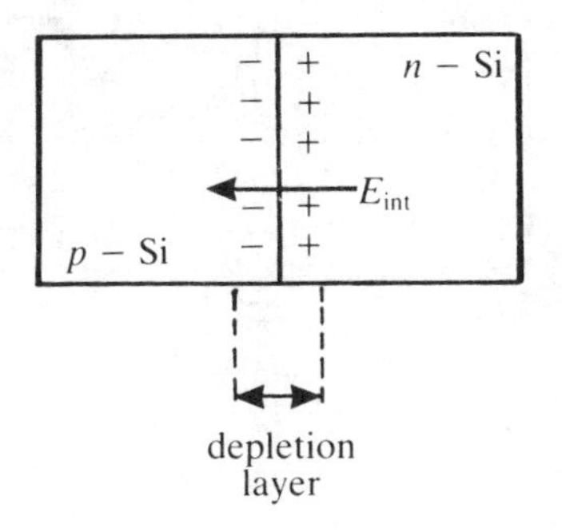

Fig. 5.4.

The region between the ions is devoid of free charge carriers and, for this reason, is known as the *depletion* layer. An internal electric field is established which acts in the wrong direction to assist further movement of electrons from *n*-Si to *p*-Si. In silicon, the pd across the depletion layer is about 0.6 to 0.7 V. This compares with 0.3 V in germanium and 1.2 V in gallium arsenide.

5.2.1 Concept of reverse bias

A voltage source is attached to the *pn* junction diode via metal contacts deposited onto the ends of the diode, as shown in Fig. 5.5a. Electrons are attracted across the *n*-

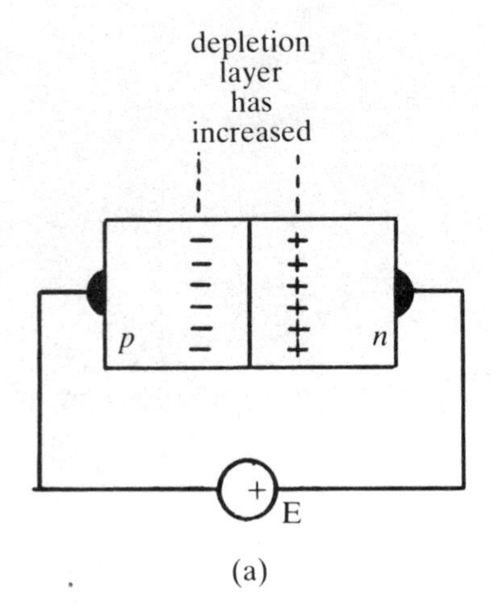

(a)

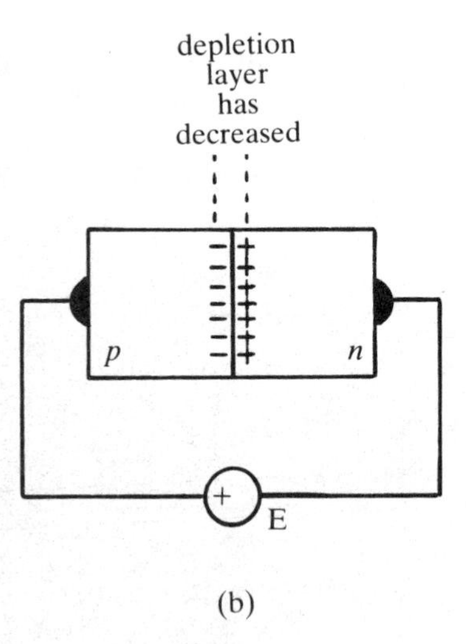

(b)

Fig. 5.5.

Si/metal junction towards the positive terminal of the source, and in so doing expose further positively-charged donor ions. Similarly, electrons flow across the metal/*p*-Si junction from the negative terminal of the voltage source and increase the concentration of acceptor ions. The overall effect is an increase in the width of the depletion layer. Reverse-biasing the diode makes it more difficult for electrons to flow across the junction from the *n*-Si to the *p*-Si. Some minority electrons will flow in the opposite direction, however, but the measured current is quite small, being ~μA or less. A more detailed discussion requires some knowledge of the contact potential between two materials.

5.2.2 Concept of forward bias

When the voltage source is reversed, as in Fig. 5.5b, electrons in the *n*-Si are repelled away from the contact towards the junction where they neutralize some of the net positive charge. The result is that the width of the depletion layer decreases. Increasing the emf of the source continuously from zero, ultimately has the effect of removing the depletion layer entirely, with the result that there is now no longer any barrier to the flow of electrons from *n*-Si to *p*-Si. Applying Kirchhoff's voltage law to the source-diode loop gives

$$E + V_{\text{dep}} = 0, \qquad (5.7)$$

where V_{dep} stands for the pd across the depletion layer. Hence E needs to be 0.7 V for free electron flow to occur across the junction.

5.3 SYMBOL FOR THE DIODE

The symbol for the pn junction diode is given in Fig. 5.6a. The arrow indicates the

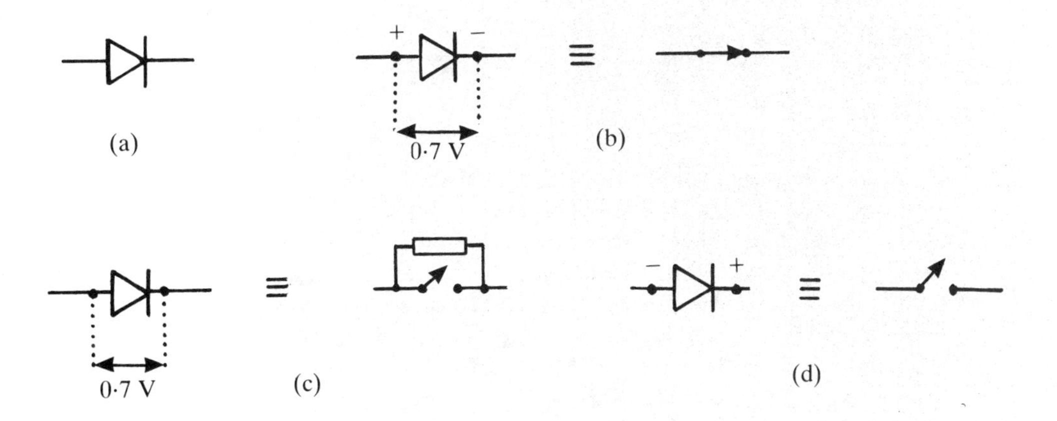

Fig. 5.6.

direction of conventional current flow (opposite to electron flow). In circuit analysis we shall equate a diode forward-biased at 0.7 V with a closed switch (Fig. 5.6b), one biased at <0.7 V with a leaky switch (Fig. 5.6c), that is, an open switch with a shunt resistor, and one under reverse-biased conditions with an open switch (Fig. 5.6d).

A complete analysis of the diode must include the *bulk* resistance of the diode, that is, the total resistance of the *n*- and *p*-regions. In the following sections, the circuit resistance can be assumed to include the bulk resistance, although it is not specifically referred to.

5.4 *I-V* CHARACTERISTICS

The shape of the current–voltage characteristic of a silicon diode can be explored by using the same basic circuit, although, for convenience, it is helpful to divide the complete characteristic into a forward and a reverse voltage region.

To measure the forward characteristic, connect up the circuit shown in Fig. 5.7.

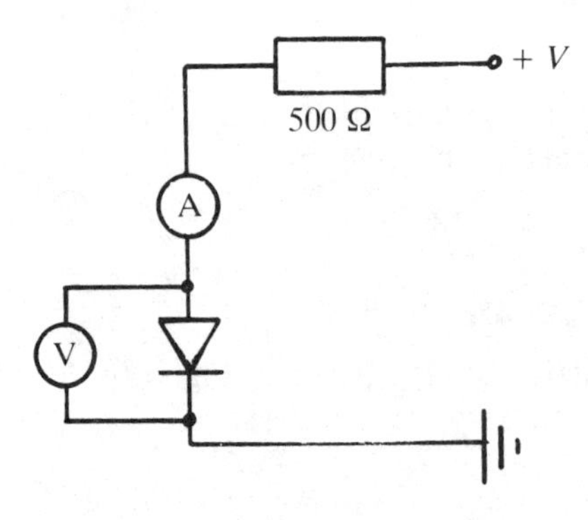

Fig. 5.7.

As the dc resistance of the diode under forward bias is quite small, about a few ohms, it is essential to connect the voltmeter across the diode only. First, scan through the forward voltage range and observe how the current changes. You should find that the current increases slowly at first, from about 1 μA, but then more rapidly — perhaps exponentially — up to about 50 mA or so. Therefore, it is advisable to measure the forward voltage for given values of current.

The ammeter actually measures the current I flowing through the diode and the voltmeter. hence it is essential to correct the experimental data in order to obtain the diode current I_d (see sections 3.7.1 and 3.7.2). From Ohm's law it can be easily shown that

$$I_d = R_v I/(R_d + R_v) \tag{5.8}$$

where R_v is the resistance of the voltmeter and R_d is the resistance of the diode.

The dc resistance of the diode is MΩ under reverse-bias. So, it is necessary to connect the voltmeter across both the diode and ammeter. The manufacturer's specifications for the diode should quote the so-called *peak-inverse voltage*. This is the maximum pd that the diode will support under reverse bias without breakdown occurring. So it is advisable to make the maximum reverse voltage about half of this figure. The current is in the microampere range. If plotted on linear graph paper the complete *I-V* characteristic should look like Fig. 5.8. However, to test whether the current under forward bias increases exponentially the current ought to be plotted on a logarithmic scale, when a straight line should be obtained.

5.4.1 Introductory theory

The forward current increases with potential difference V according to the relation

$$I = I_S[\exp(qV/nkT) - 1]$$

but for all practical purposes, because the exponential term is usually very much larger than unity,

$$I = I_s \exp(qV/kT). \tag{5.9}$$

Here, I_s is the reverse saturation current (it is $\sim\mu A$), q is the electronic charge ($= 1.6 \times 10^{-19}$ C), k is Boltzmann's constant ($= 1.38 \times 10^{-23}$ JK^{-1}), T is the absolute temperature, and n is the exponential ideality factor. n is 1 for a germanium diode and 2 for silicon, although most silicon diodes in practice give a value between 1.3 and 1.6. In order not to overcomplicate our analysis, here, and in the next chapter, it will be assumed that n is 1.

At 300 K, which is approximately room temperature, (5.9) can be simplified to

$$I = I_S \exp(40\ V) \tag{5.10}$$

Let the diode's forward pd change by ΔV and the current change by ΔI, then the transconductance g_m is the ratio $\Delta I/\Delta V$. Differentiating (5.10) gives

$$\begin{aligned} g_m &= 40\ I_S \exp(40\ V) \\ &= 40\ \mathrm{I} \end{aligned}$$

Therefore, the *dynamic* or ac resistance r of the diode in ohms is

$$\begin{aligned} r &= \Delta V/\Delta I \\ &= 1/40\ I \text{ (with } I \text{ in amperes)} \\ &= 25/I \text{ (with } I \text{ in milliamperes).} \end{aligned} \tag{5.11}$$

We shall use r in Chapter 6 when we consider small-signal amplification.

In contrast to the dynamic resistance, the dc resistance is defined by the relation

$$\text{dc RESISTANCE} = \frac{\text{TOTAL dc POTENTIAL DROP } V_d \text{ ACROSS DIODE}}{\text{DIODE CURRENT } I_d}. \tag{5.12}$$

The dashed part of the reverse characteristic is included for completeness' sake. It shows that at a certain pd the current increases in a dramatic fashion. This is the *breakdown* region, and the pd which initiates breakdown is called the *breakdown* pd V_B. For one class of diode, called the *Zener* diode, V_B is about -6 V to -10 V, whereas for the second class, the *avalanche* diode, V_B is about -50 V.

5.4.2 Effect of temperature

Changes in temperature can greatly affect the *I-V* characteristics unless the diode circuit is both designed carefully and packaged adequately. An increase in tempera-

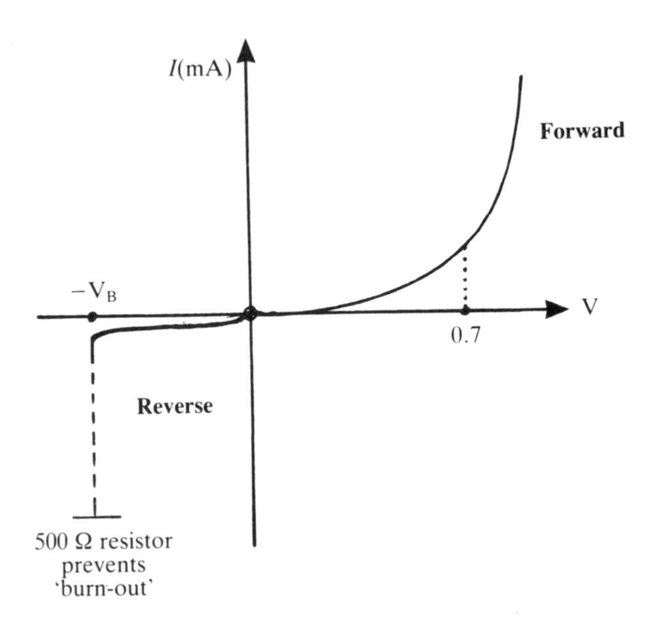

Fig. 5.8.

ture causes a reduction in the *turn-on* pd, which we know is 0.7 V in silicon. The reduction for each 1°C rise in temperature is 2 mV $°C^{-1}$. The reverse saturation current also roughly doubles for each 10 °C rise.

5.5 DIODE CIRCUITS

This section will discuss some applications of the diode in a few representative circuits. After which it is hoped that the student will be able to apply the same basic technique in understanding the principle of operation of other kinds of *diode* circuits.

5.5.1 Half-wave rectifier

Fig. 5.9 is the basic circuit to be studied. It will be beneficial to treat the diode as an

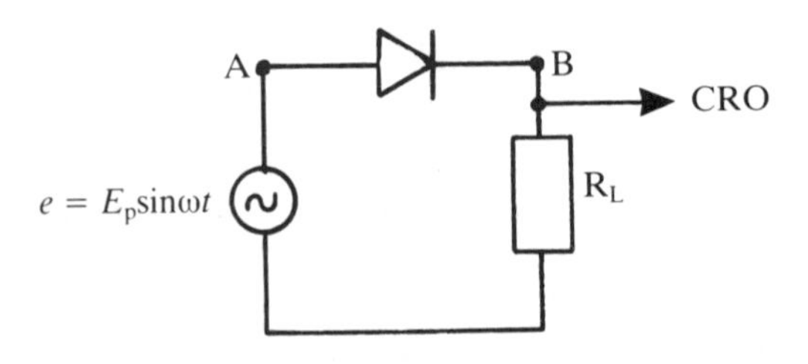

Fig. 5.9.

ideal device initially; that is, the 0.7 V potential drop across the depletion layer will be ignored. However, once the circuit principle is understood, it should be relatively straightforward to include the 0.7 V and modify the conclusions.

WORKING RULES:
(i) A forward-biased diode acts like a closed switch;
(ii) A reverse-biased diode acts like an open switch.

At the instant when the potential of node A is positive relative to earth, the diode conducts and there is a short-circuit between A and B. The potential of B rises to E_P and falls to zero in phase with that of A. If a CRO is used to probe the potential at node B, it will display a waveform identical with that at A.

On the other hand, when the potential of node A goes negative relative to earth, the diode does not conduct and there is an open circuit between A and B. No current flows at all during this negative cycle, so that the potential at B is zero during this half cycle of the voltage source. The maximum pd across the diode during this half-cycle is $-E_P$.

If the potential of node A varies sinusoidally as a function of time then v_B will vary as in Fig. 5.10. The period of v_B is identical with the ac source.

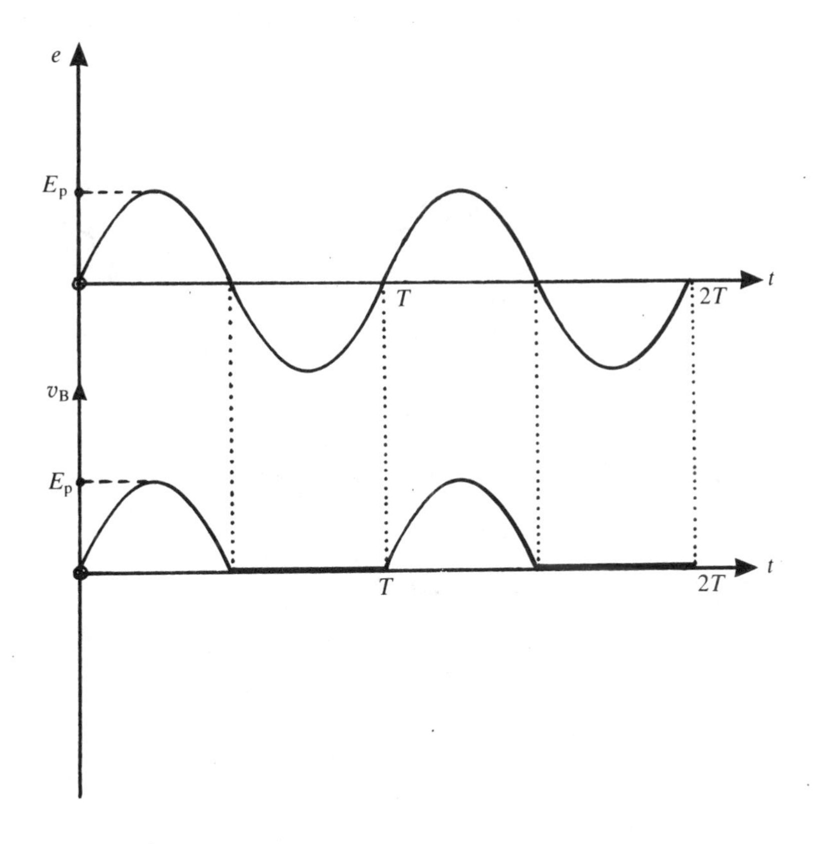

Fig. 5.10.

If we now consider an actual diode, then we can no longer neglect the pd of 0.7 V across the depletion layer. The effect that this has on the foregoing discussion is that the diode becomes forward-biased only when the potential at A reachs 0.7 V; below 0.7 V the diode is reverse-biased. The amplitude of υ_B is actually $(E_P - 0.7)$V.

Worked examples 5.1

Q1. Calculate the average value $\bar{\upsilon}_o$ of the output pd of a half-wave rectified circuit utilizing a silicon diode, if the rms value of the source voltage is 5 V.

First, we will determine an expression for $\bar{\upsilon}_o$ for the ideal diode. For this purpose the variation of υ_o with time is redrawn. By definition, the average pd over the period of the voltage waveform has a constant value — which is sometimes referred to as the dc value, υ_{dc}. We require that the area of the rectangle between υ_{dc} and the time axis, as depicted below, should be equal to the area betwen υ_o and the time axis. Mathematically, this means that

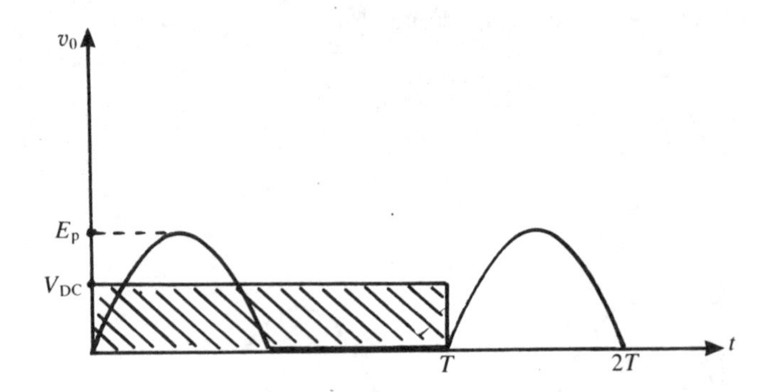

$$V_{dc}T = \int_o^T \upsilon_o \, dt = \int_o^{T/2} \upsilon_o \, dt + \int_{T/2}^T \upsilon_o \, dt.$$

The second integral is zero because υ_o is zero, hence

$$V_{dc} = (1/T) \int_o^{T/2} \upsilon_o dt = (1/T) \int_o^{T/2} E_P \sin \omega t \, dt$$
$$= E_P/\pi.$$

As the rms source E' is 5 V this means that

$$\begin{aligned} E_p &= \sqrt{2}.E' \\ &= 1.41 \times 5 \text{ V} \\ &= 7.2 \text{ V} \end{aligned}$$

and

$$\begin{aligned} V_{dc} &= 7.2/\pi \\ &= 2.3 \text{ V}. \end{aligned}$$

For a *real* diode,

$$V_{dc} = (E_P - 0.7)/\pi$$

$$= (7.2 - 0.7)/\pi$$
$$= 6.5/\pi$$
$$= 2.1 \text{ V}.$$

Q.2 Calculate the peak inverse voltage (PIV) for a silicon diode in the half-rectifier circuit of QW1.

When the diode is reverse-biased there is no output pd. The source pd falls entirely across the diode, as shown in the diagram. Using the same convention as in Chapter 3, the + and − signs indicate the high and low potential sides of the source and diode. Now apply KVL in a clockwise direction to give

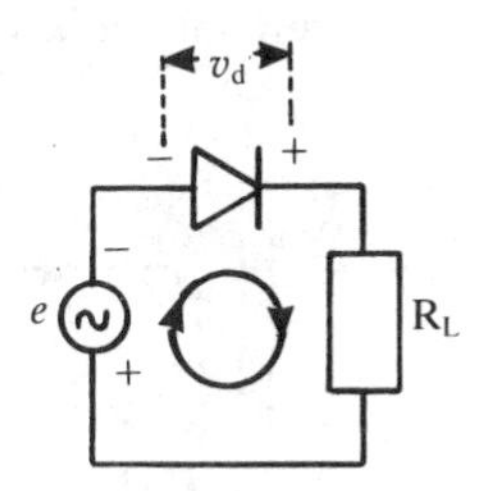

$$e - v_d = 0$$

from which

$$v_d = e$$

at every instant of time.

The maximum pd that the diode must be capable of supporting occurs when the source voltage is at its maximum value, that is E_p. hence the PIV is 7.2 V.

For the *real* diode, the PIV is $(E_P - 0.7)$ V, which equals 6.5 V.

5.5.1.1 Transformer-coupled half-wave rectifier

The function of the transformer is to either step-up or step-down the source voltage by a value which depends on the ratio of the number of turns on the secondary winding, N_2, to the number of turns on the primary, N_1. Then the instantaneous pd e_s across the secondary is given by

$$e_S = (N_2/N_1)\ e. \qquad (5.13)$$

The maximum secondary pd E_S occurs when the primary voltage waveform reaches its maximum value of E_P for the ideal diode and ($E_P - 0.7$ V) for the real diode. The results obtained in section 5.5.1 hold if E_P is replaced by E_s.

☆ ☆

AN ASIDE: The transformer can be called an *electromagnetic energy converter*. In its simplest form it consists of two coils, the primary and the secondary, wound on the same core of magnetic iron, but insulated from each other and the core. A time-varying pd across the primary produces a changing current through the primary, which, in turn, generates a changing magnetic field around it and through the core. The basic function of the core is to prevent the magnetic flux from spreading out through space; the magnetic flux is contained within the region close to, and within, the core. Some of the flux lines will thread through the secondary and induce a changing current in it. This, in turn, will induce a changing pd across the secondary. The strength of the magnetic flux depends on the number of turns in the primary, whereas the magnitude of the secondary pd depends on the number of turns in the secondary. Energy is thus transferred from the primary winding to the secondary without the need for the output of a circuit to be directly connected to the source (or input). One further advantage of the transformer is that the output is isolated from any dc component that may be present in the source. The schematic diagram of the transformer in Fig. 5.11 is very different from its actual appearance.

Consider the basic transformer circuit shown in Fig. 5.11 which also has a load

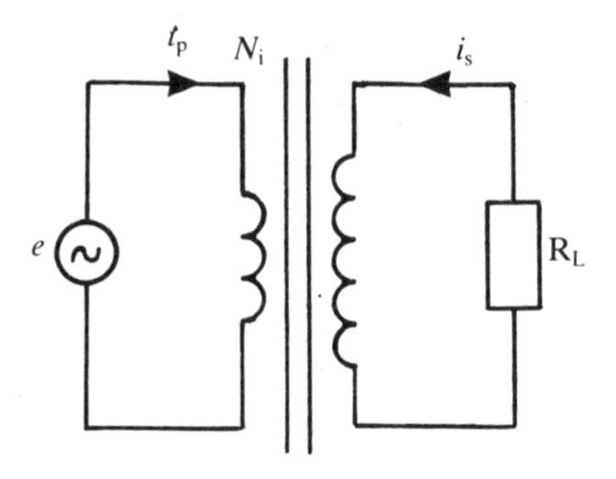

Fig. 5.11.

resistor R_L connected across the secondary. If it is assumed that the rate of loss of energy through resistive heating in both primary and secondary windings is negligible, then it is possible to write

$$e\, i_P = e_s\, i_s. \qquad (5.14)$$

That is, the power in the primary winding is equal to the power in the secondary winding. Equation (5.14) says that the transformer can change a low voltage signal at high current into a high voltage signal at low current, or vice versa. However, by Ohm's law, it follows that

$$i_s = e_s/R_L$$

so that

$$e = e_s^2/i_P R_L \qquad (5.15)$$

A resistance R_P for the primary can be defined as

$$R_P = e/i_P$$

which can now be expressed in terms of e and e_s as

$$R_P = e^2 R_L/e_s^2.$$

However, using (5.13), this reduces to

$$R_L/R_P = [N_2/N_1]^2. \tag{5.16}$$

☆ ☆

5.5.2 Full-wave rectification

5.5.2.1 Centre-tap rectifier

Fig. 5.12 illustrates the basic circuit. Note that the secondary is divided in two by a

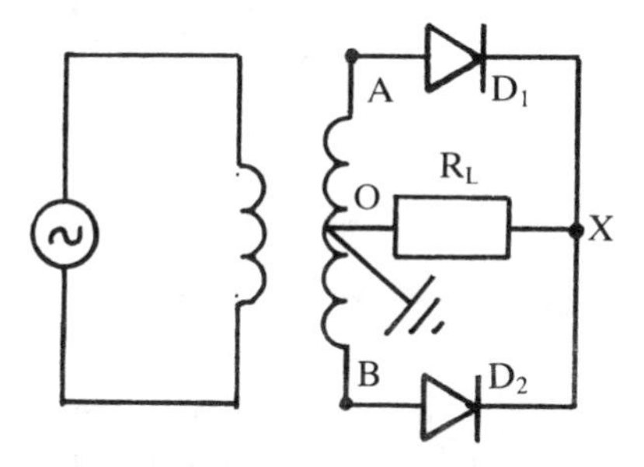

Fig. 5.12.

centre tapping which is earthed. Generally the number of turns on each half of the secondary is the same as the total number of turns on the secondary in the half-wave rectifier. We shall, therefore, let the amplitude of the ac voltage across each half equal E_s.

The action of the circuit can be best understood by considering what occurs when the ac potential at the top of the primary is first positive relative to earth and then negative. Fig. 5.13 depicts the first case, in which the + and − signs indicate the

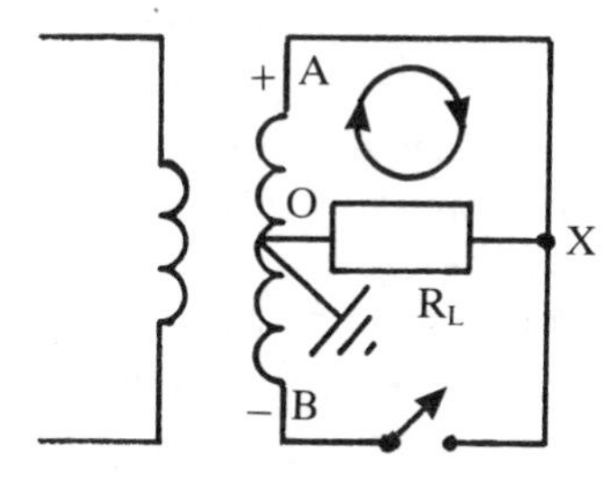

Fig. 5.13.

relative values of the potentials. Point A of the secondary is positive relative to the earthed centre tapping O. In turn, O is positive relative to point B. Thus diode D_1 is short-circuited whilst D_2 is open-circuited. Current flows through the load resistor R_L from X to O. The polarity of the ac potential at X is in phase with that at A. It attains a maximum value equal to E_s.

For the second case, the ac potential at A is now negative relative to earth, as shown in Fig. 5.14. Diode D_2 is short-circuited whilst D_1 is open-circuited. Current

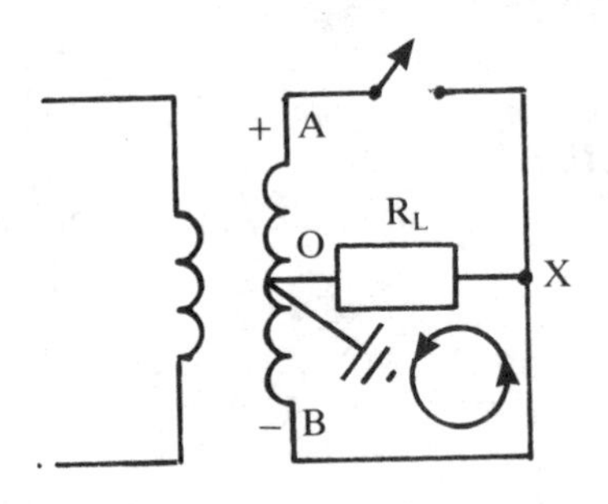

Fig. 5.14.

flows in the direction indicated — again from X to O through R_L. The polarity of the potential at X now follows that at B; the maximum potential at X again being equal to E_s.

The variation of the pd across each half of the secondary winding as a function of time is shown in Fig. 5.15a, and the variation of the potential v_x at X in Fig. 5.15b. Note that v_x is always positive because the current *always* flows through the load in the same direction. This is *full-wave* rectification, characterized by the frequency of the waveform at X being twice that of the input.

5.5.2.2 Bridge rectifier

This method sets a lower limit to the amplitude of the source voltage which can be rectified. The lower limit arises because there is a total potential drop of 1.4 V across a pair of diodes. The bridge rectifier is modelled on a Wheatstone bridge circuit, as is evident from Fig. 5.16.

When the ac potential at A is positive relative to earth, diodes D_2 and D_3 are short-circuited and current flows from X to Y; the ac potential at X is positive relative to Y. Similarly, when the ac potential at B goes positive relative to earth diodes D_4 and D_1 conduct and current again flows from X to Y.

The potential at X, v_x, varies in exactly the same way during both half-cycles of the source voltage because the current flowing through the load is always in the same direction. Thus v_x is a rectified version of the source waveform. Its time variation is identical with Fig. 5.15b; the peak value of v_x is E_P for *ideal* diodes. With real diodes the pd across the load resistor R_L is $(E_P - 1.4)$V. The frequency of the output is again double that of the input.

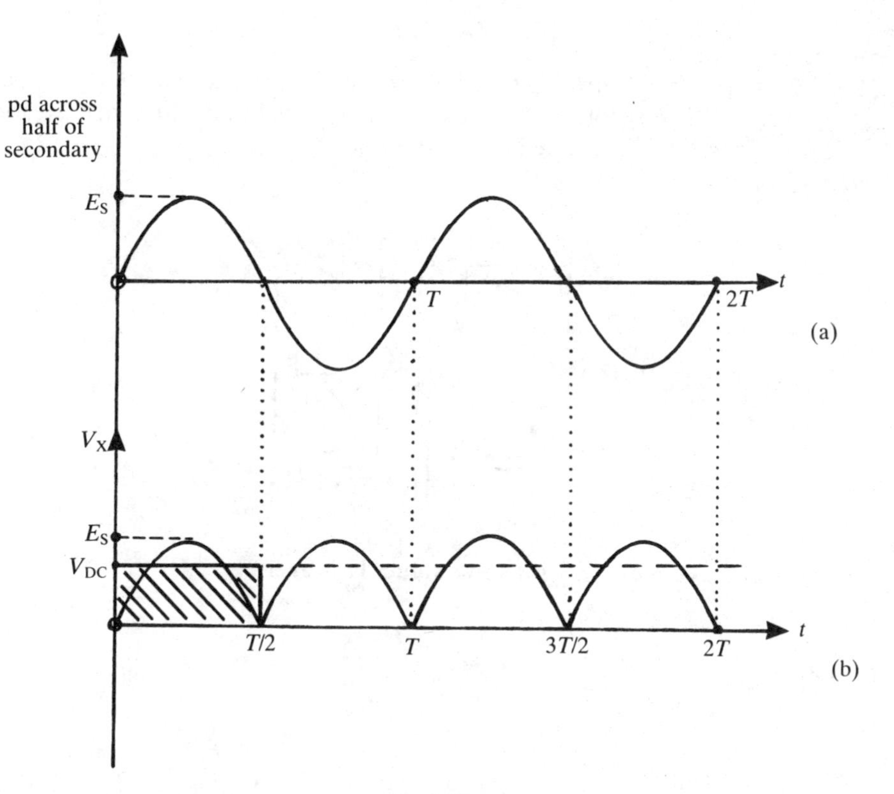

Fig. 5.15.

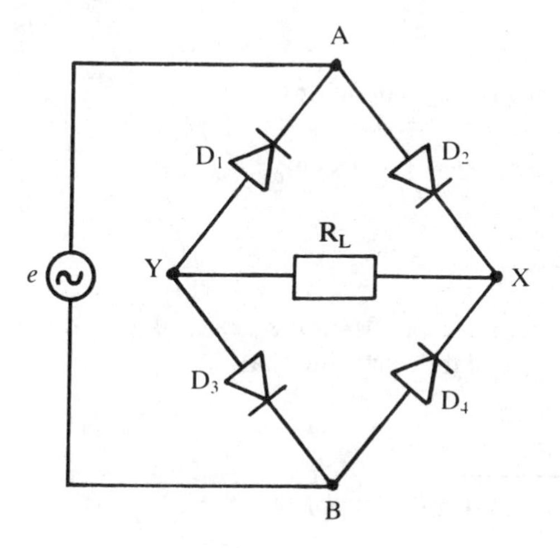

Fig. 5.16.

Worked examples 5.2

Q1. A 3:1 step-down centre-tap transformer is used to produce a fully-rectified pd across a 100Ω resistor. If the primary is connected to a 240 V mains supply, calculate the average value of the output pd.

The circuit is shown below

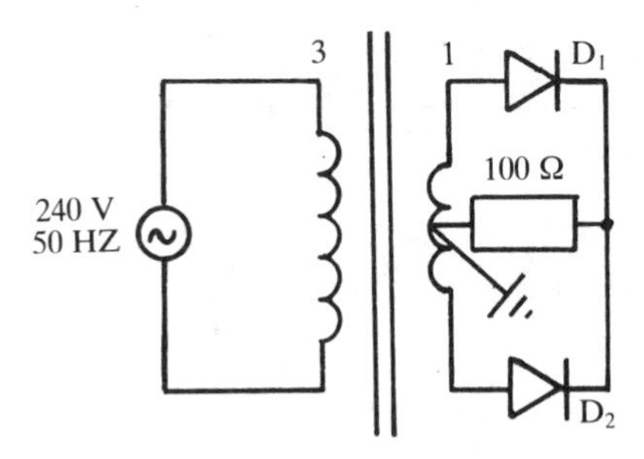

The 240 V across the primary is an rms value. However, as

$$E' = E_P/\sqrt{2}$$

we obtain

$$\begin{aligned} E_P &= E'\sqrt{2} \\ &= 240\sqrt{2}\ \text{V} \\ &= 336\ \text{V}. \end{aligned}$$

Therefore, the peak value of the pd across the secondary is 112 V. This is twice the peak value of the pd across each half of the secondary. in other words, E_s is 56 V. This is as far as we can go without obtaining an expression for the average output pd V_{dc}. So, let us use Fig. 5.15b to assist us. By definition, the area between V_{dc} and the time axis from $t = 0$ to $t = T/2$ is equal to the area under v_x over the same time interval. That is,

$$\begin{aligned} V_{dc}.\ T/2 &= \int_o^{T/2} E_s \sin \omega t\ dt \\ &= \{E_s/\omega\}\ [-\cos \omega t]_o^{T/2} \end{aligned}$$

and

$$V_{dc} = 2E_s/\pi.$$

The average output pd depends on the peak value of the pd across the secondary. Returning to the calculation, we find that

$$\begin{aligned} V_{dc} &= 112/\pi \\ &= 35\ \text{V}. \end{aligned}$$

The frequency of the rectified supply is 100 Hz.

Q2. Calculate the PIV for the centre-tap rectifier circuit of Q1.

Let us redraw the circuit for the time that diode D_1 is conducting. It is

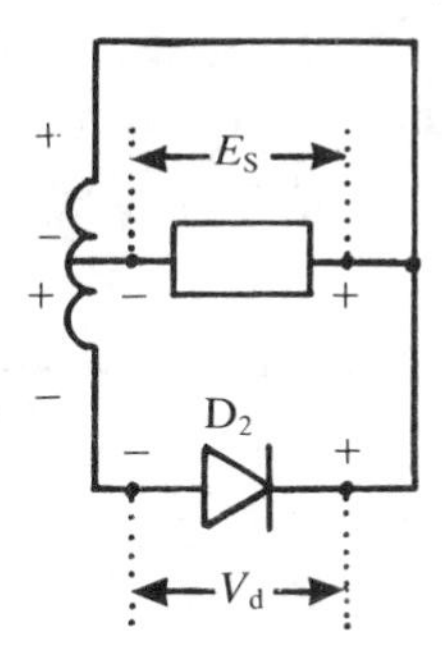

Diode D_2 is under reverse bias. At the instant when the lower half of the secondary is at its peak value of E_S, the diode D_2 must support its maximum reverse pd — call this υ_d. The pd across the load will also have reached its maximum value of E_s.

By KVL, going around the lower loop in an anticlockwise direction, we have

$$E_s - \upsilon_d + E_s = 0$$

or

$$\upsilon_d = 2\ E_s$$

So the PIV is equal to the peak value of the pd across the whole of the secondary winding. In this case it is 112 V.

Q. 3 Determine the PIV for the bridge rectifier circuit of Fig. 5.16.

Suppose that we consider the case when the ac potential at A is positive relative to earth. Diodes D_2 and D_3 are forward-biased (short-circuited) whilst diodes D_2 and D_4 are reverse-biased (open-circuited). Omitting the load, the relevant circuit is

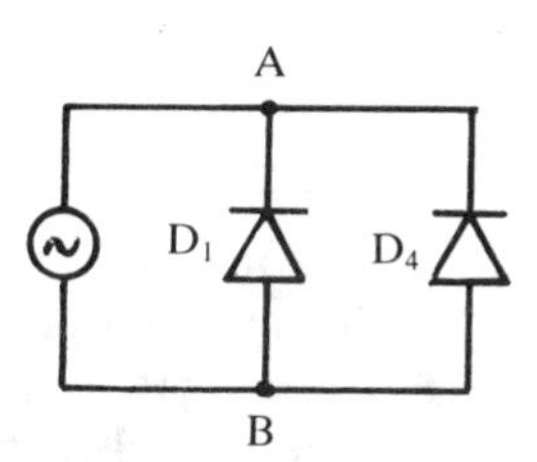

D_1 and D_4 are parallel to each other and to the source. So each must be able to support a maximum reverse pd equal to the amplitude of the source, that is, E_P. This is the PIV.

5.5.3 Capacitor-input filter

Sections 5.5.1 and 5.5.2 have described methods for producing a rectified voltage supply. That is, the conversion of a sinuisoidal ac signal into what can loosely be called a *varying* dc voltage. In both half-wave and full-wave rectification the current through a load resistor is always in the same direction irrespective of the phase of the input signal. A rectified output of this kind is suitable for use with dc motors, for example, but, generally, for most electronic applications a constant dc voltage supply is required. One way of achieving this is to use the capacitor-input filter. It filters out the time-varying component, leaving an essentially constant output.

The basic circuit is shown in Fig. 5.17. Circuit details of the full-wave rectifier are *hidden* in the black-box because they are not necessary for the ensuing discussion. We shall consider an *ideal* diode in the first instance.

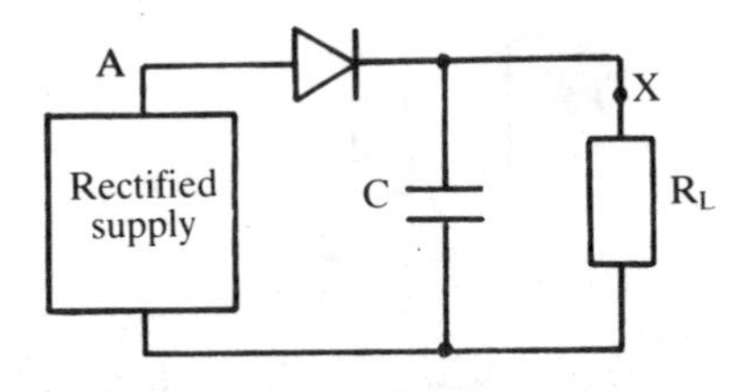

Fig. 5.17.

During the first positive quarter cycle of the rectified waveform at A, the diode conducts and behaves like a closed switch. The capacitor acquires a pd equal to E_P and v_x follows the variation in the potential at A, as can be observed by using a CRO (see Fig. 5.18a). Just after the rectified waveform reaches its positive maximum at A,

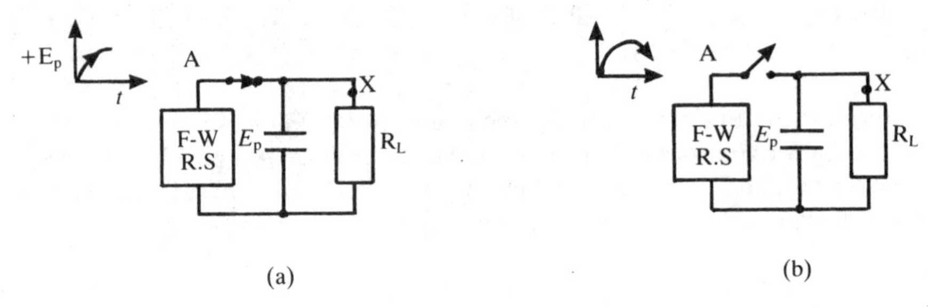

Fig. 5.18.

the diode stops conducting and behaves like an open-switch. Why? because the upper plate of the capacitor is now at a higher potential than point A (see Fig. 5.18b). The capacitor now tends to discharge through the load resistor. However, if the time constant RC is large compared with the period of the rectified supply then the capacitor will not discharge extensively. Perhaps the pd falls to $(E_P - \Delta E)$, say. Then the capacitor will recharge to its maximum value of E_P during the inverval when the diode is next forward-biased. How large should the time-constant be?

WORKING RULE: RC should be about 10 times bigger than the period T of the input voltage supply.

The variation of v_x with time will therefore look like Fig. 5.19. It is not quite constant but exhibits an ac variation or *ripple* superposed on the dc output. The ripple also has the same frequency as the rectified supply. The *ripple factor* is quoted on all power supplies and indicates the degree of ripple; the smaller the ripple factor the smoother the output is. Using rms values, the ripple factor (R.F.) is defined as

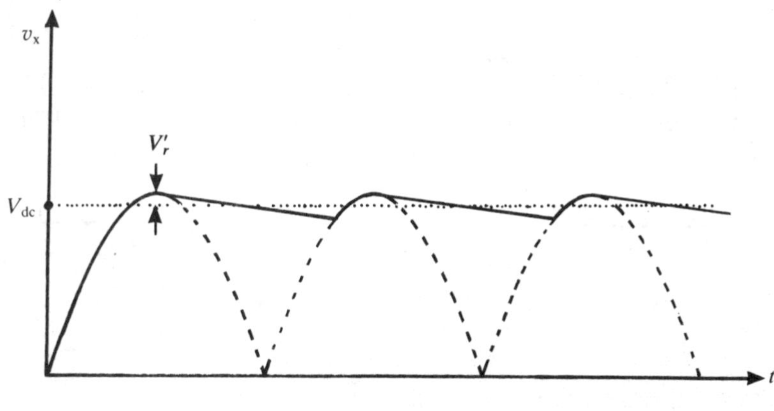

Fig. 5.19.

$$\text{R.F.} = V_r'/V_{dc} \times 100\% \quad (5.17)$$

So if V_r' is 0.05 V and V_{dc} is 20 V then R.F. is 0.25%.

5.6 CLIPPING CIRCUITS

For certain applications, for example computing, it is necessary to remove the part of the ac signal which lies either above or below a particular value. This procedure is known as *clipping*. There are *positive* clippers and *negative* clippers. The positive clipper is used to remove all of that part of the signal which lies *above* a specified voltage level whilst the negative clipper removes the signal lying *below* a specified level.

5.6.1 Positive clipper

Fig. 5.20 depicts the basic circuit. A CRO connected to X will display the resulting

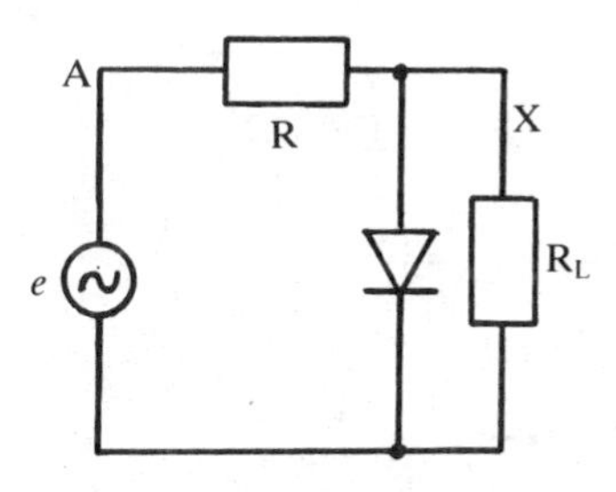

Fig. 5.20.

waveform. As soon as the potential at node A begins to rise above 0 V, an ideal diode forward-biases, as in Fig. 5.21a. Then all the current flows via the short-circuit rather than through R_L. So v_x is zero.

During the negative half-cycle at A, the diode reverse-biases, as in Fig. 5.21b. Thus, using the potential divider rule, v_x can be determined to be

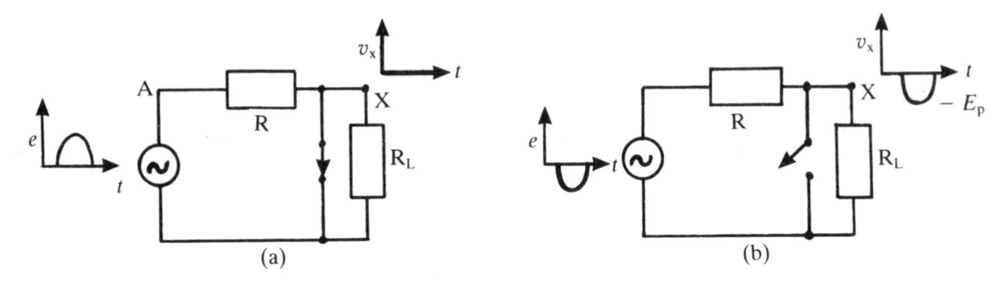

Fig. 5.21.

$$v_x = -R_L e/(R + R_L) \tag{5.18}$$

For $R_L \gg R$, v_x is equal to $-e$ with a peak value of $-E_P$. So we see that the positive half of the input signal has been clipped (see Fig. 5.22a). With a *real* diode, the output is clipped above $+0.7$ V (see Fig. 5.22b).

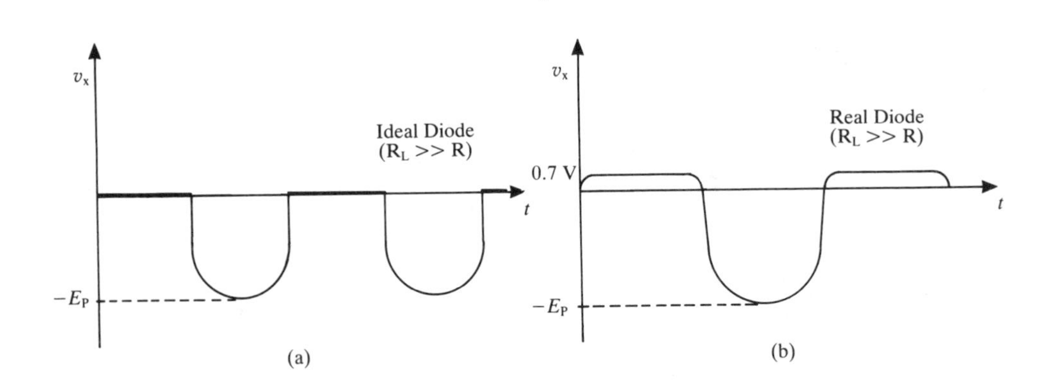

Fig. 5.22.

5.6.1.1 Biased positive clipper

It is possible to clip the input signal above $+E$ volts if a dc voltage source with emf E is put in series with the diode, as shown in Fig. 5.23a. The ideal diode will conduct only when the potential at node A is greater than $+E$. Then the pd across the load will be due to the voltage source alone, that is, $+E$. For all other values of the potential at A, v_x is given by (5.18). The complete output waveform should, therefore, resemble Fig. 5.23b. With the real diode the potential at A must reach $(E + 0.7)$ V before clipping occurs.

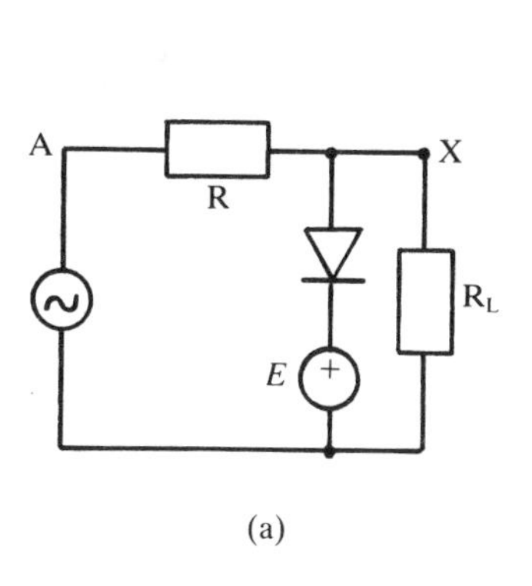

(a)

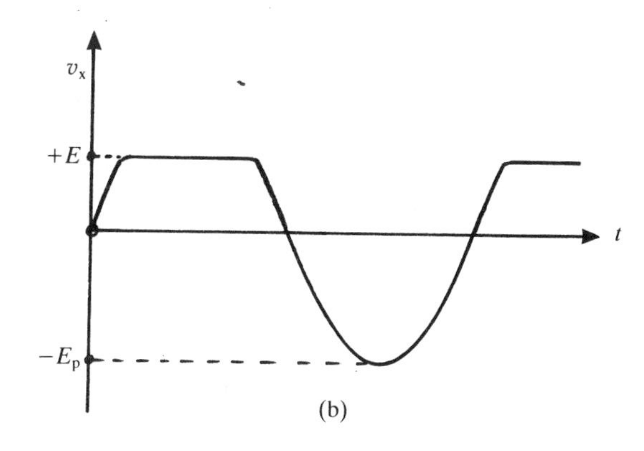

(b)

Fig. 5.23.

5.6.2 Negative clipper

The behaviour of this circuit can be understood in a similar way to the positive clipper. The only alteration that needs to be made to Fig. 5.20 is to reverse the diode. Fig. 5.24a illustrates the ouput waveform for an ideal diode and Fig. 5.24b for a real diode.

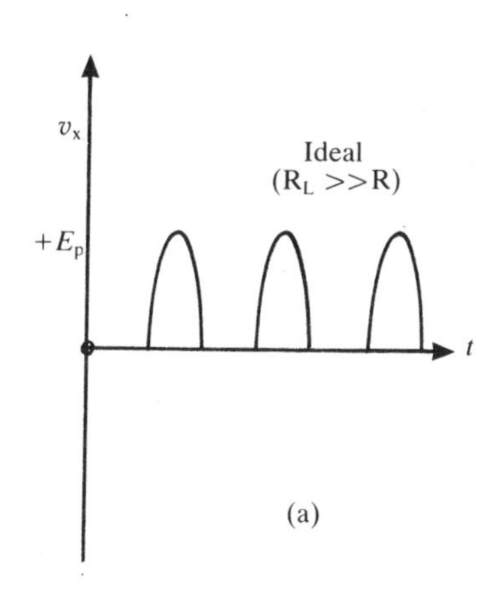

(a)

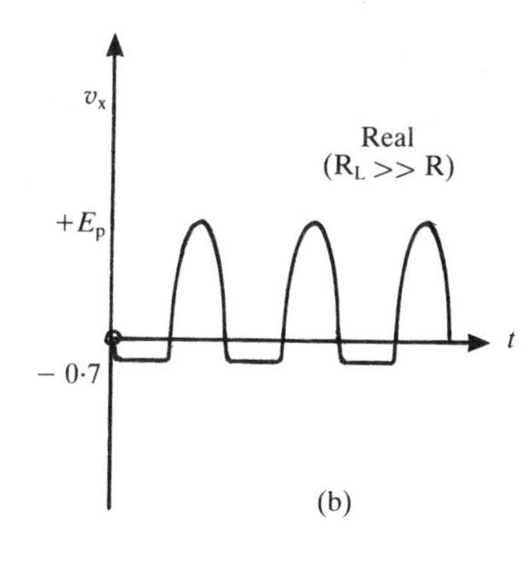

(b)

Fig. 5.24.

5.6.2.1 Negative-biased clipper

The circuit can be obtained by reversing both the diode and the voltage source in Fig. 5.23a. The *ideal* diode will conduct only when the potential at A falls below $-E$. Then v_x will be clipped below $-E$. A *real* diode conducts only when the potential at A is below $(-E - 0.7)$ V, so the input signal is clipped below this value.

5.7 THE CLAMPER

The clamping circuit either adds or substracts a dc component to an ac signal. The *positive* clamper biases the ac signal, positively, whereas the *negative* clamper biases the signal negatively.

Consider the circuit for the positive clamper, Fig. 5.25. Instead of the capacitor

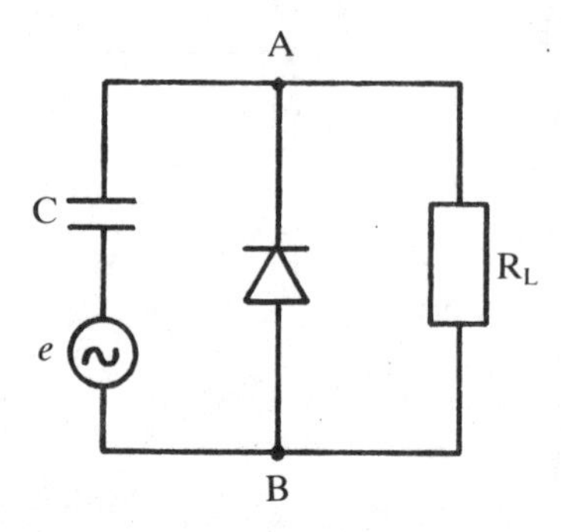

Fig. 5.25.

being in parallel with the source it is now in series with it. When node B goes positive relative to earth the *ideal* diode conducts and current flows to charge up the capacitor; the final pd across the capacitor is E_P. As soon as the potential at B falls below E_P the diode stops conducting and branch BA is on open-circuit. By choosing the time constant RC to be much longer than the period of the source the capacitor will be prevented from discharging and its pd will stay at E_P. So the pd across the load will consist of an ac component of amplitude E_P oscillating about $+E_P$. In actual fact a *real* diode will charge up only to a pd of $(E_P - 0.7)$ V, which means that the ac output signal will dip slightly below 0 V on its negative half cycle.

5.8 VOLTAGE DOUBLER

The basic voltage doubler circuit is given in Fig. 5.26. Once again we can analyse the

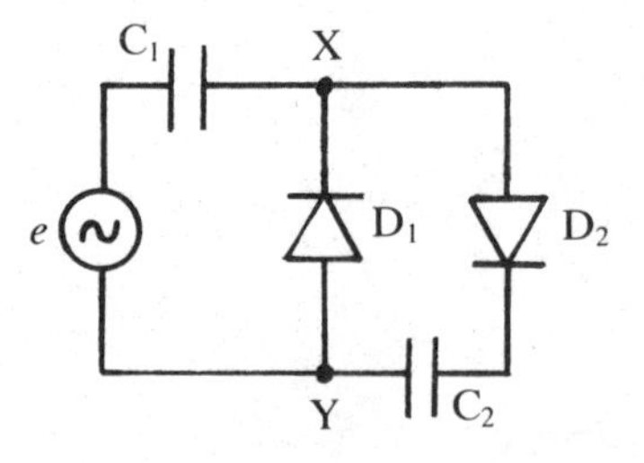

Fig. 5.26.

operation of the circuit by considering what happens in each half of the source waveform.

When the potential at Y goes positive relative to earth, the *ideal* diode D_1

conducts and capacitor C_1 charges up until the pd across it equal to E_P, see Fig. 5.27a. On the other hand, during the negative half cycle at Y, diode D_1 stops conducting and D_2 begins to conduct (see Fig. 5.27b). The source is now in series with capacitor C_1 so that C_2 charges up until the pd across it equals $2E_P$. This state of affairs does not happen immediately but will occur after several cycles have elapsed. By connecting a load resistor R_L across C_2 an output voltage of $2E_P$ can be obtained. One requirement is that R_L should be large so as to prevent C_2 discharging rapidly.

A voltage trebler may be designed by adding a further capacitor/diode stage. Although it is theoretically possible to have many capacitor/diode stages linked together, it is, in fact, rather rare to find more than four stages because voltage ripple becomes a serious problem and voltage regulation deteriorates (see section 5.10).

Worked examples 5.3

Q1. Design a diode circuit which produces the output waveform

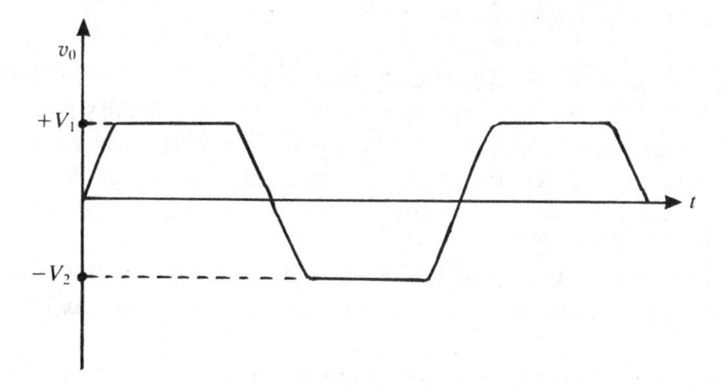

The appearance of the waveform suggests that the circuit clips the input signal above $+V_1$ and below $-V_2$. So a combination of a positive-biased and negative-biased circuit is required. The suggested circuit, without component values, is given below

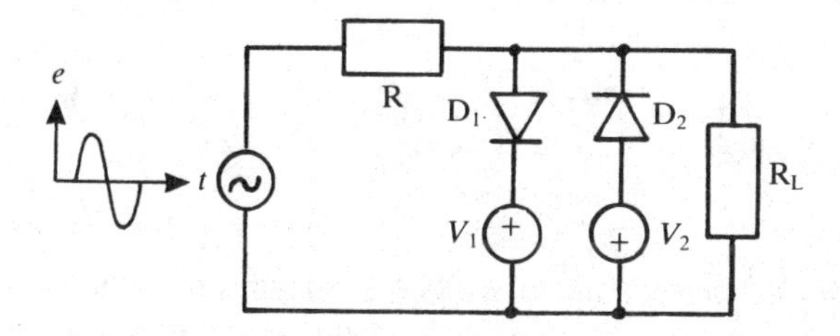

Q.2 A train of square-waves is inputted into the following circuit:

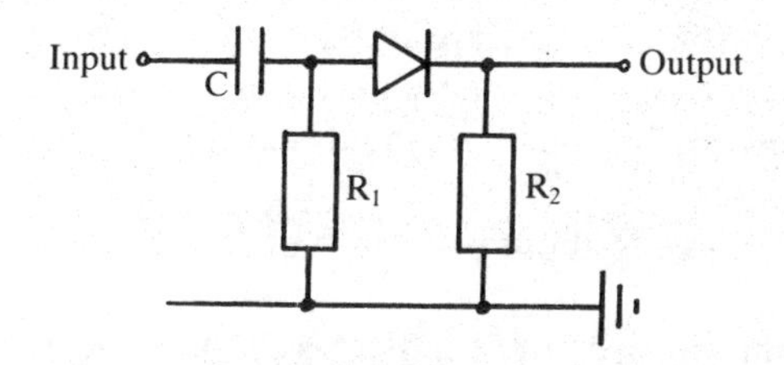

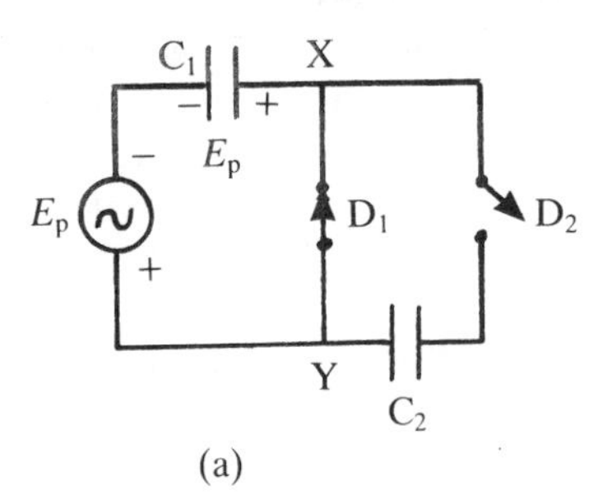

(a)

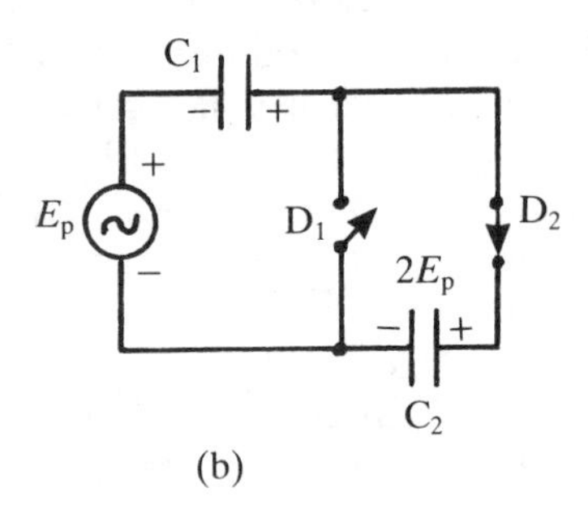

(b)

Fig. 5.27.

What form of output will be obtained?

Q2 of *Worked examples* 3.4 has already dealt with the behaviour of the first C-R stage, which is a differentiator. Its output—the input to the diode stage—consists of a series of positive and negative spikes. The diode will conduct only for the duration of each positive spike. It will be reverse-biased for the set of negative spikes. Hence the output consists of a train of positive spikes, which can, if required, be used for timing purposes.

There is just one qualification that needs to be appended to this solution. If the amplitude of the input train is less than, or equal to, 0.7 V no output will be obtained.

Q3. The following circuit is set up and is found not to work. Explain the reason for this, and suggest a remedy.

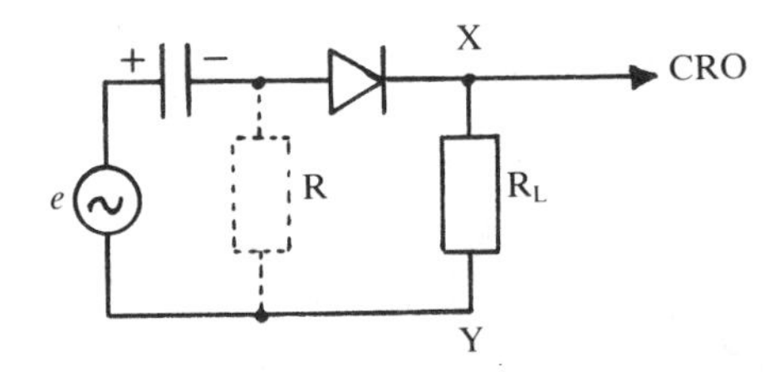

The capacitor gradually charges up to a pd of E_P during the first few cycles of the source and negatively-clamps the signal; the high and low potentials on the capacitor plates are indicated by the + and − signs. The diode is thereby prevented from conducting after the initial few cycles. This means theat an oscilloscope probing point X will not display any signal. A way around this problem is to help the capacitor to discharge during the period when the diode is not conducting. This is achieved by adding a *dc-return* resistor R (shown by the dashes in the diagram).

WORKING RULE: Make R about one-tenth of R_L.

This example is important because it indicates the kind of problems that can arise with a capacitively-coupled signal source. The idea of using capacitive-coupling is to prevent any dc component in the source from reaching the load. However, as we have seen, unwanted dc clamping can also occur.

5.9 VOLTAGE REGULATION: THE ZENER DIODE

A rectified and filtered voltage supply is to be connected to a variety of different loads. The one requirement is that the output should remain constant. It is the *voltage regulation* which informs us whether this voltage stability stipulation can be achieved. Consider Fig. 5.28. We require the rms output to be independent of the

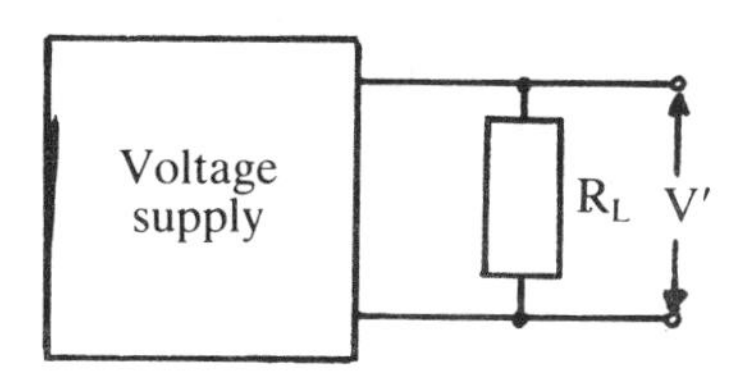

Fig. 5.28.

value of R_L. The voltage regulation $V.R$ is defined as

$$V.R = (V'_\infty - V'_o)/V'_o \times 100\% \tag{5.19}$$

where V'_∞ is the value of V' when R_L is infinite, that is, the output is open-circuited, and V'_o is the corresponding value of V' when R_L is zero, that is, there is a short-circuit. For a well-designed supply, V'_∞ should be slightly larger than V'_o so that $V.R$ is practically zero.

One way of stabilizing the voltage output is to use a Zener diode. Section 5.4.1 mentioned, very briefly, that if the breakdown pd in the reverse characteristic occurred at a value of about -10 V, then the diode was termed a Zener diode. We shall consider the action of a Zener diode through a worked example.

Worked examples 5.4

Q1. Determine the current flowing through the Zener diode in the following dc circuit if the diode has a breakdown pd of -10 V and (i) is ideal; (ii) has a resistance of 10Ω.

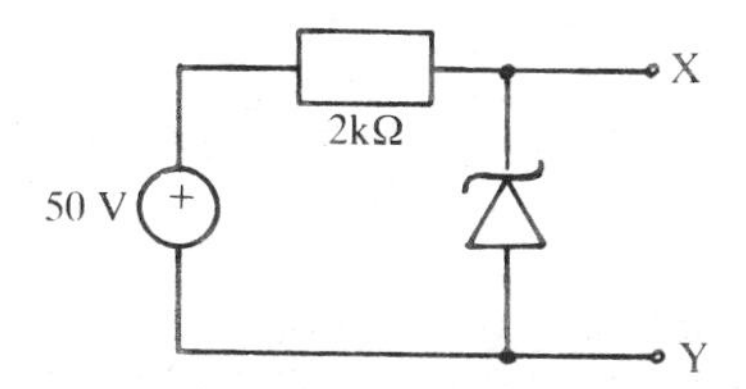

(i) First note the symbol for the Zener diode. Since the breakdown pd is -10 V, this means that the pd across XY is + 10 V. The remaining 40 V must be dropped across the 2 kΩ resistor. The current flowing through the resistor and Zener is, therefore,

$$I_Z = 40/2000 = 20 \text{ mA}.$$

The output would still remain at 10 V even if the supply was increased to 100 V. The only difference is that the Zener current would increase to 90/2000 mA = 45 mA. So long as this current is less than the *burn-out* current for the Zener there is little to worry about. The *burn-out* current can be obtained from the manufacturer's specifications.

(ii) Now we are dealing with a *real* Zener. It can, however, be thought of as an *ideal* Zener in series with a resistance of 10 Ω. The total pd across XY is now made up of the pd of 10 V across the ideal Zener plus the pd across the 10 ohms. Hence 40 V are dropped across 2010 Ω. The current through the Zener is now

$$\begin{aligned} I_Z &= 40/2010 \\ &= 19.90 \text{ mA} \end{aligned}$$

In addition the pd across XY is given by

$$\begin{aligned} V_{XY} &= 10 + 19.90 \times 10^{-3} \times 10 \\ &= 10 + 0.20 \\ &= 10.20 \text{ V} \end{aligned}$$

Repeating the calculation for a 100 V supply, we find that the Zener current is 44.78 mA and V_{XY} has increased to 10.45 V.

Q2. Load resistors, 50 kΩ and 5 kΩ in value, are connected between X and Y in the circuit of Q1. Determine the current flowing through the real Zener in each case if the supply voltage is 50 V.
The circuit now looks like:

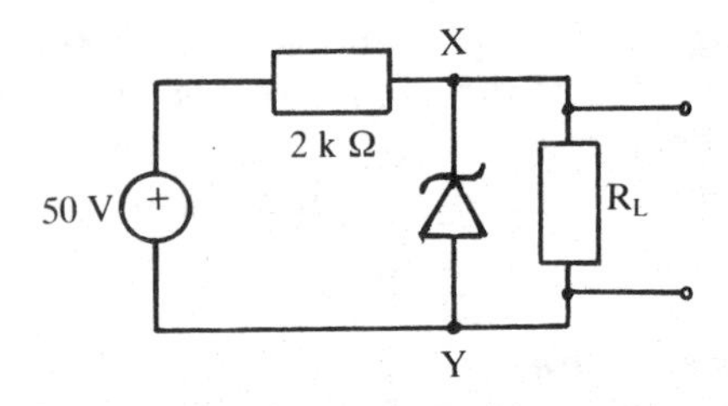

We need to Théveninize the circuit (see section 3.6). We already know the o/c Thévenin pd across XY. It is 10.20 V. The Thévenin resistance is 2k||50, i.e. approximately 10 Ω. So the Thevenin equivalent circuit is

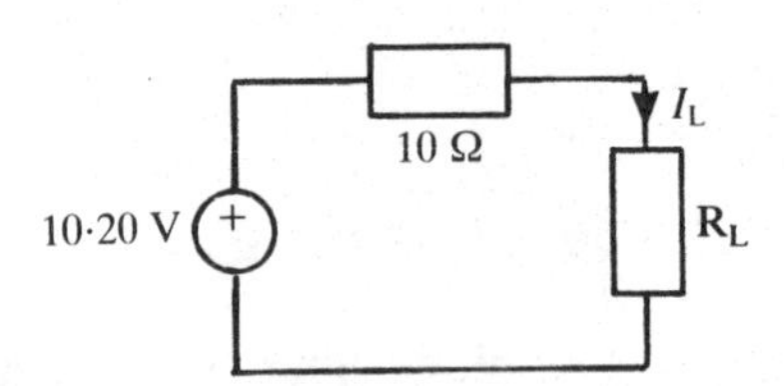

The load current I_L is given by Ohm's law as

$$I_L = 10.20/(10 + R_L).$$

Various values of R_L can now be substituted into this equation. For $R_L = 50\ \text{k}\Omega$, $I_L = 0.20$ mA. As the Zener current in the absence of the load was found to be 19.90 mA, this means that it is now (19.90-0.20) = 19.70 mA. For $R_L = 5\ \text{k}\Omega$, $I_L = 2.04$ mA and $I_Z = 17.86$ mA.

So as the resistance of the load decreases we should note that the Zener current also decreases. Therefore, there is a minimum load resistance for which the Zener current is zero. Using the above relation, this value can be calculated to be 513 Ω— when all the 19.90 mA flows through the load. Now the Zener is taken out of the breakdown region! It no longer behaves as a voltage regulator. To prevent this from happening, it is important for the load resistance to be larger than 513 Ω.

Q3. Calculate the voltage regulation for the circuit of Q2.
The pd across the load is given by the potential divider rule as

$$V_L = 10.20\ R_L/(10 + R_L)$$

So, for the 50 kΩ load, V_L is 10.20 V and for the 5 kΩ load it is 10.18 V. Therefore, by (5.19), the voltage regulation is 0.2%.

5.10 SOME OTHER DIODES

5.10.1 The photodiode

A reverse-biased diode produces a saturation current as a result of minority carriers generated through bond-breakage of the silicon lattice. The saturation current increases with increase in temperature because the degree of bond-breakage increases. Light has a similar effect to temperature.

Generally, the diode is opaque to radiation, in the visible spectrum, so that there is no detectable photoelectric effect. It can be made significant by constructing a window in the photodiode to allow radiation to fall on the junction region. The greater the intensity of the incident radiation the higher the saturation current.

5.10.2 Light-emitting diode

The light-emitting diode (LED) is available both as a discrete diode and also in displays. The seven-segment display, in particular, is used widely in digital watches and multimeters. The segment is a magnified portion of the diode surface, formed with a plastic lens.

The light is emitted when a conduction electron and a positive hole recombine; some heat is also produced. This recombination process occurs all the time in a forward-biased silicon diode, but no visible light is observed because silicon is opaque to radiations in the visible spectrum. Using various compositions of gallium, arsenic, and phosphorus, it is possible to obtain different emitted frequencies. For example, gallium arsenide emits radiation in the near infrared (λ 870 nm) whilst the addition of some phosphorus generates red light. Gallium phosphide, itself, emits green light.

The TIL221 LED has a breakdown pd of only 3 V. So it is important to connect the LED correctly in the circuit, as in Fig. 5.29. A protective rectifying diode is

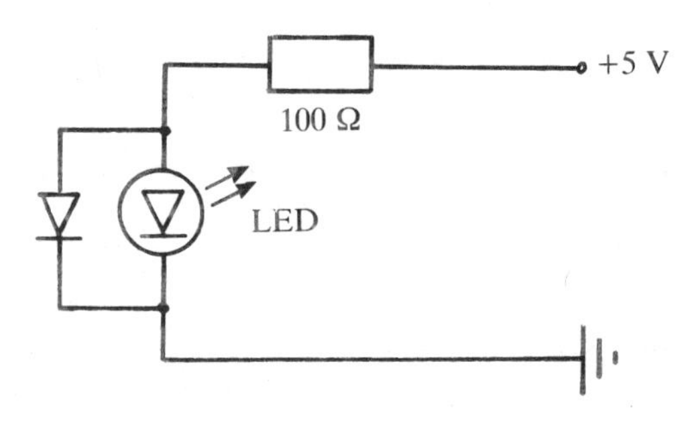

Fig. 5.29.

usually put in parallel with the LED. If the supply is incorrectly connected, for whatever reason, the diode will conduct strongly and protect the LED from being destroyed.

5.10.3 The opto-coupler

This is a LED/photodiode combination, as Fig. 5.30 illustrates. As the LED and

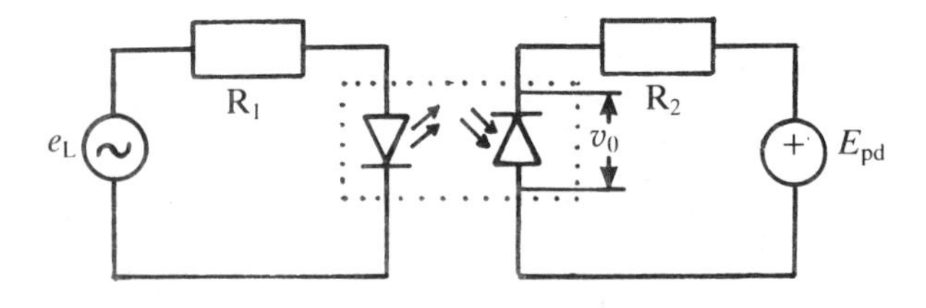

Fig. 5.30.

photodiode are isolated from each other the impedance between them can be as high as 10 GΩ, or so. This gives rise to the alternative name for this device, viz. the *opto-isolator*. As we have already observed in section 5.10.1, the saturation current flowing in the photodiode circuit depends on the intensity of the light from the LED. Driving the LED with an ac source results in the intensity of the light rising and falling systematically, with the result that the photodiode's output also varies in the same way.

6

The bipolar transistor and its circuits

Objectives

(i) To introduce the *npn* and *pnp* bipolar transistors
(ii) To identify electrons and positive holes as majority and minority carriers
(iii) How to use the bipolar transistor in its three circuit configurations; the importance of the common-emitter circuit
(iv) To obtain the collector characteristics of the *npn* transistor
(v) To discuss the effect of placing a resistor in the collector circuit; the dc load line
(vi) To discuss the significance of saturation and cut-off; use in the design of logic switches
(vii) To discuss the importance of the Q-point in transistor circuit design; negative current feedback
(viii) Why use a potential divider bias circuit?
(ix) How to analyse the small-signal amplifier at low frequencies with different degrees of rigour
(x) How to draw the dc and ac equivalent circuits.
(xi) To discuss input impedance and its role in the design of a high-quality small-signal amplifier.
(xii) Why the output impedance of a voltage source is important.
(xiii) To discuss the emitter-follower as an impedance-matching device.

6.1 BACKGROUND

The transistor is a three-terminal device used in electric equipment to control a large current with a smaller current. It has many applications, which include: amplifiers, function generators, digital switches.

The charge carriers in the *bipolar* transistor are electrons and positive holes, unlike the field-effect transistor, which is a *unpolar* device. We shall concentrate our attention on the *npn* transistor because its principle of operation concerns the behaviour of electrons, with which we are probably more familiar. (The pnp transistor must be discussed in terms of positive holes.)

The transistor is made from a single crystal of silicon and consists of three distinct regions, see Fig. 6.1. It is, therefore, a three-element semiconductor device. The

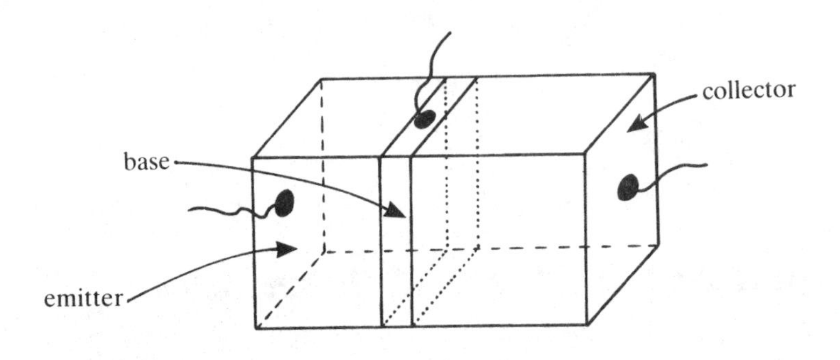

Fig. 6.1.

regions are the emitter, base, and collector. The whole transistor is about 3 mm long, with a cross-section 1 mm × 1 mm, in size. The width of the base is usually smaller than 0.01 mm. The base is lightly-doped, the emitter is heavily-doped, and the collector has a doping level between that of the base and emitter. Metal contacts, with terminating leads, allow the potentials of the three regions to be varied via external voltage sources. The whole is encapsulated in an epoxy resin to protect the device from environmental pollution, external radiation, and possible mechanical damage.

In the absence of any external voltage sources the transistor acts like two pn junction diodes placed back-to-back. So, there are two depletion layers, one at the emitter/base (E/B) junction and the other at the collector/base (C/B) junction. The E/B and C/B depletion layers are unequal in width because the doping levels of the emitter and collector are unequal. The heavily-doped emitter has a high concentration of impurity ions near the E/B junction which means that the depletion layer does not penetrate far into it; the C/B depletion layer penetrates further into the collector.

Fig. 6.2 illustrates one way of biasing the *npn* transistor. It is called the common-base (C–B) configuration. Notice that the E/B pn junction diode is forward-biased and the C/B diode is reverse-biased. The potential barrier existing at the E/B junction can be reduced, therefore, by increasing the emf of source 1. It is now easier for an electron to escape from the emitter into the base. Following the discussion in section 5.2.2, there will be no barrier to electron flow when the emf of source 1 is 0.7 V.

What can happen to electrons when they arrive at the base? There are two possibilities: (i) the electrons recombine with positive holes; (ii) as the base-width is quite small, the electrons diffuse across it and enter the collector. (i) Is less likely than (ii) because the concentration of holes in the base is low. The reverse-biased C/B junction region actually assists the electrons to enter the collector. In fact, about 95%, or more, of the electrons arriving at the base flow into the collector.

In terms of conventional currents, also indicated on Fig. 6.2.

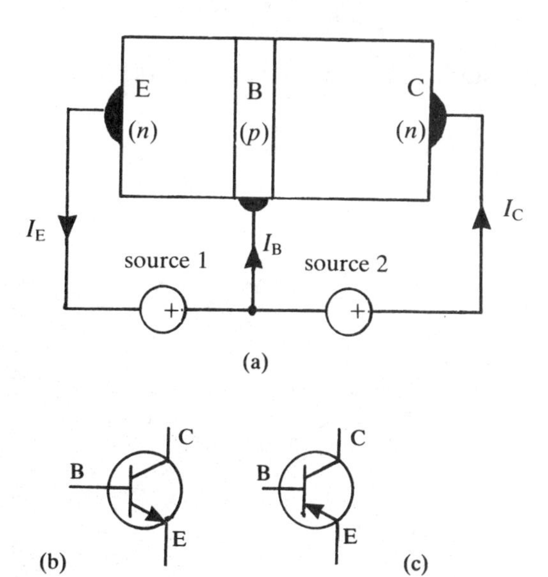

Fig. 6.2.

$$I_E = I_B + I_C \tag{6.1}$$

The symbol for the *npn* bipolar transistor is shown in Fig. 6.2b; the arrow from base (B) to emitter (E) indicates the direction of conventional current flow. The pnp transistor is shown in Fig. 6.2c; conventional current now flows in the *same* direction as the positive holes. NB. An *npn* transistor will not conduct if the base is negative with respect to the emitter, for then both junction regions are reverse-biased; a current will flow only if the base is positive relative to the emitter.

The ratio:

$$\frac{\text{collector current } I_C}{\text{emitter current } I_E} \tag{6.2}$$

is known as the *forward-current transfer* ratio, and is given the symbol α_{dc}. The subscript dc is used to indicate that dc currents are being measured. α_{dc} is less than 1 because I_C is less than I_E by the amount of the base current I_B. Normally, α_{dc} lies in the range 0.95 to 0.99; the higher the value of α_{dc} the better the quality of the transistor. As α_{dc} lies very close to 1, we will generally assume, especially in calculations, that I_C is equal to I_E.

The ratio:

$$\frac{\text{collector current } I_C}{\text{base current } I_B} \tag{6.3}$$

is known as the *forward-current gain* β_{dc} or h_{FE}. In this book we shall use only the symbol β_{dc}. β_{dc} can lie between 20 and 1000. It indicates how much control the base current has over the collector current.

Worked examples 6.1

Q1. A bipolar transistor has a β_{dc} of 50. Calculate the base current if the collector current is 5 mA.

By (6.3),

$$5\,\text{mA}/I_B = 50 \ ,$$

therefore,

$$I_B = 5/50\,\text{mA}$$
$$= 0.1\,\text{mA} \ .$$

Q2. The base current of the transistor in Q1 is now decreased by 0.02 mA. Calculate the new collector current.

Now I_B is 0.08 mA. Therefore,

$$I_C = 50 \times 0.08\,\text{mA}$$
$$= 4\,\text{mA} \ .$$

This apparently trivial example indicates that a relatively small decrease of 0.02 mA in the base current produces a much larger decrease of 1 mA in the collector current. This is a 20% drop in the value of I_C. This calculation underlies the statement made earlier, *viz*. that the value of β_{dc} tells us the degree of control that the base current has over the collector current. Changes in the base current alter the resistance between emitter and collector; a high base current produces a high collector current with the result that the transistor has a low resistance, and vice versa. This is the origin of the word *transistor* or *transfer-resistor*. If the transistor conducts heavily its resistance is very low and the transistor acts as a closed switch, whereas if very little current flows it acts as an open switch.

6.2 SOME RELEVANT THEORY

Equation (6.1) says that the dc emitter current is the sum of the dc base current and the dc collector current. Let us restate it here

$$I_E = I_B + I_C \ . \tag{6.1}$$

An accurate description of the bipolar transistor takes account of the fact that I_C actually consists of two component currents: I_{C1} and I_{C0}, such that

$$I_C = I_{C1} + I_{C0} \tag{6.4}$$

I_{C1} is the current arising from majority electrons leaving the emitter and reaching the collector. It is this current which is related to I_E through (6.2). I_{C0} is a current resulting from the flow of minority electrons in the base and minority holes in the collector crossing the C/B junction. The minority carriers are generated through bond-breakage of the silicon crystal lattice (see sections 5.1.3.1 and 5.1.3.2).

Rewrite (6.4) as

$$I_C = \alpha_{dc} I_E + I_{C0} \ . \tag{6.5}$$

Now eliminate I_E from (6.1) and (6.5) to give

$$I_C = [\alpha_{dc}/(1-\alpha_{dc})] I_B + I_{C0}/(1-\alpha_{dc}) \ . \tag{6.6}$$

I_{C0} is very much smaller than I_C, so that

$$\begin{aligned}\alpha_{dc}/(1-\alpha_{dc}) &\simeq (I_C/I_E)/(1-I_C/I_E) \\ &\simeq I_C/(I_E - I_C) \\ &\simeq I_C/I_B \\ &\simeq \beta_{dc} \ .\end{aligned}$$

Hence (6.6) becomes

$$I_C = \beta_{dc} I_B + I_{C0}/(1-\alpha_{dc}) \ . \tag{6.7}$$

6.3 CIRCUIT CONFIGURATIONS

There are three ways of connecting a bipolar transistor into a circuit. These are called the common-emitter (C-E), common-base (C-B), and common-collector (C-C) configurations; the C-B configuration was discussed in section 6.1. The relation (6.7) is valid for all circuit configurations because it was not necessary to invoke any of them in deriving it.

Fig. 6.3 shows the C-B circuit configuration with a switch in the emitter-base

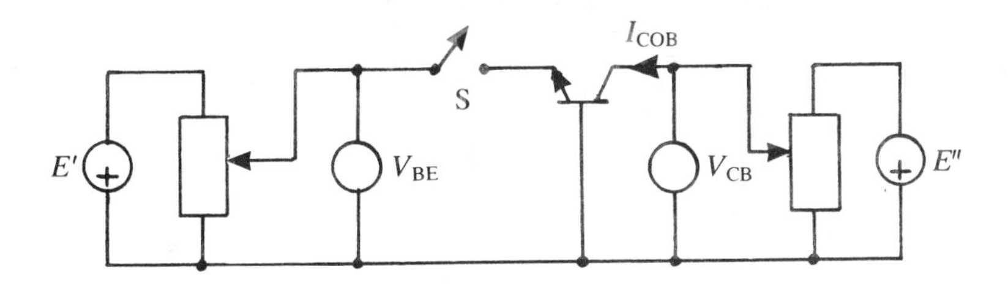

Fig. 6.3.

loop. With switch S open $I_E = 0$, but a current still flows in the collector-base loop owing to minority carriers generated through bond-breakage. We shall call this current I_{COB}. Then, when switch S is closed, $I_E > 0$ and (6.7) becomes

$$I_C = \beta_{dc} I_B + I_{COB}/(1-\alpha_{dc}) \tag{6.8}$$

I_C can be measured for various values of V_{CB} but with I_E constant. The results are depicted schematically in Fig. 6.4a. The curve for $I_E = 0$ is called the *cut-off* curve. The scale on the I_C axis has been exaggerated to represent I_{COB} on the same figure.

Now look at the C-E circuit configuration in Fig. 6.5. With no base current,

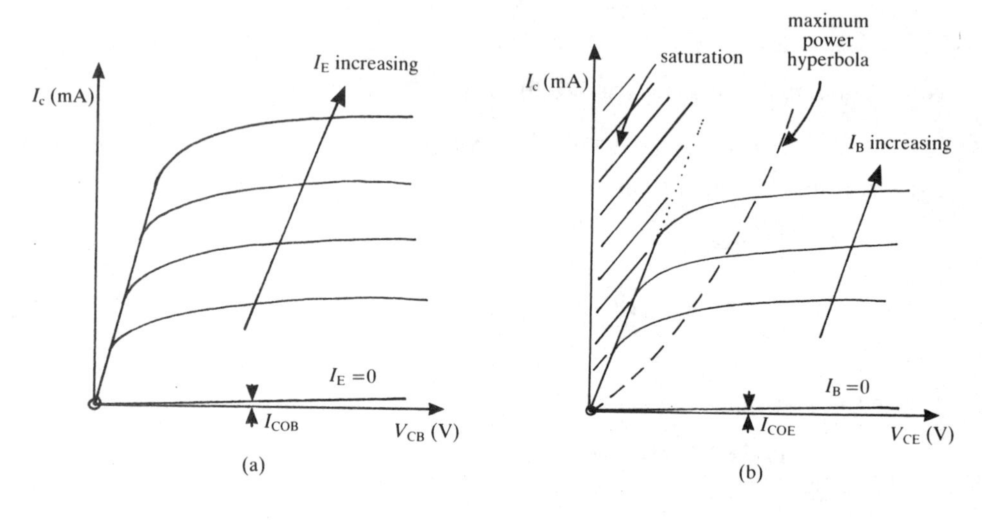

Fig. 6.4.

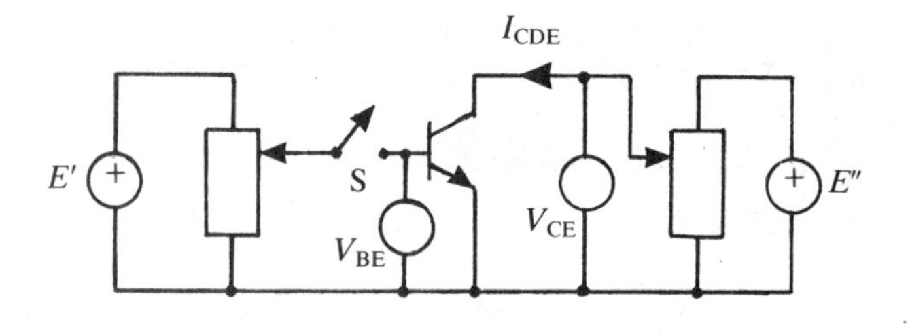

Fig. 6.5.

$I_B = 0$, with switch S open, but a minority current I_{COE} flows in the collector-emitter loop. However, by putting $I_B = 0$ in (6.8), we obtain

$$I_C = I_{COB}/(1 - \alpha_{dc}) \tag{6.9}$$

which must be identical with I_{COE}. Thus it is possible to relate the minority flowing in the C-E and C-B configurations. Therefore,

$$I_C = \beta_{dc} I_B + I_{COE} \ . \tag{6.10}$$

Experimental plots of I_C vs V_{CE} at various fixed values of I_B look like Fig. 6.4b. Once again, the scale on the I_C axis is exaggerated to include the cut-off characteristic.

6.4 MAXIMUM POWER HYPERBOLA

The emitter-base junction requires only 0.7 V to forward-bias it. Then electrons, unimpeded by any potential barrier, pour into the base and thence into the collector.

V_{CE} needs to be greater then 0.7 V to forward bias the E/B junction and to reverse bias the C/B junction. However, V_{CE} cannot be increased indefinitely because, as we learnt in section 5.4, the C/B diode will breakdown. For this reason, all transistors have a *maximum power rating*, given by

$$P_{MAX} = V_{CE} . I_C \ . \tag{6.11}$$

It can be found in the manufacturer's specifications. Once P_{MAX} is known, a set of values of I_C and V_{CE} may be calculated and plotted on the collector characteristics of Fig. 6.4b. The curve shown dashed is known as the *maximum power hyperbola*. The transistor should be used only in the region to the left of this hyperbola.

6.5 DC LOAD LINE: SATURATION

Now let us see how the introduction of a resistor in the collector line affects the collector current. The relavant C-E circuit is in Fig. 6.6. This resistor is referred to as

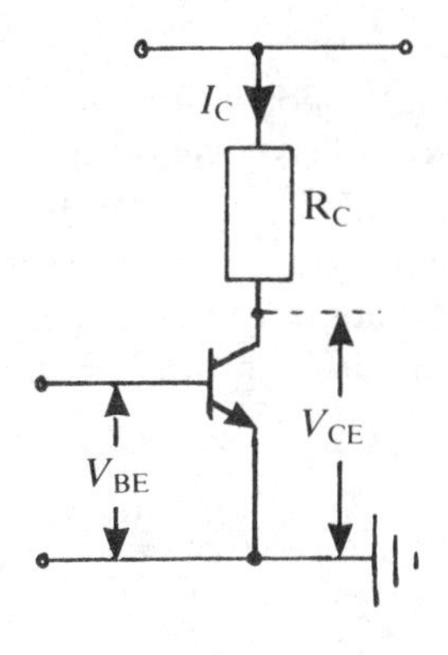

Fig. 6.6.

either the collector resistor or the load resistor. On applying Ohm's law we obtain

$$V_{CC} = I_C R_C + V_{CE}$$

or

$$I_C = -(1/R_C)V_{CE} + V_{CC}/R_C \ . \tag{6.12}$$

Now it is obvious that a graph of I_C vs V_{CE} is a straight line with a slope of $-1/R_C$, as Fig. 6.7 illustrates. This straight line is known as the *dc load line*. To draw the dc load line it is only necessary to know the end points. The maximum value of I_C — the *saturation* current — is equal to V_{CC}/R_C, which occurs for $V_{CE} = 0$. The value of V_{CE} at the point where the load line intersects the V_{CE} axis is V_{CC}. This is the cut-off point. Actually, the cut-off point is slightly less than V_{CC}. It is the point of intersection of the dc load line with the cut-off characteristic in Fig. 6.4b, but because the latter is practically coincident with the V_{CE} axis it is always taken to be V_{CC}.

The saturation region is also indicated in Fig. 6.4b. It lies to the left of the

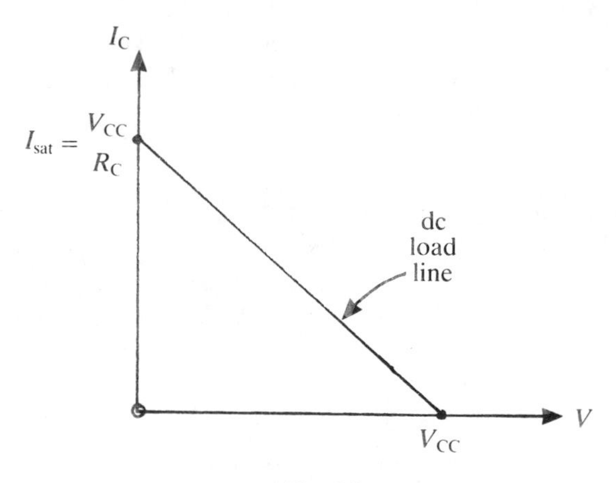

Fig. 6.7.

collector characteristics. In this region both the E/B and the C/B junctions are forward-biased; the transistor conducts heavily and acts as a short-circuit. Note that the saturation current depends on V_{CC} and R_C only.

In circuits based on saturated logic, the transistor switches between cut-off and saturation; the logic circuit designer must ensure that switching occurs as rapidly as possible. The basic switching circuit is shown in Fig. 6.8. It is also known as a *logic*

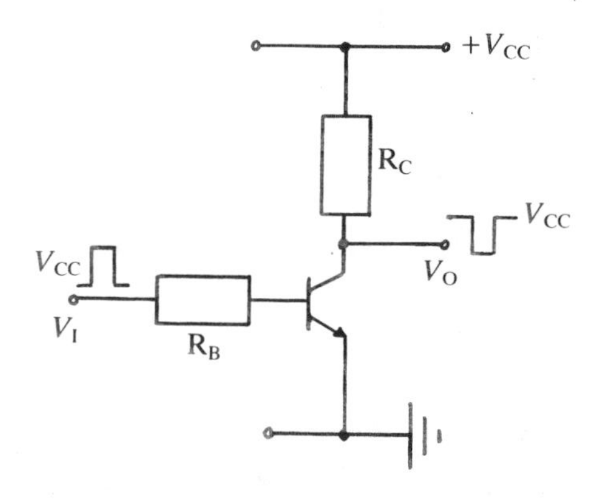

Fig. 6.8.

inverter because the polarity of the output (V_O) is in antiphase with the polarity of the input (V_I).

6.6 QUIESCENT OPERATING POINT

Suppose that under dc conditions a base current I_B flows in the transistor of Fig. 6.6. The collector characteristic for the transistor itself and the dc load line for the circuit are redrawn in Fig. 6.9. The point of intersection Q indicates the collector current

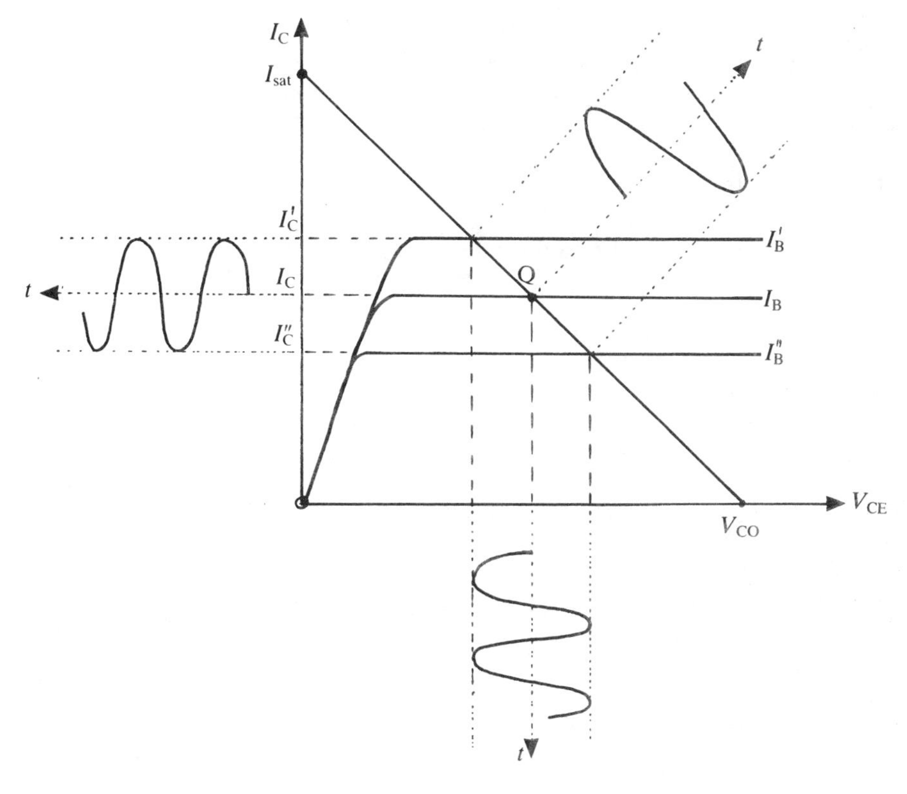

Fig. 6.9.

flowing down the resistor–transistor network for the given base current. Now, let the base current vary between I'_B ($>I_B$) and I''_B ($<I_B$). The collector characteristics for these base currents are also drawn in Fig. 6.9. Moving along the dc load line, we see that the collector current becomes I'_C and I''_C, respectively. The collector current oscillates between these values as the base current oscillates between its extreme values. This can be achieved by using an ac voltage signal applied to the base. So long as the variation in base current is not so large as to drive the transistor into saturation or cut-off, the collector current will appear undistorted and have a regular ac waveform. Everything hinges on choosing an appropriate value for I_C, and, hence, V_{CE}, under dc conditions. The point on the dc load line with these coordinates is known as the *quiescent* operating point (Q-point, for short). It is possible to fix it roughly midway along the dc load line. In any case the amplitude of the ac base signal must not be allowed to increase too drastically, or else signal distortion will still occur. This kind of amplification is referred to as *small-signal amplification*.

Worked examples 6.2

Q1. Determine the dc load line and the Q-point for the following circuit. Is the Q-point satisfactory?

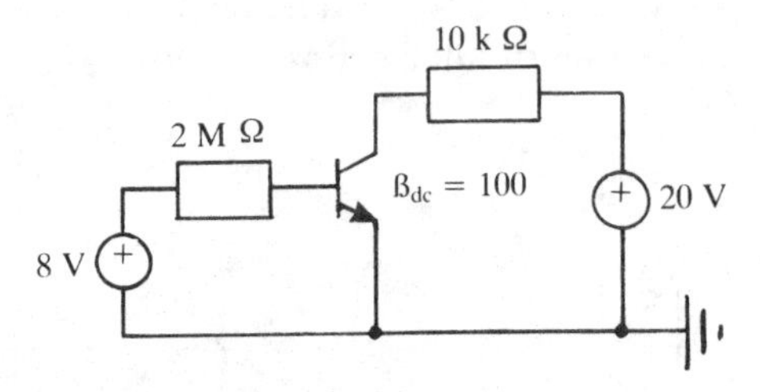

As the base-emitter pd is 0.7 V, the pd across the 2 MΩ base resistor is 7.3 V and the dc base current I_B is 7.3/2 μA, that is, 3.65 μA. The dc collector current I_C is 3.65 × 100 μA, or 0.37 mA. From the collector-emitter loop.

$$\begin{aligned} V_{CE} &= 20 - (0.37 \times 10) \\ &= 16.3\,V\ . \end{aligned}$$

So the Q-point is at (16.3 V, 0.37 mA).

For the dc load line,

$$\begin{aligned} I_{sat} &= V_{CC}/R_L \\ &= 20/10\,\text{mA} \\ &= 2\,\text{mA} \end{aligned}$$

and

$$\begin{aligned} V_{CO} &= V_{CC} \\ &= 20\,\text{V} \end{aligned}$$

So, the dc load line, with Q-point included, looks like

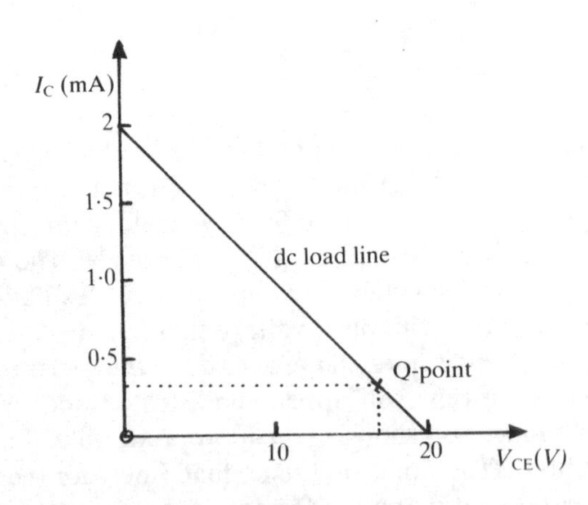

The Q-point is not well-placed. If the amplitude of any ac signal applied to the base of the transistor is too large, the transistor will be driven into cut-off. This is an example of bad circuit design. What is also noticeable, and will be commented on further in Q3, is that the Q-point depends on β_{dc}.

Q2. Draw the dc load line for the following circuit. Mark in the Q-point

The input capacitor is incorporated to prevent any dc component that may be outputted by either the source or a previous circuit stage from upsetting the Q-point of the circuit. The circuit is similar to that in Q1, except that the base and collector potentials are derived from the same supply V_{CC}. I_B is (10 − 0.7)/300 mA or 31 μA, which means that I_C is 3.1 mA. In turn, V_{CE} is (10 − 3.1) or 6.9 V.

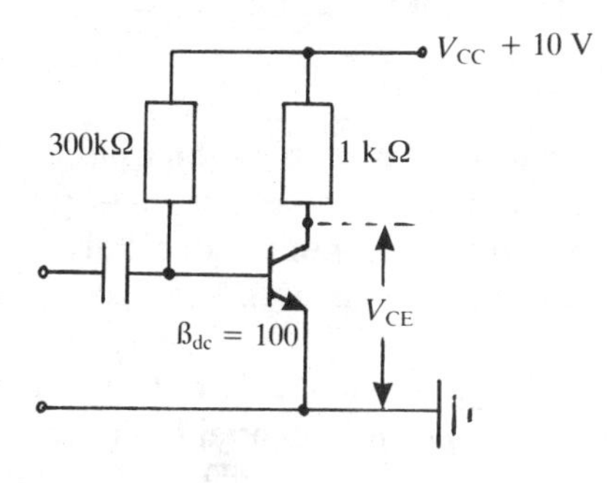

For the load line, the extreme points are:

$$V_{CO} = 10\,V$$

$$I_{SAT} = 10/1\,mA$$
$$= 10\,mA$$

The dc load line with Q-point are drawn below, together with the collector characteristic for the base current of 31 mA.

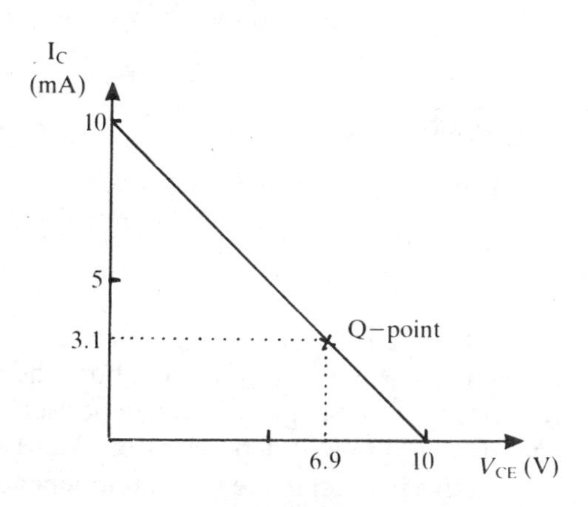

Q3. Determine the Q-point of the following circuit.

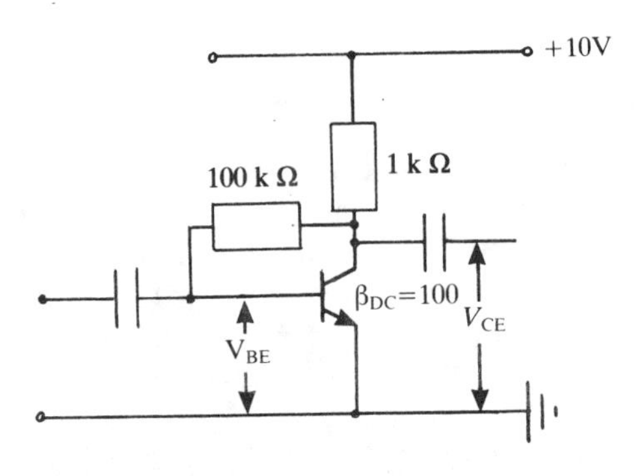

Because of manufacturing differences between transistors of the same type, β_{dc} may vary significantly, with the result that the Q-point may also vary. Even variations in ambient temperature can alter the Q-point. So, ideally, what is needed is a circuit which will make the Q-point independent of the transistor. The above circuit is a first shot at doing this.

As V_{BE} is fixed at 0.7 V, if β_{dc} were to increase, I_C would increase and V_{CE} would decrease. This means that the collector potential V_C decreases, I_B decreases, and I_C decreases. The total process has the effect of stabilizing the collector current. This is an example of *negative current feedback*.

I_B is given by

$$I_B = (V_{CE} - V_{BE})/R_B \quad . \tag{1}$$

But what is V_{CE}? From the load-transistor network, we obtain

$$\begin{aligned} V_{CE} &= V_{CC} - I_C R_C \\ &= V_{CC} - \beta_{dc} I_B R_C \quad . \end{aligned} \tag{2}$$

Substituting from (1) into (2), we have

$$V_{CE} = V_{CC} - \beta_{dc} R_C V_{CE}/R_B + 0.7\beta_{dc} R_C/R_B \quad ,$$

from which

$$V_{CE} = \frac{[V_{CC} + 0.7\beta_{dc} R_C/R_B]}{1 + \beta_{dc} R_C/R_B} \quad . \tag{3}$$

As $\beta_{dc} R_C/R_B = 1$, numerical substitution into (3) gives

$$\begin{aligned} V_{CE} &= [10 + 0.7]/2 \\ &= 5.35\text{ V} \quad . \end{aligned}$$

I_C can be calculated to be 4.65 mA.

It is important to recognize that the second term in both the numerator and the denominator of (3) are too large to be neglected in comparison with the first term. Hence V_{CE} and, through (1), I_C are still dependent on β_{dc}. The circuit does not successfully achieve its objective in making the Q-point independent of the value of

β_{dc}. This can be done by using the *potential divider* circuit, which is the subject of the next few sections.

6.7 STABILIZATION OF THE Q-POINT: THE POTENTIAL-DIVIDER BIAS CIRCUIT

Fig. 6.10 shows the universal method used for stabilising the Q-point. There are two

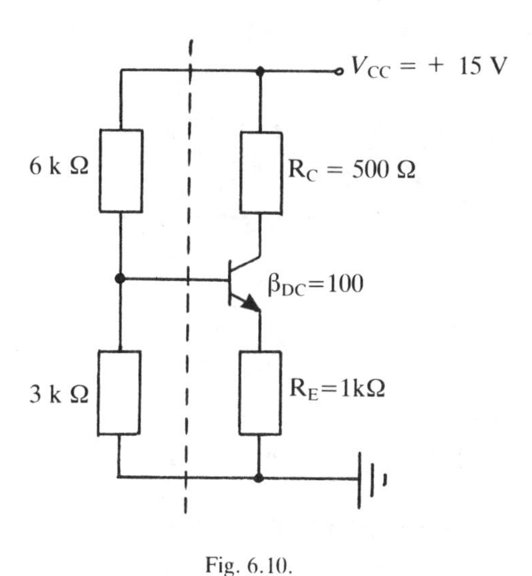

Fig. 6.10.

aspects of this circuit that need commenting on: (i) a potential divider network sets the dc base potential; (ii) there is a resistor between emitter and earth. The reason for the latter will emerge shortly. The circuit can be quite simply analysed if it is first Théveninized. Treat the transistor network to the right of the dashed line as a load. Then the Thévenin open-circuit pd is

$$V_{TH} = (3/9) \times 15\ \text{V}$$

This is 5 V. The Thévenin equivalent resistance R_{TH} is $6\,\text{k}\Omega \| 3\,\text{k}$, $= 2\,\text{k}\Omega$.

Now Fig. 6.10 can be reduced to the circuit in Fig. 6.11. On applying KVL (clockwise) to the base-emitter loop, we obtain

$$-V_{TH} + I_B R_{TH} + V_{BE} + I_E R_E\ . \tag{6.13}$$

Using

$$I_B = I_C/\beta_{dc} \simeq I_E/\beta_{dc}$$

in (6.13), and transposing terms, gives

$$I_E = (V_{TH} - V_{BE})/(R_E + R_{TH}/\beta_{dc})\ . \tag{6.14}$$

It is difficult to make any deductions from (6.14) unless the numerical values are used. Once this is done, we obtain

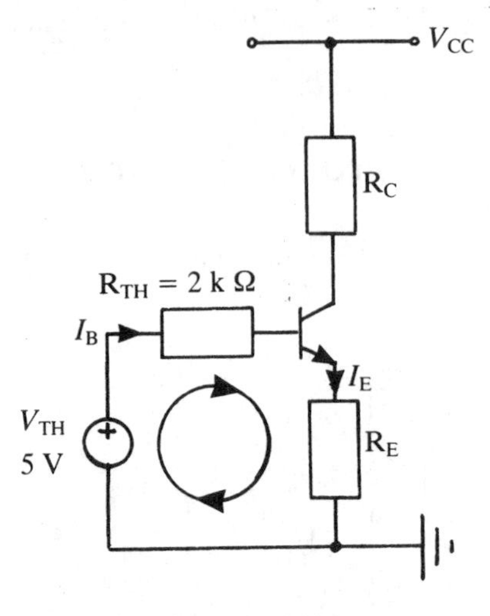

Fig. 6.11.

$$I_E = (5 - 0.7)/(1 + 2/100)\,\text{mA} \ .$$

Now it can be seen that the second term in the denominator is small compared with the first. This means that I_E (and I_C) is essentially independent of the β_{dc} of the transistor. It also follows that V_{CE} is independent of β_{dc}. We have found, therefore, a method of stabilising the Q-point of the circuit. The necessary condition is

$$R_E \gg R_{TH}/\beta_{dc} \ . \qquad (6.15)$$

The universal nature of the potential divider biased circuit of Fig. 6.10 emerges when the possible effects of temperature-variations are explored. Without going into details, it turns out that (6.15) also achieves the optimum thermal stability. The reasons for introducing an emitter resistor of the appropriate value into the circuit of Fig. 6.10 are that manufacturing differences between transistors of the same type now become unimportant, as do slight changes in temperature.

6.8 SMALL-SIGNAL AMPLIFICATION AT LOW FREQUENCIES

Before applying an ac signal to the base of the transistor, the Q-point must first be satisfactorily established (approximately midway along the dc load-line) using dc voltage sources. From now on, it will be assumed that this has been done. Further, unless stated to the contrary, the β referred in what follows is β_{ac} (that is, i_c/i_b). The term *small-signal* concerns the application of a relatively small amplitude signal to the base of the transistor to ensure that the variation in the collector potential is undistorted. The basic *base-driven* form of small-signal transistor amplifier is shown in Fig. 6.12, with the associated dc circuitry omitted. R_L is a load resistor across which the output pd is developed. Under forward bias, and at room temperature, the emitter current is given by (5.10) and the dynamic resistance r of the base-emitter diode is given by (5.11), *viz*.

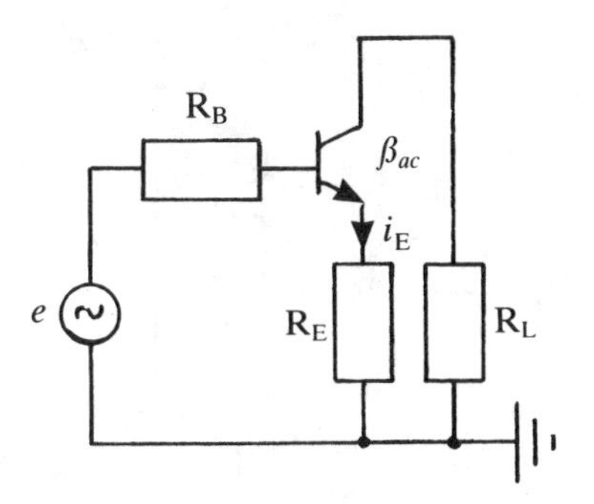

Fig. 6.12.

$$r = \Delta V_{BE}/\Delta I_E = v_{BE}/i_e = 25/I_E \qquad (6.16)$$

with I_E in mA. r is the change in the dc base-emitter pd, when the ac signal is applied to the base, divided by the change in the dc emitter current. As far as the ac signal is concerned, r is the resistance of the base-emitter diode. In addition, the collector characteristics in Fig. 6.4b clearly show the presence of a constant current region once V_{CE} goes above a few tenths of a volt. Taken together, both pieces of information allow us to picture a transistor as an ideal current source in series with a resistor of value r. This is the *Ebers-Moll* model which we shall use to study ac signal amplification (Fig. 6.13).

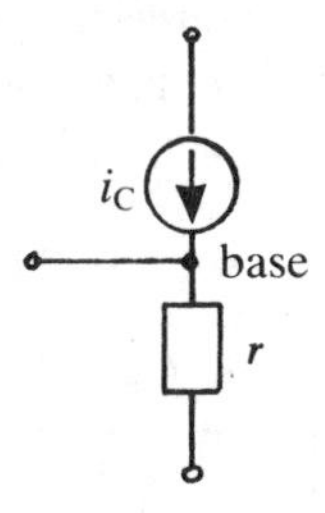

Fig. 6.13.

Fig. 6.12 may now be redrawn to obtain an expression for the ac voltage gain of the circuit. This is depicted in Fig. 6.14. First, note that the ac potentials at A and B are in phase with each other whereas the potentials at E and C are in antiphase. At the instant depicted in Fig. 6.14, i_E flows *down* to earth but i_C flows *up from* earth; that is, the AC potential at E lies above 0 V whilst that at C lies below 0 V.

Applying KVL to the base-emitter loop in Fig. 6.14 gives

$$-e + (i_E/\beta)R_B + (r + R_E)i_E = 0$$

and

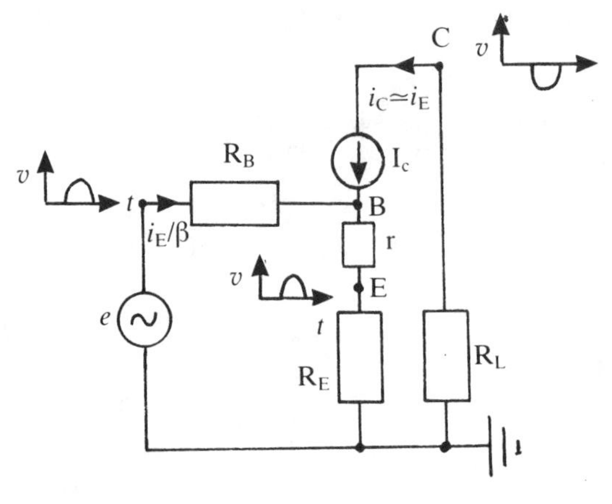

Fig. 6.14.

$$i_E = e/[r + R_E + R_B/\beta] \quad . \tag{6.17}$$

The ac voltage gain A is defined by

$$A = \text{pd across } R_L/\text{pd between base and earth} \tag{6.18}$$

i.e.

$$A = i_C R_L / i_E(r + R_E)$$

$$= R_L/(r + R_E) \quad . \tag{6.19}$$

In a *swamped* amplifier, $R_E \gg r$ and A reduces to R_L/R_E, whereas in the *common-emitter* amplifier, A is R_L/r. In the latter circuit, the emitter is now connected directly to earth. This can be done by placing a bypass capacitor across R_E.

Worked examples 6.3

Q1. Find the dynamic resistance and the ac voltage gain in the following circuit in the absence of capacitor C_2.

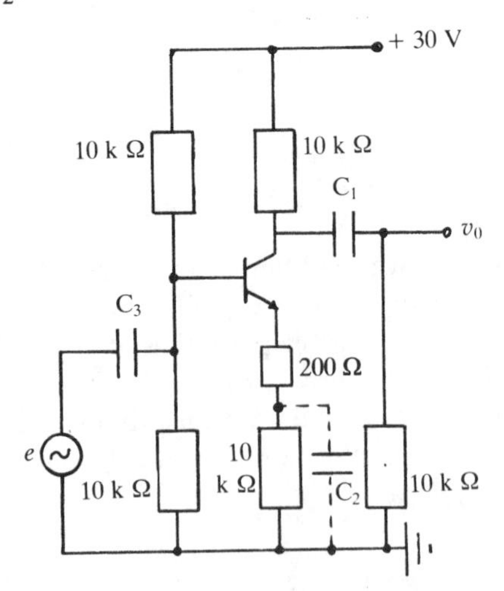

By (5.11) and (6.16), r is given by

$$r = 25/I_E(\text{mA})$$

Although it is possible to determine I_E directly from the above circuit, it is probably much safer to redraw the dc part of the circuit. This is known as the *dc equivalent* circuit. C_1 and C_3 are coupling capacitors and C_2 is a bypass capacitor. So let us say a few words about the role of these capacitors in a circuit.

A capacitor which connects one unearthed point (A) in a circuit to another (B) is called a *coupling* capacitor. Look at the following figure.

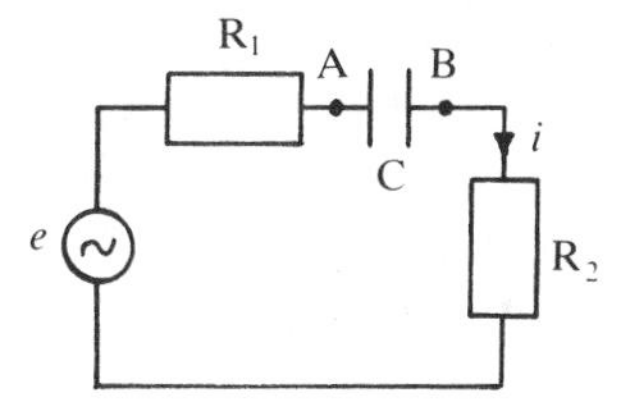

The instantaneous ac current i is given by

$$i = e/\sqrt{(R^2 + X_C^2)}$$

where R is the total resistance in the network and X_C is the capacitive reactance. i will take its largest value for $X_c \ll R$. So how large does the capacitor have to be before this inequality can be satisfied? X_C will have its largest value at the lowest frequency at which the circuit is to be used. Hence is is crucial for this frequency to be ascertained. For, then, the time constant RC of the circuit can be equated with the period ($= 1/f_{min}$) of the ac voltage source, and the capacitance C determined.

The bypass capacitor couples an unearthed point (A) in a circuit to earth, as below.

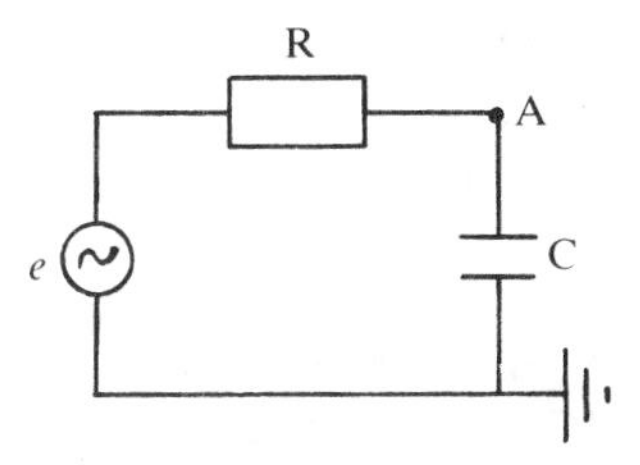

The instantaneous current i is still given by $e/\sqrt{(R^2 + X_c^2)}$. The circuit will act as though C is absent if $X_c \ll R$, and then point A becomes an ac earth. The capacitance can be determined once the lowest source frequency to be used is known. It should be assumed that capacitors included in circuits like the one given with this question satisfy the $X_C \ll$ R inequality, even though their values are not stated.

The third role of a capacitor has been discussed fully in Chapter 4 and in Q2: *Worked examples* 6.2. The capacitive reactance is infinite under dc conditions. Hence a capacitor can be used to isolate one part of a circuit from another or separate the stages of a cascaded amplifier. Thus C_1, enables the Q-points of the amplifier stages to be set independently of one another, whilst C_3 removes any dc component from the source voltage.

Returning to the solution to the problem, the dc equivalent circuit is:

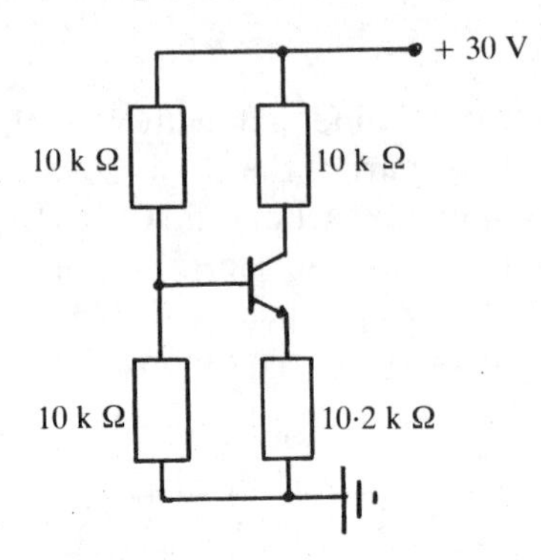

The base potential is 15 V (potential divider rule) and the emitter potential is 14.3 V. Hence, I_E is 14.3/10.2 kΩ, = 1.40 mA. Now

$$r = 25/1.40$$
$$= 18\,\Omega.$$

To calculate the ac voltage gain it is necessary to reduce the circuit to the base-driven form by drawing the ac equivalent circuit. This can be obtained by reducing the dc voltage source to zero. The required ac equivalent circuit is:

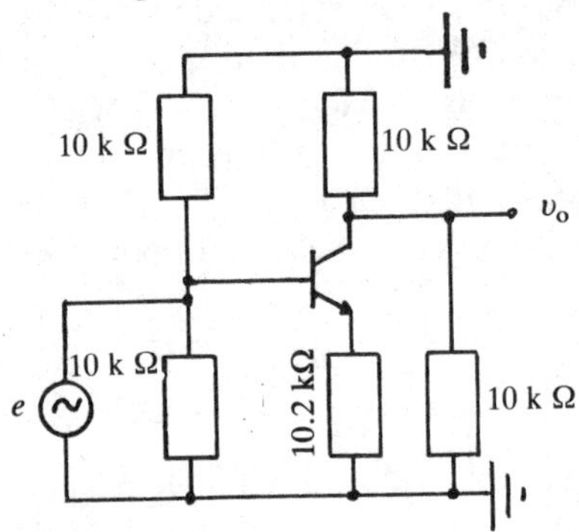

This can be further reduced by putting the two 10 kΩ resistors in the potential divider chain in parallel. The combined load resistor is also equal to 10 kΩ||10 kΩ. We obtain, therefore,

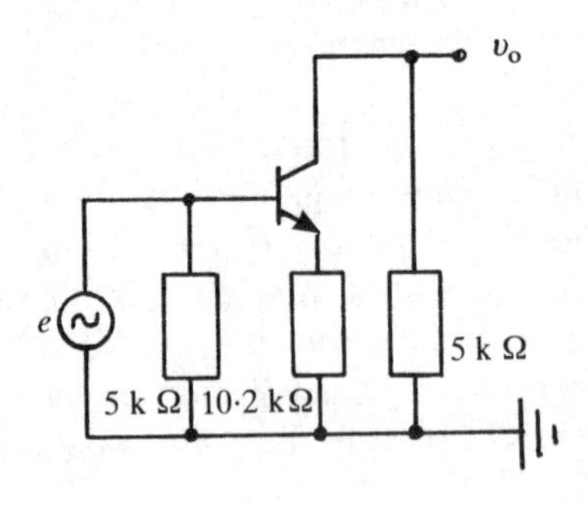

or

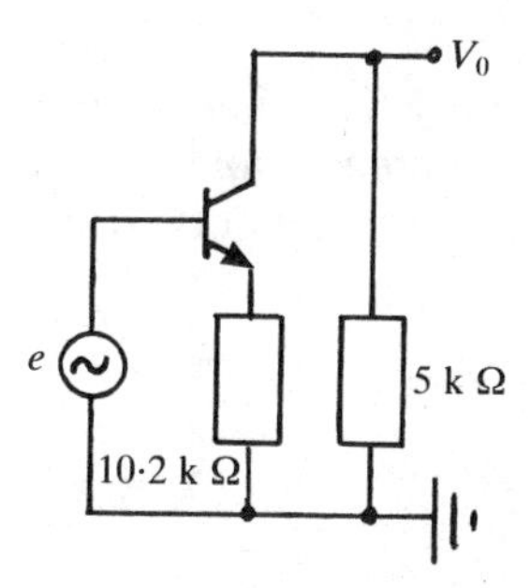

The ac voltage gain can be determined by using (6.19). It is

$$\begin{aligned} A &= R_L/(R_E + r) \\ &= 5000/10218 \\ &= 0.5 \end{aligned}$$

— a very poor amplifier indeed!

Q2. Determine the ac voltage gain of the amplifying circuit of Q1 in the presence of the bypass capacitor C_2.

The dc equivalent circuit is unchanged. Hence I_E and r retain their values. The ac equivalent circuit must be modified, however, because ac current flow through C_2 in preference to the 10 kΩ resistor.

The final ac equivalent circuit is similar to that of Q1 except that the emitter resistance is now 200 Ω. As $(r + R_E)$ is 218 Ω, the voltage gain A is 5000/218 = 23 — a more realistic figure!

6.8.1 Input impedance Z_I of the amplifier

It is necessary now to discuss the input impedance of the transistor amplifier. A knowledge of the input impedance allows that ac voltage gain to be calculated relative to the source itself.

Fig. 6.15 depicts a source connected directly to a black box which contains either

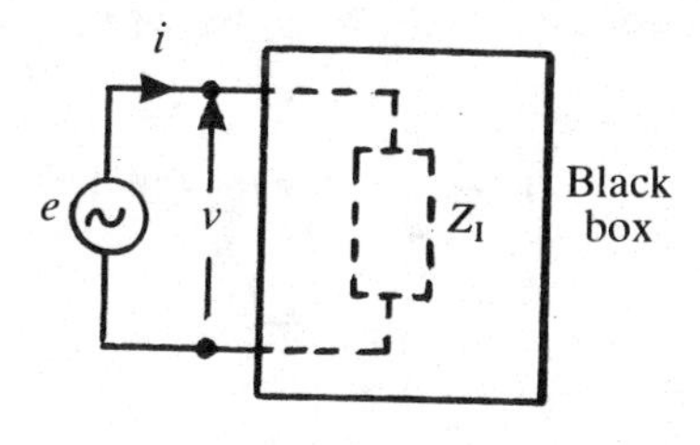

Fig. 6.15.

a device which does not require to be specified, or an amplifier. v is the pd between the input terminals of the black box and i is the input current. Z_I may now be defined as

$$Z_I = v/i = V'/I' \tag{6.20}$$

where V' and I' represent rms values. Looking into the input side of the black box, it seems as though the current i is limited by an impedance Z_I (shown dashed in Fig. 6.15). Fig. 6.16a shows the complete amplifier circuit, whereas Fig. 6.16b is the ac

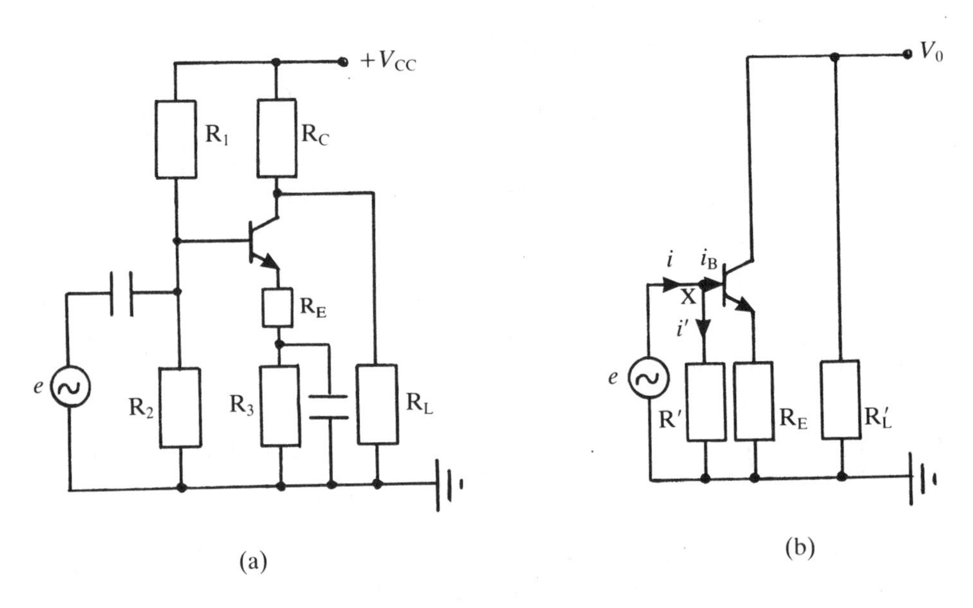

Fig. 6.16.

equivalent circuit. R'_L is $R_C \| R_L$. At each instant of time the input current i divides at node X into a part i_B which flows on to the base of the transistor and a part i' which flows through R' $(= R_1 \| R_2)$. To the source, it seems as though the input impedance Z_I of the amplifier consists of R' in parallel with an impedance Z_{IB} between the base of the transistor and earth.

What is Z_{IB}? With the help of Fig. 6.17, it may be observed that

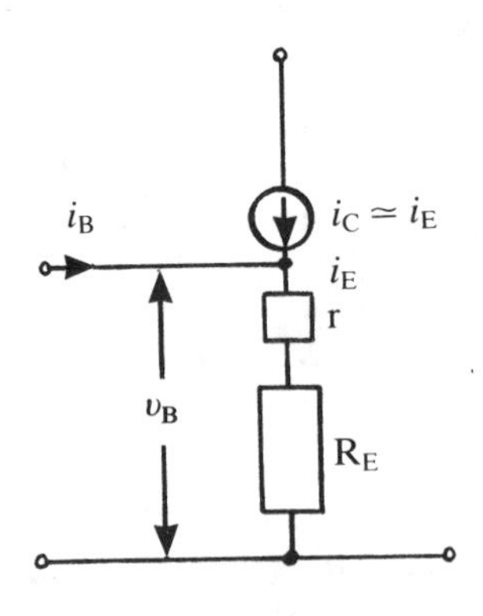

Fig. 6.17.

$$Z_{IB} = v_B/i_B$$
$$= i_e(r + R_E)/(i_E/\beta)$$
$$= \beta(r + R_E) \quad (6.21)$$

Therefore, the input impedance of the amplifier is given by

$$Z_I = R' \,\|\, \beta(r + R_E) \ . \quad (6.22)$$

Before proceeding with some worked examples, let us first learn how the input impedance can be measured experimentally.

6.8.1.1 *How to measure Z_I*

The general circuit used to measure Z_I is given in Fig. 6.18. The primes indicate that

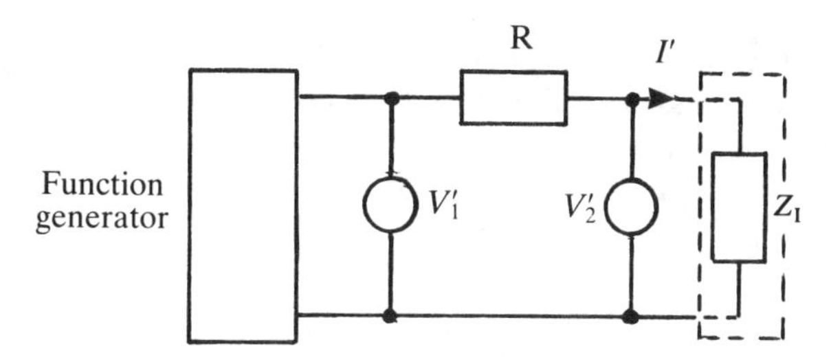

Fig. 6.18.

rms values are being used. Both voltmeters have a high resistance so that very little current flows through them. So, by Ohm's law,

$$I' = (V_1' - V_2')/R$$

and

$$Z_I = V_2'/I'$$
$$= V_2'R/(V_1' - V_2') \ . \quad (6.23)$$

In the case of an amplifier, there is an alternative method that can be used. Fig. 6.19

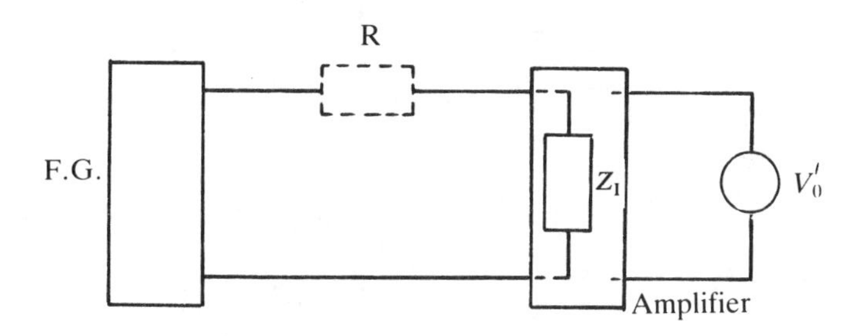

Fig. 6.19.

gives the necessary circuit. A function generator is initially connected directly to the input of the amplifier and the amplifier's output pd V'_o is noted for a given input pd, E', the rms output of the function generator. So we can write

$$(V'_o)_1 = AE' \tag{6.24}$$

where A is the voltage gain of the amplifier at the frequency used. Now insert R into the circuit. By the potential divider rule, the input to the amplifier is now $E'Z_I/(R + Z_I)$. After amplification, we obtain

$$(V'_o)_2 = AE'Z_I/(R + Z_I) \tag{6.25}$$

where $(V'_o)_2$ is the new rms output. Dividing (6.25) by (6.24) and transposing terms gives

$$Z_I = R/[\{(V'_o)_1/(V'_o)_2\} - 1] \tag{6.26}$$

6.8.2 Output impedance Z_s of a voltage source

So far, the source has been assumed to be *ideal*; that is, it has no impedance. This is obviously not the case in practice. A *real* source must be represented as an ideal source with an impedance Z_s in series, as in Fig. 6.20.

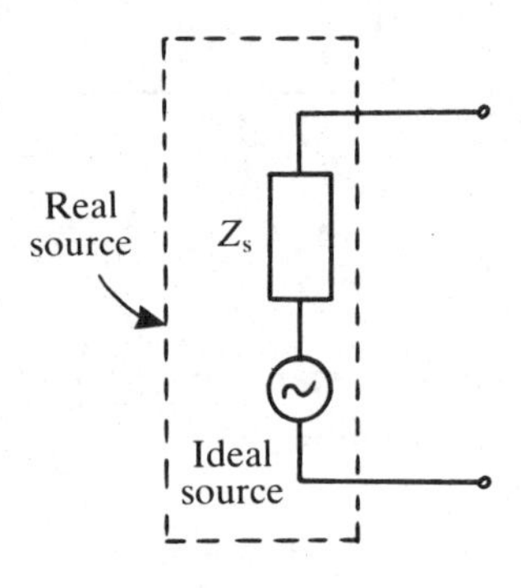

Fig. 6.20.

The base-driven circuit of Fig. 6.16b must be modified to that shown in Fig. 6.21.

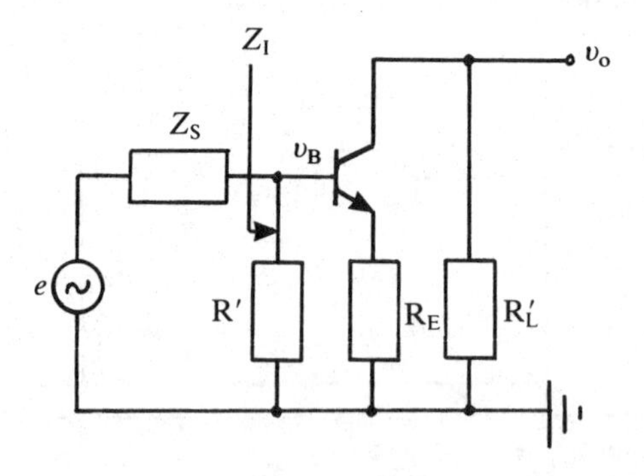

Fig. 6.21.

The potential at the base of the transistor is different from e. It depends on the input impedance of the amplifier. Using the potential divider rule, the potential v_B can be calculated to be

$$v_B = eZ_I/(Z_I + Z_s) \ . \tag{6.27}$$

However, by making $Z_I \gg Z_s$, v_B can be equated with e. This is a vitally important design consideration when constructing an amplifier.

6.8.2.1 How to measure Z_s

The ideal way to measure Z_s is to short-circuit the output terminals in Fig. 6.20 and note the current i_{sc} flowing. Then

$$\begin{aligned} Z_s &= \frac{\text{open-circuit emf}}{\text{short-circuit current}} \\ &= e/i_{sc} \\ &= E'/I'_{sc} \ . \end{aligned} \tag{6.28}$$

However, there is one major problem. A zero resistance current meter would have to be used to record I'_{sc}! This is clearly impracticable. The usual way to measure Z_{sc} is to use the circuit of Fig. 6.22. A high-resistance voltmeter is placed between the

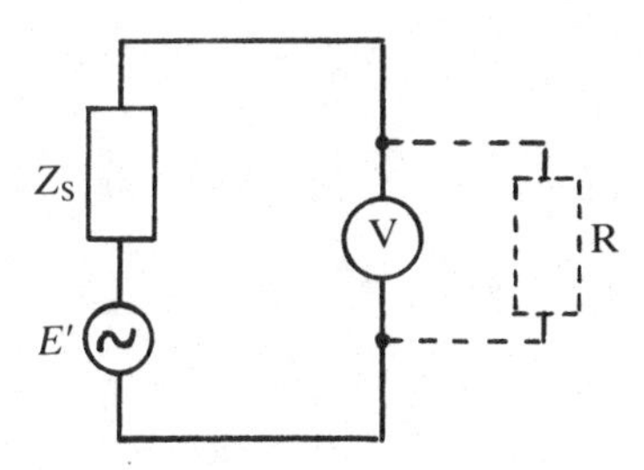

Fig. 6.22.

terminals of the *real* source and records the open-circuit emf (because no current flows through it). A resistance R (shown dashed) is then added to the circuit, and the voltmeter records the pd V'_R across it. The difference $(E' - V'_R)$ is, therefore, the rms pd across Z_s. The current flowing through Z_s is V'_R/R (for the same reason as before, *viz.* that no current is able to flow through the voltmeter). Hence the output impedance of the voltage source is given by

$$\begin{aligned} Z_s &= (E' - V'_R)/(V'_R/R) \\ &= R[(E'/V'_R) - 1] \end{aligned} \tag{6.29}$$

6.9 THE EMITTER-FOLLOWER

As we have seen, a well-designed amplifier has an input impedance which is large compared with the output impedance of the source. The same result holds for a cascade amplifier; the input impedance of stage N must be much larger than the output impedance of stage $N-1$. But suppose that this is not the case. A way of overcoming this difficulty is to insert a *buffer* amplifier between the two stages. The buffer does not contribute to the amplification process; the ac voltage gain is $\leqslant 1$. However, it does have: (i) a high input impedance which avoids substantial attenuation of the output signal from amplifier stage $N-1$; (ii) a low output impedance compared with the input impedance of amplifier stage N. One such buffer, which uses a bipolar transistor is known as the *emitter-follower*.

The basic circuit of the emitter-follower is shown in Fig. 6.23a, with the AC

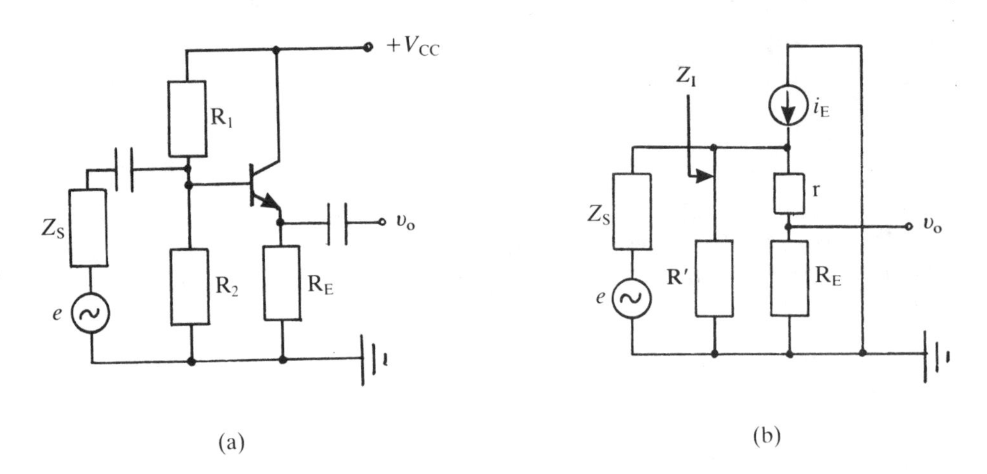

Fig. 6.23.

equivalent circuit in Fig. 6.23b. Note that: (i) the output is taken from the emitter of the transistor, and (ii) the collector is directly connected to the high voltage rail. For this reason it is also known as the *common-collector amplifier*.

The ac voltage gain can be worked out fairly easily. The pd between the base of the transistor and earth is given by

$$v_B = i_E(r + R_E)$$

and the output pd by

$$v_o = i_E R_E \ .$$

So, the ac voltage gain A is

$$A = R_E/(r + R_E) \tag{6.30}$$

A is less than 1, although, by making $R_E \gg r$, A will tend to the value 1. Unlike the ac collector potential, the emitter potential is in phase with the base potential at every instant of time.

For the emitter-follower to function as an impedance-matching unit its output impedance must be low compared with the input impedance of the next amplifier stage. In fact, as we shall see, the emitter-follower effectively reduces the output impedance of the previous amplifier stage (which, of course, may be the source) by the factor β_{ac}.

If we assume that the emitter-follower is connected directly to the source, then Fig. 6.24a depicts the relevant AC equivalent circuit. (The potential divider resistor

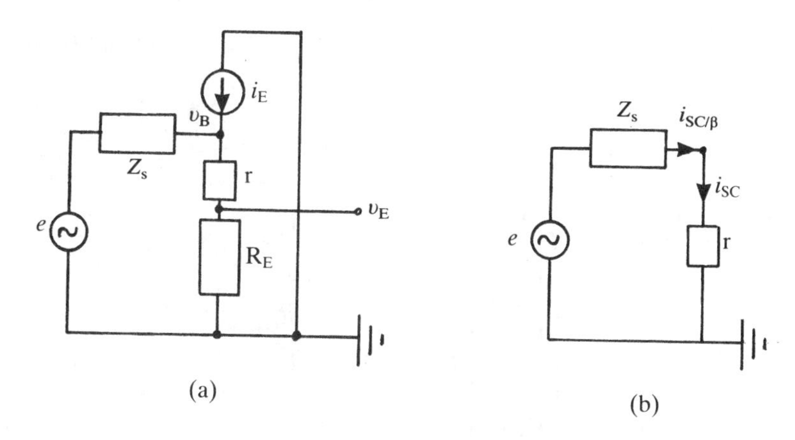

Fig. 6.24.

has been omitted to keep the circuit as basic as possible.) We require to obtain a relation for the output impedance Z_o between the output lead and earth. Using the technique adopted for the measurement of the source impedance in section 6.8.2.1, it should be noted that the open-circuit emf is numerically equal to the ac emitter potential υ_E and the short-circuit current is i_{sc}. Hence, by (6.28),

$$Z_o = \frac{\text{open-circuit emf}}{\text{short-circuit current}}$$

$$= \upsilon_E / i_{sc} \ . \tag{6.31}$$

However, by (6.30), υ_E is practically equal to υ_B. Further, by working with small base currents, the pd across Z_s is negligible and υ_E can be identified with e. Hence

$$Z_o = e/i_{sc} \ . \tag{6.32}$$

Now we need to determine i_{sc}.

Applying KVL to the relevant part of the circuit shown in Fig. 6.24b, we obtain

$$e - (i_{sc}/\beta).Z_s - i_{sc}.r = 0$$

from which

$$i_{sc} = e/[(Z_s/\beta) + r] \ . \tag{6.38}$$

Substitution into (6.32) gives

$$Z_o = (Z_s/\beta) + r \ . \tag{6.34}$$

Equation (6.34) tells us that the purpose of the emitter-follower is to decrease the output impedance of the source (or the previous amplifier stage) by the factor β_{ac}. Now, so long as Z_o is small compared with the input impedance of the next stage, the input signal will not be substantially attenuated.

Worked examples 6.4

Q1. Find the ac voltage gain of the 2-stage amplifier which has an *ideal* voltage source:

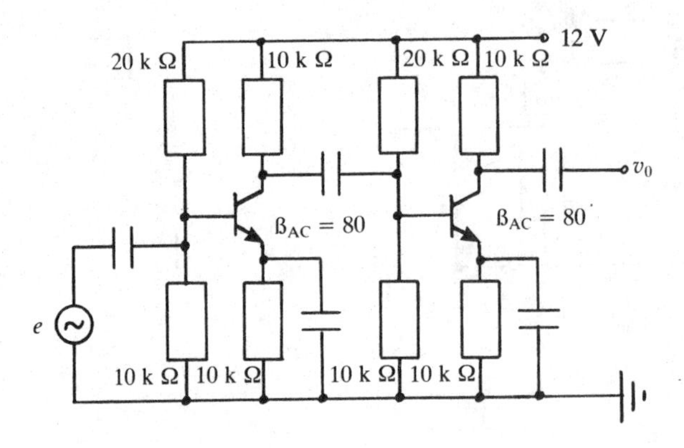

The capacitor separates the two stages and, is so doing, allows the Q-points to be set independently.

DC calculations:
The dc equivalent circuit will not be drawn, as it should be possible to extract the essential information from the above circuit. However, if any difficulty is experienced then it is advisable to draw the dc equivalent circuit before proceeding further.

For each stage we have

$$V_B = (10/30) \times 12\,\text{V} \quad \text{(potential divider rule)}$$
$$= 4\,\text{V}$$

and

$$V_E = (4 - 0.7)\,\text{V}$$
$$= 3.3\,\text{V} \ .$$

Therefore,

$$I_E = 3.3/10\,\text{k}\Omega$$
$$= 0.33\,\text{mA} \ .$$

The dynamic resistance r is $25/0.33 = 75\,\Omega$.

AC calculations:
The ac equivalent circuit is more complicated to visualise. It is:

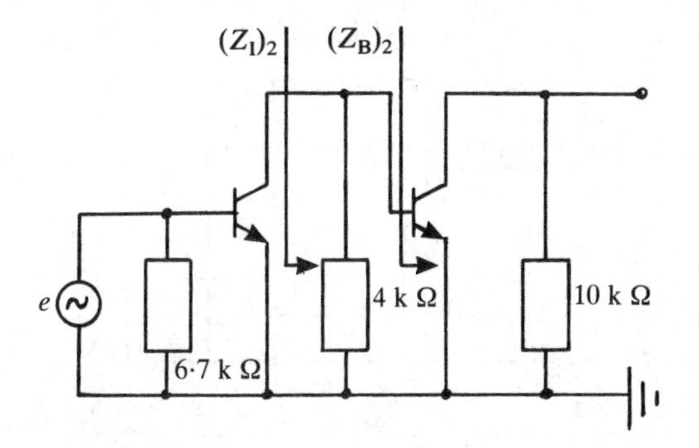

The 4 kΩ resistor arises from the parallel combination of the 10 kΩ collector resistance of stage 1 with the 20 kΩ and 10 kΩ resistances in the potential divider chain of stage 2; that is, 10 kΩ||20 kΩ||10 kΩ. The input impedance between base and earth for each transistor is given by

$$(Z_{IB})_1 = (Z_{IB})_2 = \beta r$$
$$= 80 \times 75\,\Omega$$
$$= 6\,\text{k}\Omega\ .$$

As far as stage 1 is concerned its load resistance is 4 kΩ||(Z_{IB}); that is, 4 kΩ||6 kΩ or 2.4 kΩ. This is equal to the input impedance of stage 2. Now the ac voltage gain of stage 1 can be calculated to be

$$A_1 = (Z_{IB})_2/r$$
$$= 2.4\,\text{k}\Omega/75$$
$$= 32$$

and the ac voltage gain of stage 2 is

$$A_2 = R_L/r$$
$$= 10000/75$$
$$= 133\ .$$

Therefore, the overall ac voltage gain A is

$$A = A_1 \times A_2$$
$$= 32 \times 133$$
$$= 4300\ .$$

Q2. An ac source with an output impedance of (a) 1 kΩ and (b) 4 kΩ produces a voltage signal of amplitude 1 mV. It is connected to the first stage of the amplifier in Q1. Calculate the amplitude of the output signal.

(a) The circuit is:

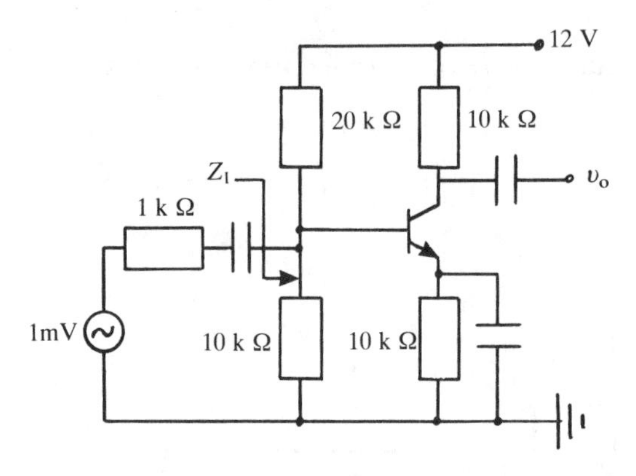

The dc values of V_B, V_E, and I_E as well as the dynamic resistance r and Z_{IB} have been previously calculated in Q1. Therefore, the input impedance is

$$\begin{aligned} Z_I &= 20\,k\Omega \,\|\, 10\,k\Omega \,\|\, 6\,k\Omega \\ &= 3.16\,k\Omega. \end{aligned}$$

Using the potential divider rule, the amplitude of the voltage signal arriving at the base of the transistor is

$$\begin{aligned} v_B &= 3.16/(1+3.16) \times 1\,mV \\ v_B &= 3.16/(1+3.16) \times 1\,mV \\ &= 0.76\,mV \end{aligned}$$

The ac voltage gain is identical with that of stage 2 in Q1, viz. 133.3. Therefore, the amplitude of the output signal v_o is 0.76×133 or 101 mV.

(b) The circuit is identical with part (a) except that the output impedance of the source is increased to 4 kΩ. v_B is now equal to

$$3.16/(4+3.16) \times 1\,mV \text{ or } 0.44\,mV$$

As can be seen, there is a substantial attenuation in the amplitude of the source signal. v_o is only 59 mV.

Part (b) reinforces the point made in section 6.8.2 that the source impedance must be much smaller than the input impedance of the amplifier to avoid excessive attenuation of the source signal occurring.

Q3. An emitter-follower is placed between the voltage source of Q2(b) and the amplifier. Calculate the output voltage and assess whether the introduction of the emitter-follower has resulted in an improvement in the output.

The complete circuit is shown below.

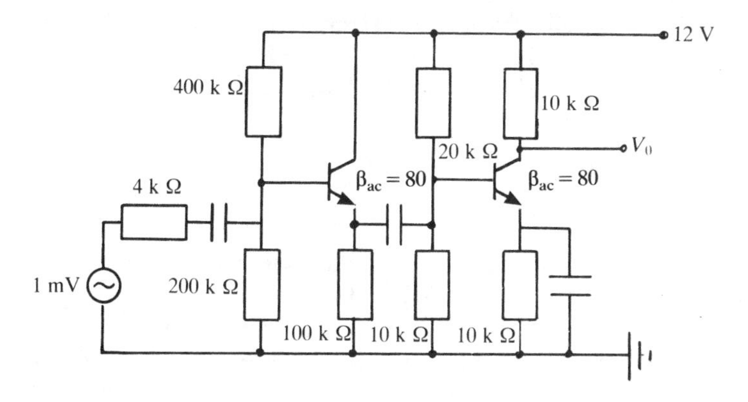

The ac voltage gain A_2 of the amplifier was calculated to be 133 in Q1. This means that in the absence of any attenuation, that is, using an ideal voltage source, the output voltage from the amplifier will be 133 mV. However, for a source with an output impedance of 4 kΩ the output voltage falls to 59 mV, as Q1(b) showed. Following section 6.9, it was proved that an important function of an emitter-follower is to reduce the source impedance by the factor β_{ac}. Hence, it is expected that placing an emitter-follower between source and amplifier will have the effect of enhancing the output voltage. Let us now calculate its actual value with the assistance of the following ac equivalent circuit.

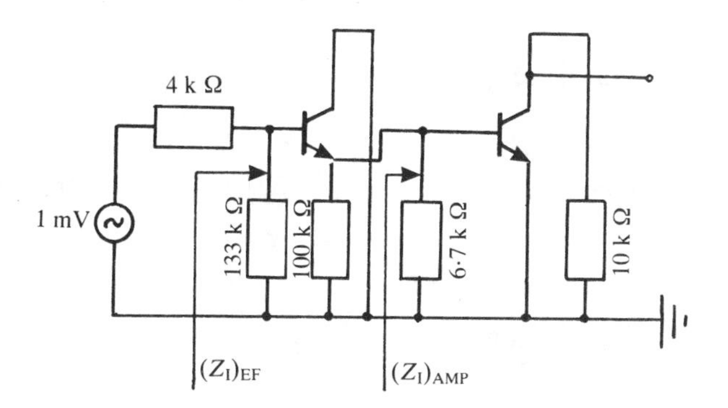

r and Z_{IB} of the amplifier have already been calculated as 75 Ω and 6 kΩ, respectively. Therefore, the input impedance $(Z_I)_{AMP}$ of the amplifier is $6.7\,k\Omega \| 6\,k\Omega, = 3.2\,k\Omega$.

For the emitter-follower, the essential dc values are:

$$V_B = 4\,V; \quad V_E = 3.3\,V\,; \quad I_E = 3.3/100\,k\Omega = 0.033\,mA\,;$$
$$r = 25/0.033 = 758\,\Omega\,; \quad (Z_{IB})_{EF} = \beta(r + R_E)\,.$$

But what value has R_E? R_E is not just 100 kΩ! Another look at the ac equivalent circuit shows that the emitter resistance of the emitter-follower is in parallel with the input impedance of the amplifier. Therefore, the value of R_E to use here is $100\,k\Omega \| 3.2\,k\Omega, = 3.1\,k\Omega$.

So, $(Z_{IB})_{EF}$ is $80 \times (758 + 3.1\,k\Omega) = 309\,k\Omega$, and the input impedance $(Z_I)_{EF}$ is

$133\,\text{k}\Omega \| 309\,\text{k}\Omega = 93\,\text{k}\Omega$. Now, we are in a position to calculate the base potential of the emitter-follower as $(93/97) \times 1\,\text{mV} = 0.96\,\text{mV}$.

The ac voltage gain A_{EF} of the emitter-follower is

$$\begin{aligned} A_{EF} &= (R_E)_{TOT}/[r + (R_E)_{TOT}] \\ &= 3.1/3.86 \\ &= 0.80\ . \end{aligned}$$

Therefore, the overall ac voltage gain of the emitter-follower and amplifier is $0.80 \times 133 = 107$. The output voltage is 107 mV.

The effect of the emitter-follower is to improve the output voltage of the amplifier by about a factor of 2 times. Its function into the circuit is obviously very desirable.

Q4. Obtain an expression for the ac power gain of the emitter-follower.

Power is the rate at which the emitter-follower does work on a load. Of course, the load may include the input impedance of any subsequent amplifier stage. Refer to the emitter-follower circuit in Fig. 6.23.

The output power is $i_E^2 R_E$, and the input power to the base of the transistor is $i_B^2 Z_{IB}$.

From (6.21), the ac power gain G is given by

$$\begin{aligned} G &= i_E^2 R_E / i_B^2 \beta (r + R_E) \\ &= \beta^2 i_B^2 R_E / \beta i_B^2 (r + R_E) \\ &= \beta R_E / (r + R_E) \end{aligned}$$

This result is the product of the ac current gain β_{ac} and the ac voltage gain A.

7

The field-effect transistor

Objectives

(i) To distinguish between the JFET and the MOSFET; channel shape and pinch-off
(ii) To discuss the form of the drain characteristics and the transconductance curve
(iii) To meet the concept of self-biasing and midpoint biasing in the design of a JFET amplifier
(iv) To introduce the common-source circuit as an analogue to the common-emitter bipolar circuit
(v) The JFET source-follower an alternative to the emitter-follower
(vi) To discuss the problem of electrical noise
(vii) How to use a JFET for amplifying dc voltages
(viii) How to use the MOSFET in logic applications.

Unlike the bipolar transistor in which both electrons and positive holes contribute to the total current, the field-effect transistor is a unipolar device; that is, only one type of carrier determines the current.

There are two species of field-effect transistor: the *junction* field-effect transistor (JFET), and the *metal-oxide semiconductor* (MOSFET) or *insulated gate* (IGFET) field-effect transistor. These can be further broken down into two sub-species: the *n*-channel and the *p*-channel device; electrons are the mobile charge carriers in the *n*-channel device, and positive holes in the *p*-channel device. As an added complication the MOSFET conducts in two modes: *enhancement* and *depletion*. In this introductory text we shall discuss only the *n*-channel JFET and the *n*-channel enhancement-mode E-MOSFET.

7.1 THE *n*-CHANNEL JFET

7.1.1 Structure

The basic structure of this device is depicted in Fig. 7.1a, with its symbol in Fig. 7.1b. The JFET consists of a bar of *n*-type silicon with a region of very heavily doped *p*-type silicon embedded into it. It is usual to indicate the high degree of doping by writing a

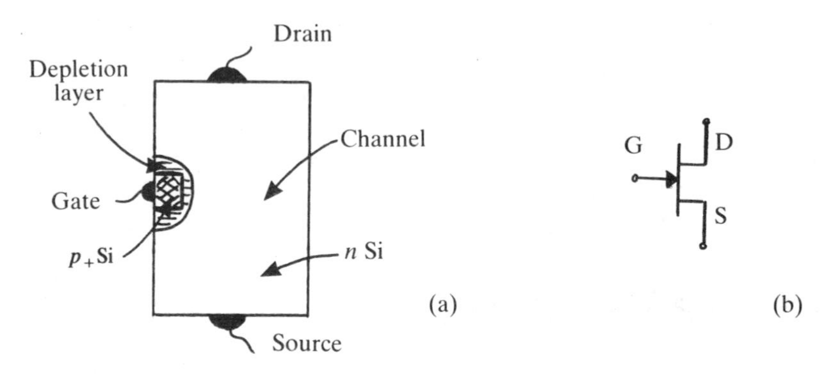

Fig. 7.1.

plus (+) sign as a superscript to either n or p. Here, we need to write p^+. A *pn* junction is produced with the depletion layer existing almost entirely in the n-Si bar. The channel is the region created between the depletion layer and the opposite surface of the bar. Some JFETs are created slightly differently: two p^+ Si regions are fabricated opposite each other. Now the channel lies between the two depletion layers so produced. Metallic (ohmic) electrodes are deposited on each end of the n-Si bar and also on the p^+Si region. These are known as the *source*, *drain* and *gate*. Electrons are supplied by the source, removed at the drain, and the rate of flow is controlled by the gate.

7.1.2 Drain characteristics

Fig. 7.2a shows a circuit that can be used to study the drain characteristics depicted in Fig. 7.2b. As can be seen, $I_D = 0$ for $V_{DS} = 0$ and the channel is fully open. Then as

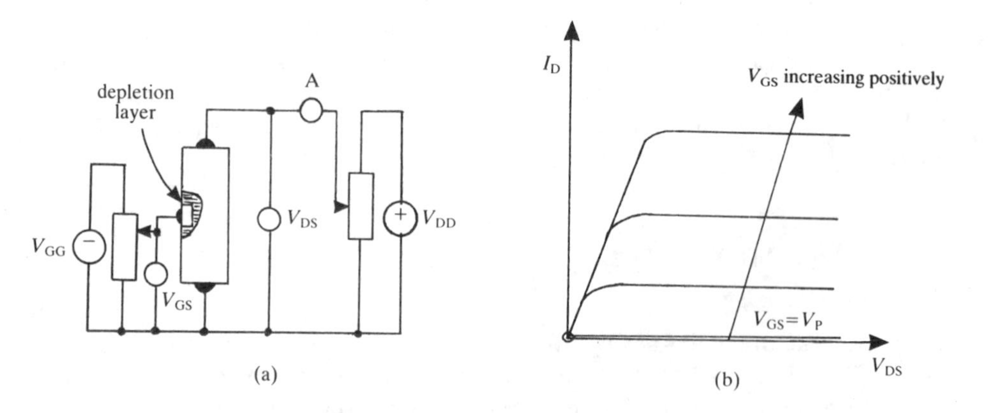

Fig. 7.2.

V_{DS} increases, the *n*-Si bar behaves like a semiconductor resistor and I_D rises almost linearly. This is the *ohmic* or *non-saturation* region. The *pn* junction reverse-biases, and the conducting part of the channel begins to narrow. The potential increases more positively on moving from source to drain, which means that the junction is under a larger reverse bias nearest the drain; the width of the depletion layer is non-uniform, being largest at its drain end. The channel becomes *pinched-off* at a specific value of V_{DS} and I_D saturates. For $V_{GS} < 0$, the depletion layer reduces the width of the channel even for $V_{DS} = 0$. So, pinch-off occurs at a smaller volume of V_{DS} than previously. Of course, at $V_{GS} = V_P$ the channel is completely pinched-off, and $I_D = 0$ at all $V_{DS} > 0$.

7.1.3 Transconductance

The transconductance curve is another useful characteristic to consider (see Fig. 7.3). It shows how I_D varies with V_{GS} at a constant value of V_{DS}. The curve is

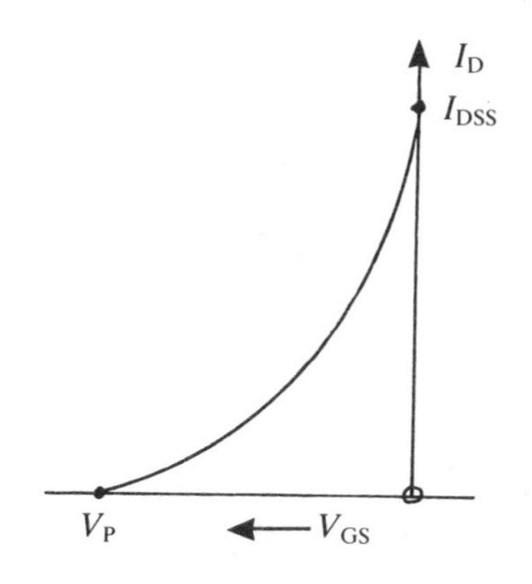

Fig. 7.3.

parabolic and obeys the relation

$$I_D = I_{DSS}\{1 - (V_{GS}/V_P)\}^2 \qquad (7.1)$$

where I_{DSS} stands for drain to source current with shorted-gate (that is, $V_{GS} = 0$). It is the maximum value that I_D can have. At the other end of the curve, $I_D = 0$ because the channel is pinched-off before any positive value of V_{DS} is applied. In the 2N5951 JFET, V_P is -3.5 V.

The transconductance g_m is given by

$$g_m = \Delta I_D / \Delta V_{GS} = i_D / v_{GS} \qquad (7.2)$$

UNIT: Siemens (S) or mho.

The transconductance is sometimes given the symbol g_{mo} in the neighbourhood of I_{DSS}.

7.1.4 Gate leakage current

As the gate-source region of the JFET is reverse biased, only a very small reverse current flows through the depletion layer. As a result, the input impedance of the JFET is exceptionally high, typically > 10–100 MΩ. For this reason, the JFET is preferred to the bipolar transistor in those applications for which a large input impedance is desirable.

7.2 DESIGNING A JFET AMPLIFER

7.2.1 Self-bias concept

This section describes a method for setting the Q-point without the need for a voltage supply in the gate line. Look at Fig. 7.4. Owing to the high input impedance of the

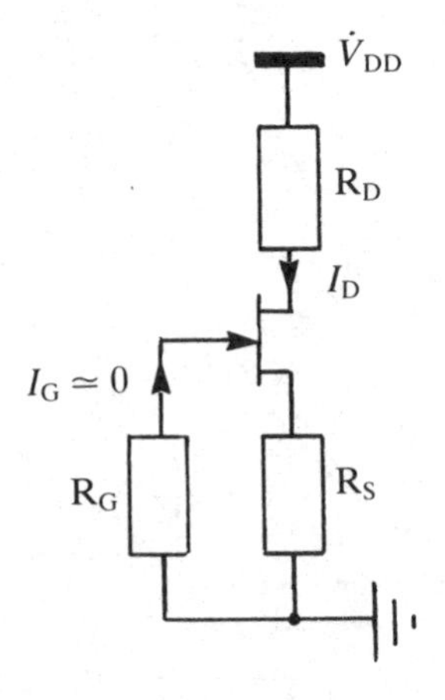

Fig. 7.4.

gate region, the gate current $I_G \simeq 0$. This means that the gate is practically at 0 V, as there is a negligible pd across resistor R_G. The value of R_G is not critical. However, as $I_G R_G$ is very much smaller than $I_D R_S$ and I_G is $\sim \mu$A, R_G needs to lie between 1 and 10 MΩ.

Now,

$$V_{DS} = V_{DD} - I_D(R_D + R_S) \tag{7.3}$$

and, applying KVL to the gate-source loop,

$$V_{GS} + I_D R_S = 0$$

or

$$V_{GS} = -I_D R_S \,, \tag{7.4}$$

Equation (7.4) tells us that although there is no gate voltage supply, the gate still acquires a negative potential relative to the source, of magnitude $I_D R_S$.

In the absence of any resistor network, the drain current I_D varies along the transconductance curve in Fig. 7.3. However, in the presence of resistors (Fig. 7.4), I_D now varies linearly with V_{GS}. Why? Rearranging (7.4) to

$$I_D = -V_{GS}/R_S \tag{7.5}$$

gives the answer; (7.5) is the equation of a straight line passing through the origin and having the gradient $-1/R_S$. It is the dc load line of the JFET circuit. Fig. 7.5 depicts

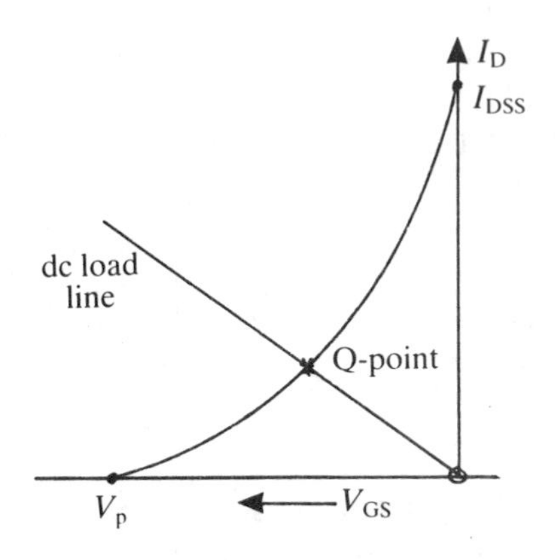

Fig. 7.5.

the Q-point — the point of intersection of the load line and the transconductance curve, for the *same* drain current must flow through the JFET and the resistor network.

Fig. 7.4 exhibits negative current feedback. What happens if I_D were to increase slightly, for whatever reason? By (7.4) or (7.5), V_{GS} would become more negative and the depletion layer would increase in size, thus narrowing the width of the channel and decreasing I_D. So the effect of self-bias is to offset any change in the drain current.

There is a problem in using the self-bias procedure for setting the Q-point. Like bipolar transistors, slight differences in their fabrication ensure that the drain and transconductance characteristics of JFETs of the same type are not completely identical. Therefore, the coordinates of the Q-point may change. The technique described in the next section makes the Q-point independent of the JFET being used.

7.2.2 Current–source bias

The principle behind this technique is simply to make I_D independent of V_{GS}. The relevant circuit is shown in Fig. 7.6. The FET is in series with a bipolar transistor. The emitter current I_E of the bipolar is given by

$$\begin{aligned} I_E &= \{V_{DD}R_1/(R_1 + R_2) - 0.7\}/R_E \\ &\simeq I_C \\ &= I_D \ . \end{aligned} \tag{7.6}$$

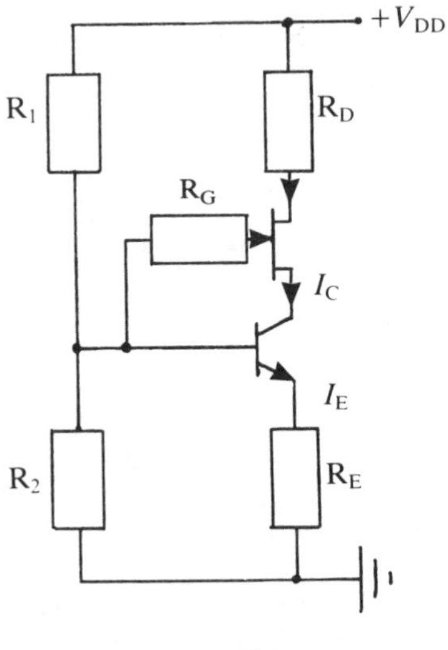

Fig. 7.6.

Note that I_D depends only on V_{DD} and the resistors R_1, R_2, and R_E and not on V_{DS} or V_{GS}. The only precaution required is that I_C must be less than I_{DSS}, the maximum value of the drain current. Although it is true that V_{BE} will vary slightly from one bipolar to another of the same type, the change should not produce a significant shift in either I_D or the Q-point.

Worked examples 7.1

Q1. A JFET is to be used in the *midpoint* bias condition. Obtain an expression for the source resistance R_S in terms of V_P and I_{DSS}.

The appearance of V_P and I_{DSS} in the desired expression for R_S suggests that it is necessary to discuss I_D via the transconductance curve for the JFET. The midpoint bias condition indicates that at the Q-point the drain current is equal to $I_{DSS}/2$. Once the corresponding value of V_{GS} has been determined from (7.1), the value of R_S can be obtained from (7.5).

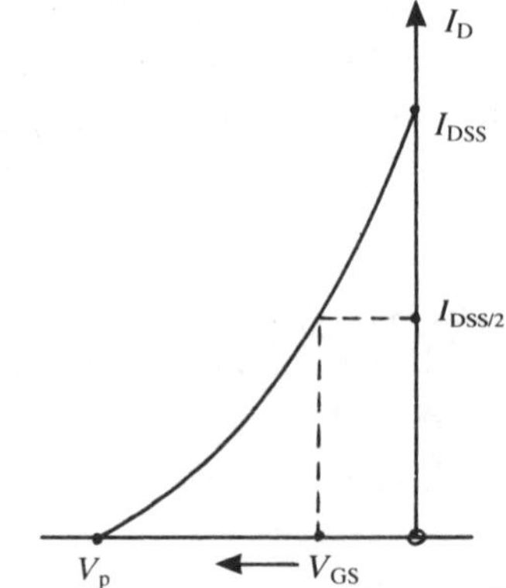

The equation of the transconductance curve is stated in (7.1). It is

$$I_D = I_{DSS}\{1-(V_{GS}/V_P)\}^2 \ . \quad (1)$$

For $I_D = I_{DSS}/2$, we have

$$\tfrac{1}{2} = \{1-(V_{GS}/V_P)\}^2 \ . \quad (2)$$

This can be solved exactly to give

$$V_{GS} = V_P(1-0.71)$$
$$= 0.29\,V_P \ . \quad (3)$$

Or, as is more usual, observe that (2) will practically be satisfied if

$$V_{GS}/V_P = \tfrac{1}{4} \ , \quad (4)$$

for then,

$$(1-\tfrac{1}{4})^2 = \tfrac{9}{16}$$

which is close to $\frac{1}{2}$.
From (4),

$$V_{GS} = V_P/4 \ .$$

Substituting for V_{GS} in (7.5), we obtain

$$V_P/4 = -I_D R_S$$
$$= -I_{DSS}R_S/2$$

from which

$$R_S = -V_P/2\,I_{DSS} \ . \quad (5)$$

The negative sign will disappear when numerical values are substituted into (5), for V_P is less then 0 V.

If V_{GS} is varied by $\pm v_{GS}$, I_D will vary by $\pm i_D$ and an ac output is generated. However, because of the parabolic form of the transconductance curve, it is essential for the variations in V_{GS} to be small if a sinusoidal output is required.

WORKING RULE: i_D should be about $\frac{1}{10}$th of I_D.

The output is said to suffer from *square law* distortion once v_{GS} gets too large. This deficiency is put to advantage in frequency mixers for eliminating spurious signals.

Q2. The value of the transductance g_{mo} near $I_{DSS} = 4$ mA is 3000 μS. Determine the value of the source resistance R_S in the circuit of Fig. 7.4 under midpoint bias. Calculate any other circuit parameters if $V_{DD} = 20$ V and $R_o = 4$ kΩ.

From the wording of the question, it should be possible to determine R_S from a knowledge of g_{mo} alone. Let us investigate this aspect of the question further. g_{mo} is defined as $\Delta I_D/\Delta V_{GS}$ evaluated close to I_{DSS}. So differentiate (7.1) with respect to V_{GS} to give

$$g_m = 2I_{DSS}(-1/V_P)\{1 - V_{GS}/V_P\}$$

and hence obtain g_{mo} by putting $V_{GS} = 0$, that is,

$$\begin{aligned} g_{mo} &= \Delta I_D/\Delta V_{GS}|_{V_{GS}\to 0} \\ &= -2I_{DSS}/V_P \ . \end{aligned} \tag{1}$$

With the help of equation (5) in Q1, this expression becomes

$$g_{mo} = 1/R_S \ , \tag{2}$$

therefore,

$$\begin{aligned} R_S &= 1/4 \times 10^{-3} \\ &= 250\,\Omega \ . \end{aligned}$$

Using (7.5) with $I_D = I_{DSS}/2$ at midpoint bias, we have

$$\begin{aligned} V_{GS} &= -I_{DSS}R_S/2 \\ &= -2 \times 10^{-3} \times 250 \\ &= -0.5\,\text{V} \ . \end{aligned}$$

(7.3) allows us to calculate V_{DS} as

$$\begin{aligned} V_{DS} &= 20 - [2 \times 10^{-3} \times 4.25 \times 10^3] \\ &= 20 - 8.5 \\ &= 11.5\,\text{V} \ . \end{aligned}$$

7.3 COMMON SOURCE AMPLIFIER

The basic circuit is similar to the CE bipolar transistor amplifier, see Fig. 7.7, as the source is common to the input and output sides of the amplifier. For the moment ignore the swamping resistor (shown dashed). The output signal is in antiphase with the input signal. Why? As V_{GS} increases by e, for example, I_D increases by i_D. The pd across R_D increases by i_DR_D and, therefore, the output decreases by $v_D = i_DR_D$. So let us write

$$\begin{aligned} v_0 &= -i_DR_D \\ &= -g_m v_{GS} R_D \\ &= -g_m e R_D \end{aligned} \tag{7.7}$$

and the ac voltage gain A is

$$A = -g_m R_D \ . \tag{7.8}$$

Remember that if the amplifier is cascaded, then R_D must include the input impedance of the next stage.

The ac equivalent circuit is given in Fig. 7.8; the input and output stages are separated, for convenience.

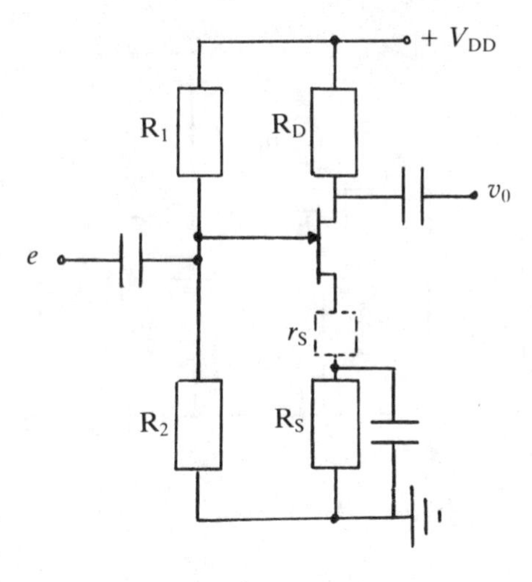

Fig. 7.7.

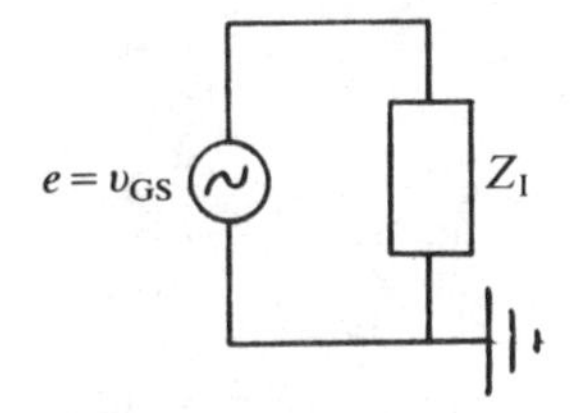

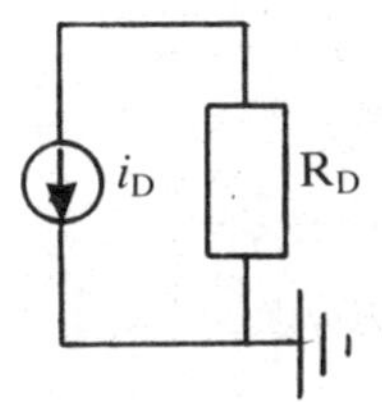

Fig. 7.8.

The input impedance Z_I is given by

$$Z_I = [1/R_1 + 1/R_2 + 1/R_{GS}]^{-1} \tag{7.9}$$

in which R_{GS} is the resistance of the gate-source depletion layer. As R_{GS} is usually very much higher than either R_1 or R_2, $1/R_{GS}$ can be neglected in (7.9), which reduces to $R_1 \| R_2$.

What is the effect on the ac voltage gain if the swamping resistor r_s (shown dashed in Fig. 7.7) is now included? The first point to note is that the source is no longer at 0 V. The ac equivalent circuit now looks as shown in Fig. 7.9. R_{GS}, the resistance of the gate-source depletion layer, is no longer in parallel with R_1 and R_2.

Apply KVL to the input-source loop at some time t. It gives

$$e - i_D r_S - v_{GS} = 0$$

from which

$$e = (g_m r_s + 1) v_{GS} \ . \tag{7.10}$$

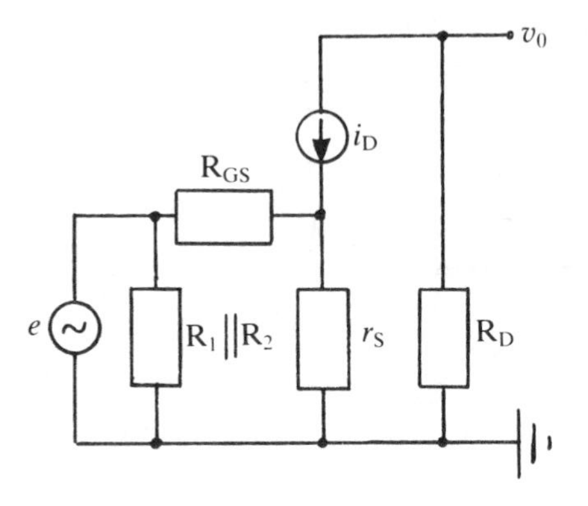

Fig. 7.9.

As

$$v_o = -i_D R_D$$
$$= -g_m v_{GS} R_D \ ,$$

the ac voltage gain is found as

$$A = v_o/e$$
$$= -g_m R_D/(1 + g_m r_s)$$
$$= -R_D/(r_s + 1/g_m) \ . \qquad (7.11)$$

The effect of the swamping resistor is to: (i) reduce the ac voltage gain; (ii) make changes in the value of g_m ineffectual for $r_s \gg 1/g_m$, when $A = -R_D/r_S$.

7.4 SOME JFET APPLICATIONS

7.4.1 A buffer amplifier

The function of the buffer amplifier was discussed in section 6.9 when we introduced the emitter follower. The source follower is the FET analogue. It will be shown that the ac voltage gain is $\leqslant 1$ and that the output impedance is R_S. As have been mentioned previously, the input impedance is the resistance of the gate-source depletion layer.

Fig. 7.10a is the schematic of a typical source follower, and Fig. 7.10b is the ac equivalent circuit. Fig. 7.10b indicates that an increase in the dc gate potential by v_I results in the drain current increasing by i_D and the output potential increasing by $i_o R_S$; v_o is in phase with v_I. Hence,

$$v_o = i_D R_S \ . \qquad (7.12)$$

Also, using

$$v_{GS} = e - v_o$$

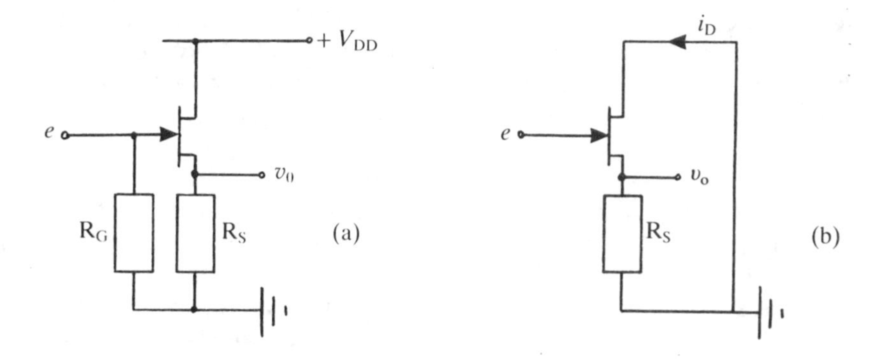

Fig. 7.10.

with (7.2), leads to

$$i_D = g_m(e - v_o) \quad . \tag{7.13}$$

Substituting (7.13) into (7.12) gives

$$v_o = g_m(e - v_o)R_S$$

and, hence

$$v_o(1 + g_m R_S) = g_m R_S e \quad . \tag{7.14}$$

The ac voltage gain is, therefore, given by

$$A = g_m R_S/(1 + g_m R_S) \tag{7.15}$$

which is <1, but can tend to the values 1 for $g_m R_S \gg 1$.

The output impedance is v_o/i_D, which is R_S. It is usually about a few hundred ohms, which may not be low enough for all applications.

7.4.2 Low-noise amplifier

Whenever weak signals are to be amplified, it is very important for the input stage of the amplifier to be *noise-free*. FETs are superior to bipolars in this respect; the input impedance of a JFET is higher than that of the bipolar because the depletion region has a smaller concentration of free charges.

Electrical noise is a major problem in electronics because it generates an unwanted disturbance which is superimposed on the *input* signal to be amplified. In some instances the noise signal can completely mask the amplified input signal. A JFET placed at the input stage of the amplifying system can greatly minimise the incidence of noise.

7.4.2.1 Brief introduction to electrical noise

Electromagnetic noise, such as rf pick-up, is generated by electric motors, power lines, ignition systems, etc. Such systems generate electromagnetic radiation which, in turn, induces an unwanted noise signal in an electronic circuit. Circuits used in the neighbourhood of such disturbances must be shielded by an earthed Faraday cage, and all connecting cables should be screened and kept as short as possible.

Voltage ripple may be categorized as noise, as it is an unwanted ac voltage superimposed on the steady dc voltage produced by a power supply. It can appear as *mains pick-up*.

Mechanical vibration can alter the separation of the plates of a capacitor, for example. This, in turn, changes the steady-state charge distribution on the plates and, hence, the pd between them. The disturbance comes under the general heading of *microphonics*.

Consider a resistor lying on the table in front of you. The conduction electrons in it move in various directions with different speeds; the electons acquire their kinetic energy thermally, that is from their surroundings—this, as we saw in section 5.1.1, is the mechanism which initiates bond-breakage in a semiconductor. At some instants, it is possible to find more electrons moving up the resistor than in any other direction. When this occurs, a small positive pd, say, will be generated. At other instants, more electrons will tend to move down the resistor. Then, a small negative pd will be produced. These noise signals, when amplified and viewed on an oscilloscope screen, appear as a random oscillatory disturbance. This is *thermal* or *Johnson* noise. The rms noise voltage V'_N is given by

$$V'_N = \sqrt{(4kTBR)} \tag{7.16}$$

where k is Boltzmann's constant, T is the absolute temperature, B is the bandwidth of the noise signal which can be about 10^{12} Hz, and R is the resistance of the component under test. V'_N may be about several μV.

$1/f$, or *flicker*, noise depends on the way the device is fabricated as well as on the nature of the contacts. The amplitude of the noise signal varies inversely with the frequency of the input signal.

7.4.3 FET chopper for the amplification of dc voltages

There is a general problem in using a cascaded RC-coupled amplifier for amplifying dc voltages. It concerns the presence of the interstage capacitor. This capacitor prevents the flow of dc current and, thereby, allows the Q-point of each stage to be set independently; ac signals pass without attenuation. So the system is not suitable for amplifying dc. One way around this problem is to convert the dc voltage into a train of rectangular pulses, amplifying them by using the RC-coupled amplifier, and then extract the amplified dc voltage, using a peak detector. The complete system is shown schematically in Fig. 7.11.

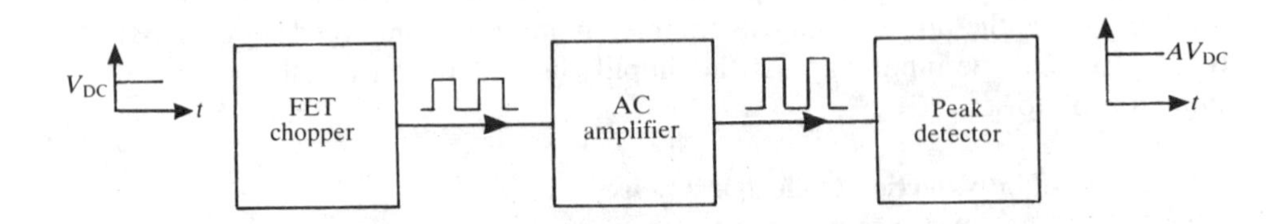

Fig. 7.11.

The *chopper* JFET circuit is drawn in Fig. 7.12a. A train of square waves of negative polarity and amplitude V_P is applied to the gate of the JFET. The JFET is ON for $V_{GS} = 0$ and OFF for $V_{GS} = -V_P$. During the ON state, dc current flows through R_D and the JFET to earth, and node X is pulled down to 0 V. During the OFF state, when no dc current flows, node X is pulled-up to $+V_{DC}$. The dc voltage is chopped into a train of rectangular pulses of amplitude V_{DC} and having the same frequency as the signal applied to the gate. The output from a conventional RC-coupled amplifier may then be passed to the diode peak detector of Fig. 7.12b to obtain the amplified dc voltage. In order for the pd across the capacitor plates to remain at AV_{DC}, the time constant $R_L C$ of the network must be large compared with the period of the inputted signal.

7.5 *n*-CHANNEL ENHANCEMENT MODE MOSFET

7.5.1 Structure

The structure of the MOSFET is very different from that of the JFET. The gate electrode, of either metal or polycrystalline silicon, is deposited on a layer of insulator, such as silicon dioxide, which, in its turn, is deposited on a substrate of *p*-type silicon. Fig. 7.13a illustrates the overall structure together with the appropriate potentials of the source, drain, and gate.

The presence of the silicon oxide insulating layer gives the MOSFET a very high input impedance, varying from 10^8 to $10^{14}\,\Omega$. This, of course, can be a big advantage in certain applications, but because its time constant is fairly long, free charges are not removed quickly; the switching speed is relatively low (~ 100 ns). Fig. 7.13b gives the symbols for the *n*-channel MOSFET; the arrow indicates that carriers are attracted towards the gate.

☆ ☆

AN ASIDE: The *p*-channel MOSFET has an *n*-type silicon substrate with the polarities of the voltage supplies in Fig. 7.13 inverted.

☆ ☆

7.5.2 Operation

Referring to Fig. 7.13, we should observe that as the potential of the gate is raised above 0 V, ($V_{GS} > 0$), minority electrons in the substrate are attracted towards the region beneath the gate. Eventually, the gate's potential reaches a value, called the *threshold* gate potential V_T, at which the concentration of electrons $[e^-]$ in this region equals the concentration of positive holes $[h]$. A further increase in the gate potential creates an *inversion* layer because now $[e^-]$ is greater than $[h]$.

The term *inversion* is used to indicate that the *normal* charge condition of the substrate near the insulating layer has been inverted electrically. The *n*-type inversion layer extends from source to drain, behaving as a channel along which electrons can drift when $V_{DS} > 0$.

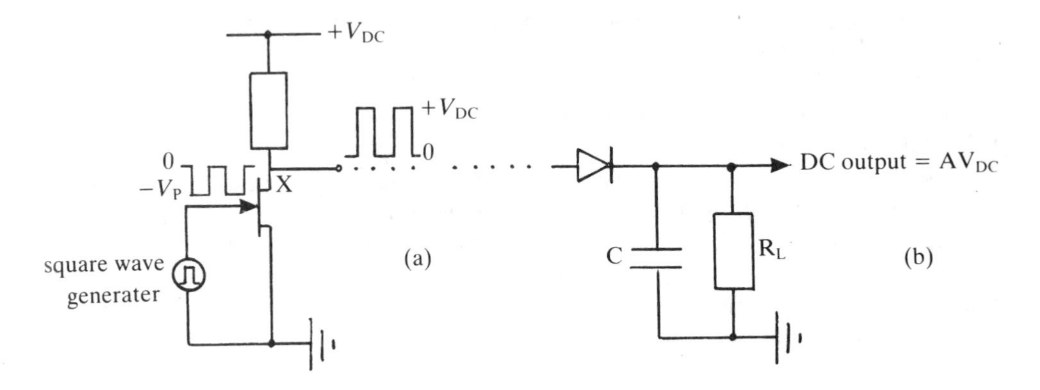

Fig. 7.12.

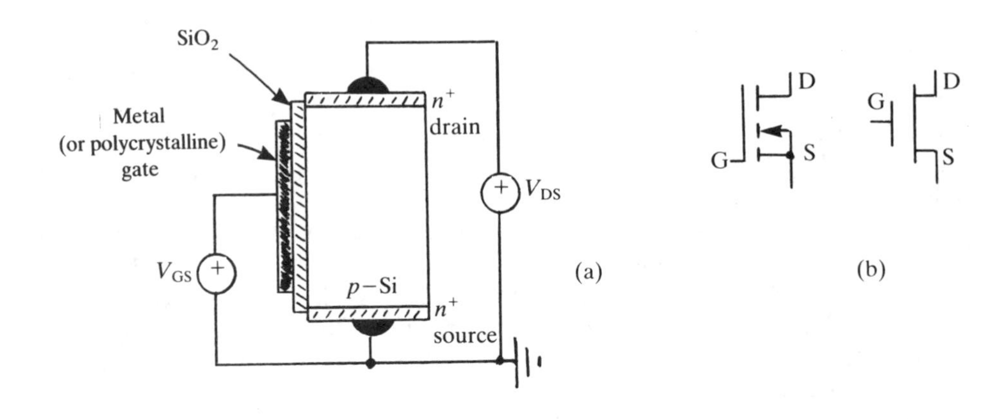

Fig. 7.13.

☆ ☆

AN ASIDE: In a p-channel MOSFET, positive holes drift from source to drain.

☆ ☆

7.5.2.1 Shape of the channel

(i) $V_{DS} = 0$

The potential at every point in the inversion layer is at 0 V. So with $V_{GS} > V_T$, the electron concentration $[e^-]$ will be the same everywhere beneath the gate and the channel will have a uniform width between source and drain, as Fig. 7.14a illustrates. The gate (+ve) and channel (−ve) act as the electrodes of a parallel-plate capacitor.

(ii) $0 < V_{DS} < V_{GS}$

Now the gate-drain pd is less than the gate-source pd. This means that there will be a smaller concentration of electrons at the drain end of the channel than at the source end. The width of the channel becomes triangular in shape, see Fig. 7.14b. The drain current flow is *ohmic* in character and is controlled by V_{DS}.

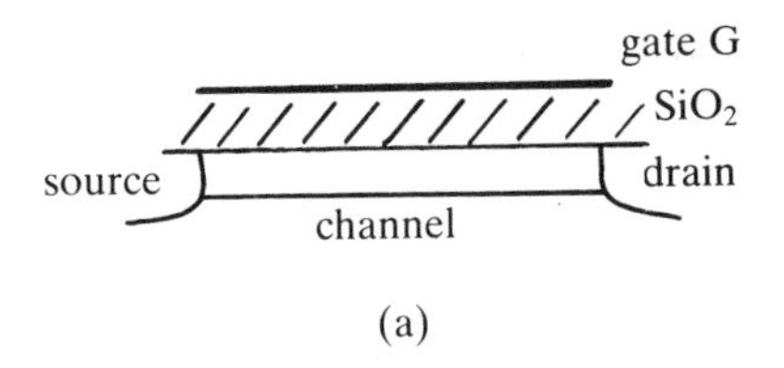

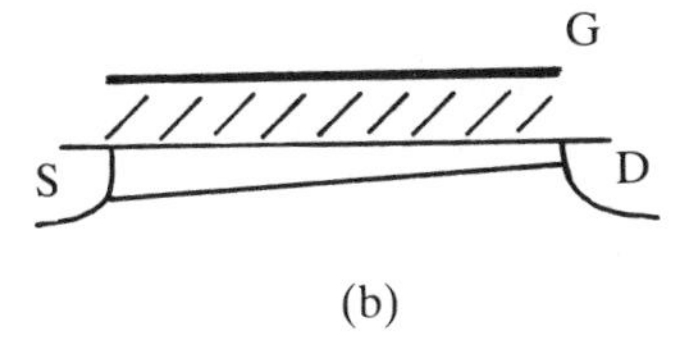

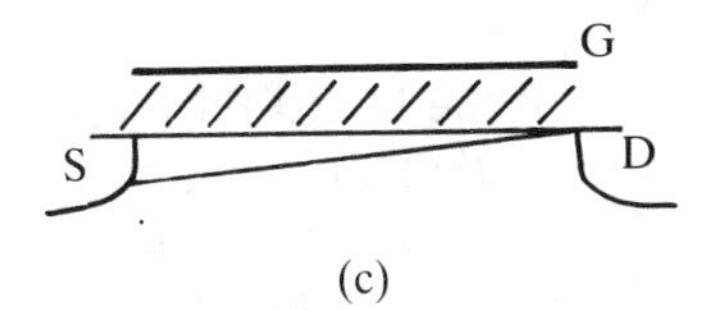

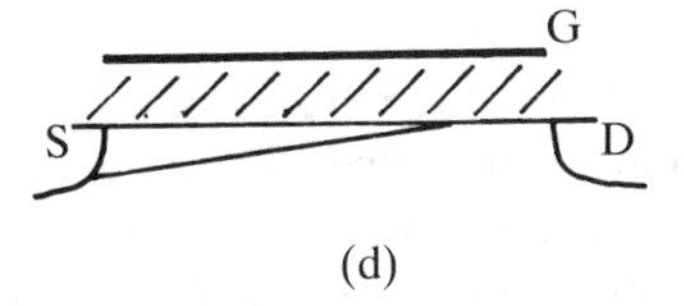

Fig. 7.14.

(iii) $V_{DS} \geqslant V_{GS}$
For $V_{DS} = V_{GS}$, the gate-drain pd is zero and the width of the channel is zero at the drain (Fig. 7.14c); the drain current is sometimes referred to by $(I_D)_{ON}$. Increasing V_{DS} above V_{GS} means that there is now some point between source and drain at which the potential is the same as the gate. The channel is *pinched-off* (Fig. 7.14d) and the drain current saturates. Fig. 7.15a shows the drain characteristics and Fig. 7.15b the transconductance curve.

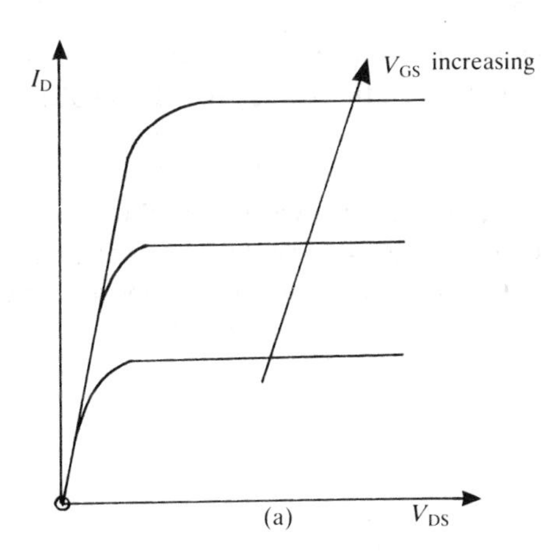

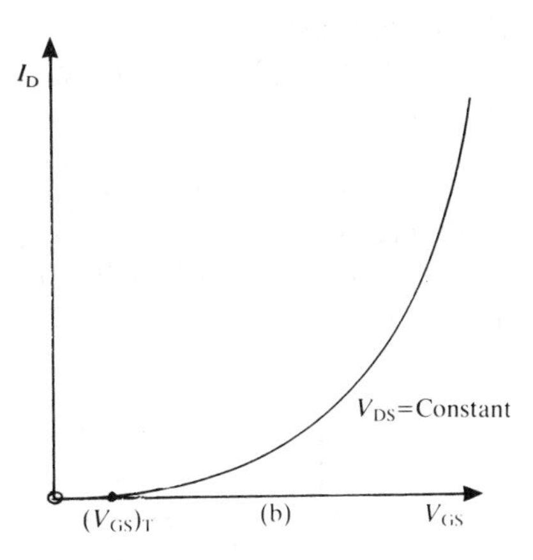

Fig. 7.15.

Worked examples 7.2
Q1. Set up a self-biasing circuit for the *n*-channel E-MOSFET. Determine the value of the drain resistor required if $(I_D)_{ON}$ is 8 mA at $V_{DS} = V_{GS} = 5$ V and $V_{DD} = 20$ V.

The required circuit is shown below.

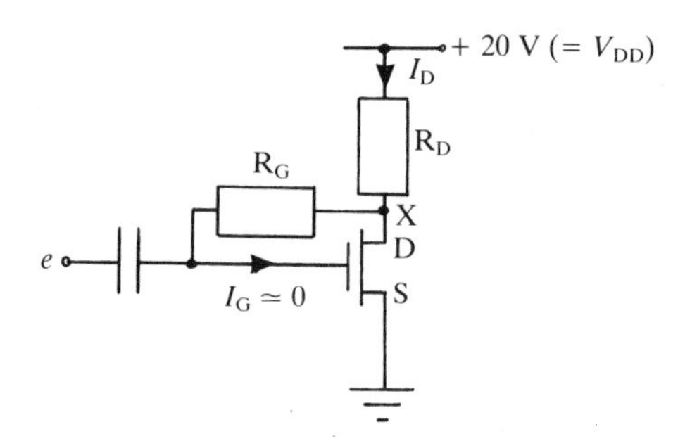

Note that the gate current is negligibly small because of the MOSFET's very high input impedance; $V_G \simeq V_X$ and $V_{GS} \simeq V_{DS}$. So we see that the gate acquires a positive potential relative to the source even in the absence of a voltage supply. Once again, observe that if I_D increases, V_X decreases, V_G decreases, and I_D decreases again. This is an example of negative current feedback.

Now,

$$V_{DD} = V_{DS} + I_D R_D$$

so that

$$\begin{aligned} R_D &= (V_{DD} - V_{DS})/I_D \\ &= (20 - 5)/8\,\text{mA} \\ &= 15/8\,\text{mA} \\ &= 1.9\,\text{k}\Omega. \end{aligned}$$

Remember that the amplitude of any ac signal applied to the gate must not be too large; otherwise, V_{GS} will fall below V_T during the negative cycle, and the MOSFET switches off.

Q2. Design a MOSFET potential divider amplifier circuit with a maximum drain current of 10 mA and a maximum drain-source pd of 24 V. The drain characteristics are drawn below. The Q-point of the circuit is (10 V, 6 mA) for a gate-source pd of 4 V. V_T is 1.5 V.

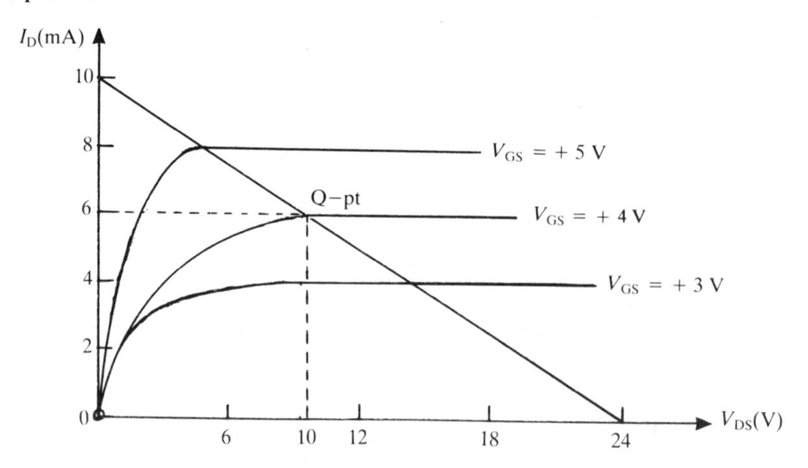

As the Q-point is now known, the dc load line of the circuit can be superimposed on the drain characteristics. On doing this, it can be seen that applying an ac signal of 1 V amplitude to the gate results in I_D varying between 4 mA and 8 mA. V_T is given as 1.5 V, which indicates that the MOSFET will conduct normally at all times.

The desired circuit has the same form as the common-source amplifier in Fig. 7.7, *viz*.

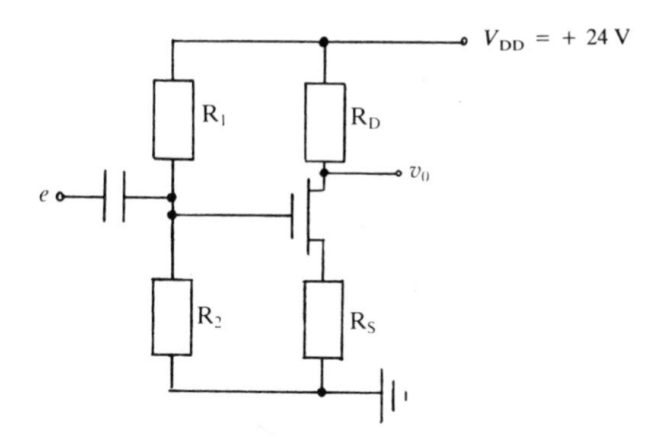

R_s and R_D may be found if the dc circuit is Theveninized to:

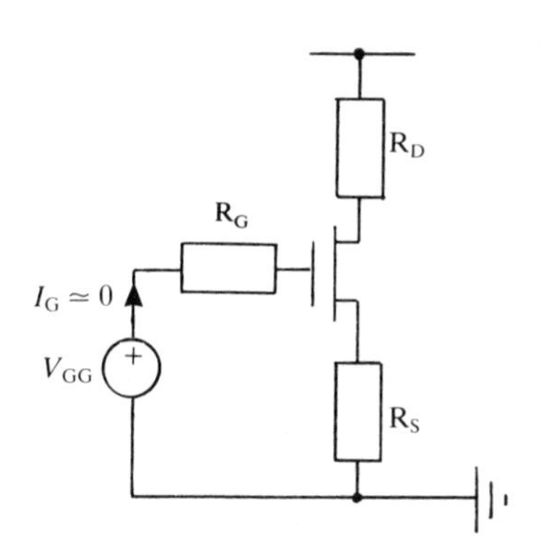

Here,

$$V_{GG} = R_2 V_{DD}/(R_1 + R_2) \tag{1}$$

and

$$R_G = R_1 \| R_2 \ . \tag{2}$$

Also, for the gate-source loop,

$$V_{DS} = V_{DD} - I_D(R_D + R_S) \ . \tag{4}$$

Because the input impedance of the MOSFET is very high, let us choose high values for R_1 and R_2 — 100 MΩ, say — to snsure that a small gate current flows.

Numerical substitution into (1) gives

$$V_{GG} = \tfrac{1}{2} \times 24\ V$$

$$= 12\,\text{V} \ ,$$

and into (3) and (4) gives

$$R_S = (12–4)/6\,\text{mA}$$
$$= 1.3\,\text{k}\Omega$$

and

$$R_o = (24 - 10 - 8)/6\,\text{mA}$$
$$= 1\,\text{k}\Omega \ .$$

7.6 SOME MOSFET APPLICATIONS

MOSFETs have certain advantages over bipolar transistors in digital circuits: (i) they have a simpler construction; (ii) they are smaller and, hence, a larger density of devices can be fabricated with a given slice of silicon; (iii) the power consumption is lower. The main disadvantage is a lower switching speed.

There are three kinds of logic circuits which use MOSFETs: an NMOS circuit uses only *n*-channel devices; a PMOS circuit uses only *p*-channel devices; a CMOS circuit uses both *p*-channel and *n*-channel devices — the letter C stands for *complementary*. Typical switching rates and average propagation delay per logic element are: 5–10 MHz and 50 ns (NMOS); 2 MHz and 300 ns (PMOS); 5–10 MHz and 70 ns (CMOS).

The following sections briefly consider particular logic functions performed using E-MOSFETs.

7.6.1 NMOS logic inverter

The basic circuit is shown in Fig. 7.16; Q_1 acts as an *active* load resistor to Q_2. The

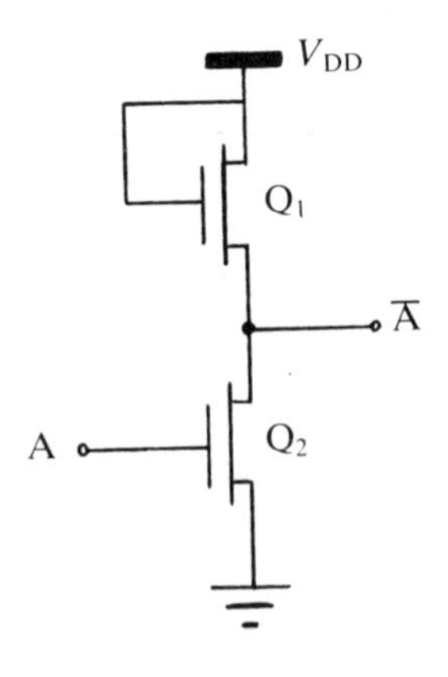

Fig. 7.16.

effect of strapping the gate of Q_1 to the drain ensures that Q_1 is permanently conducting (ON). If the input potential at A (V_A) is less then the threshold potential of Q_2, then Q_2 will not conduct (OFF). The output then rises until it practically equals V_{DD}. This is the binary '1' state. Alternatively, if V_A is greater than the threshold value of Q_2 then Q_2 permanently conducts and the output is *pulled-down* to earth. This is the binary '0' state.

Typically, a V_{GS} of 2.5 V on the input side marks the transition from '0' to '1', whereas on the output side a pd less than 0.5 V is equivalent to '0' and greater than 4.5 V is '1'. V_{DD} is about 5 V.

For the inverter to operate in the manner described it is essential for the resistance of the inversion layer in Q_2 to be less than the ON resistance of Q_1; a 1:20 ratio is adequate. If this condition is not fulfilled the output will stay at binary '1'.

7.6.2 NMOS NOR circuit

Fig. 7.17 shows the NOR circuit with two inputs, A and B. With either A or B, or

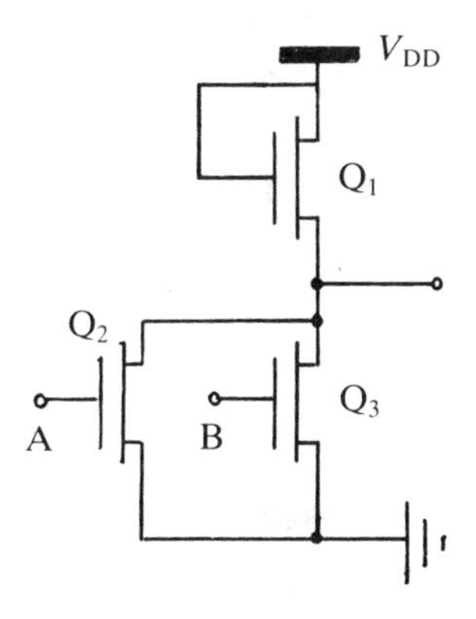

Fig. 7.17.

both, at binary '1' the output is '0'. Only when both inputs are at '0' will the output be '1'. The truth table for the NOR function, contained within Table 2.4, is thus satisfied.

7.6.3 NMOS NAND circuit

Both Q_2 and Q_3 in Fig. 7.18 need to be ON for the output to be '0'. However, if either one or both are OFF then the output will be '1'. These results satisfy the NAND logic function.

7.6.4 CMOS logic circuits

7.6.4.1 The inverter

The circuit is given in Fig. 7.19. Note the polarities of the source in Q_1 and Q_2. It is also important to realize that when Q_1 is ON, Q_2 will be OFF, and vice versa. This latter fact means that current cannot flow directly from the supply, V_{DD}, to earth. When A is '0', Q_2 is OFF and Q_1 is ON. Therefore, the output is at '1'. On the other

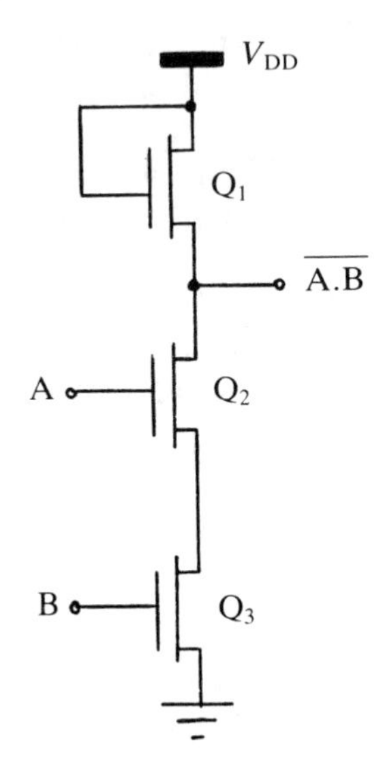

Fig. 7.18.

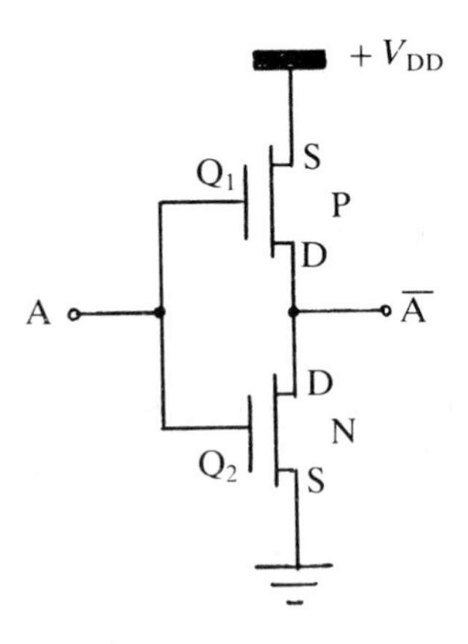

Fig. 7.19.

hand, when A is '1', Q_2 is now ON and Q_1 is OFF. The output is '0'. This is the inverting action.

Switching A continuously between '0' and '1' results in some current flowing between the supply and earth. This will occur, however, only during the small time interval when both Q_1 and Q_2 are momentarily ON together, that is during part of the rise time and decay time of the input pulse. This explains why the power consumption in CMOS circuits is very low. Typically, the power consumption at low input frequencies is about 10 nW per gate, rising to 1 mW at 1 MHz.

7.6.4.2 CMOS NOR logic gate

Fig. 7.20 displays the circuit. As Q_1 and Q_2 are in series, the output will be '1' only

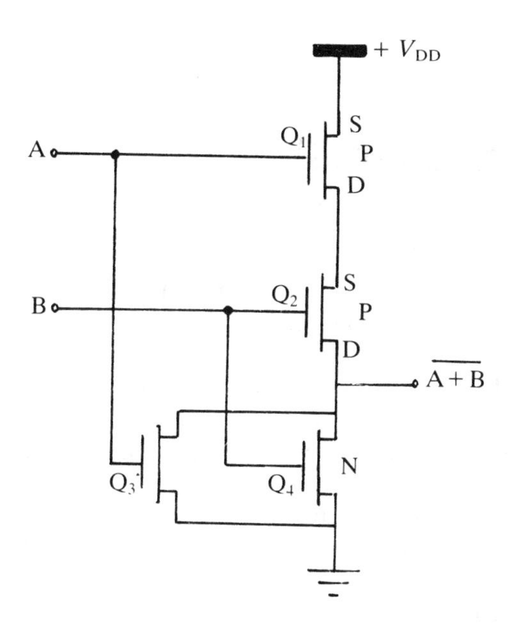

Fig. 7.20.

when they are both ON, that is when A and B are at '0'. All other input combinations will result in the output staying at '0'.

8

The operational amplifier

Objectives

(i) To introduce the '741' as a typical op-amp; blackbox treatment
(ii) To discuss the open-loop and closed-loop voltage gain of a general amplifier; positive and negative voltage feedback
(iii) How to use the Nyquist diagram to assess the stability of an amplifier
(iv) To use the ideal characteristics of an op-amp for designing circuits which perform (a) mathematical operations; (b) charge, current and voltage amplification
(v) To discuss the op-amp in real life
(vi) To extend the bandwidth by increasing the degree of feedback.

8.1 GENERAL COMMENTS

The operational amplifier (op-map for short) is probably, without exception, the most versatile electronic system available to the circuit designer. Generally, when one thinks of an op-amp it is the '741' which springs to mind. The '741' op-amp, however, is only one of many op-amps that are readily available on the market.

The name *operational amplifier* comes from one of its original applications, viz. carrying-out specific mathematical operations, such as addition, multiplication, differentiation, integration, in an analogue computer. Nowadays, all *stable*, *high-gain* amplifiers are referred to as op-amps.

In its simplest form, the op-amp has six connections — see Fig. 8.1. Note the

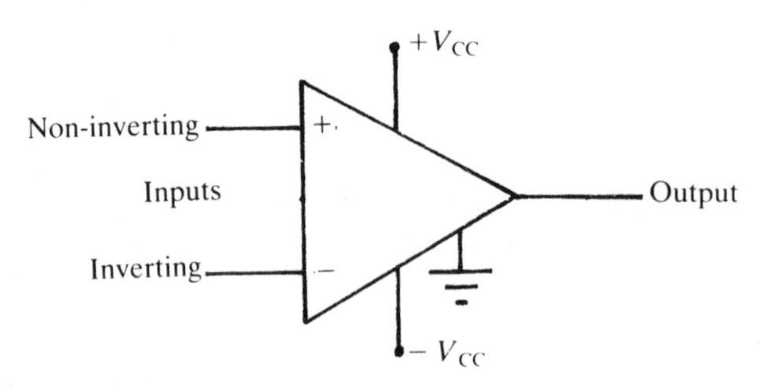

Fig. 8.1.

universal symbol used for an amplifier. The dc supples are needed for setting the Q-points in the transistors in the integrated circuit (IC) chip. They are typically ± 15 V for the '741' op-amp. As there can be as many as 20 or so transistors within a typical op-amp, it is sensible to treat it as a *black box*. It is the transfer characteristics, which relate a parameter at the output to a parameter at the input (for example output voltage/input voltage) which become of great importance.

The input marked − in Fig. 8.1 is called the *inverting* input and the one marked + is called the *non-inverting* input. It is the pd between these inputs which is amplified. Applying a signal to the − input generates an output signal in *antiphase* with the input. On the other hand, applying the signal to the + input generates an *in-phase* output signal.

The op-amp can be divided into an input stage and an output stage, and it is often helpful to do this when analysing its various circuits. The equivalent circuit is shown in Fig. 8.2. The open-loop voltage gain for the '741' is quite high, being about

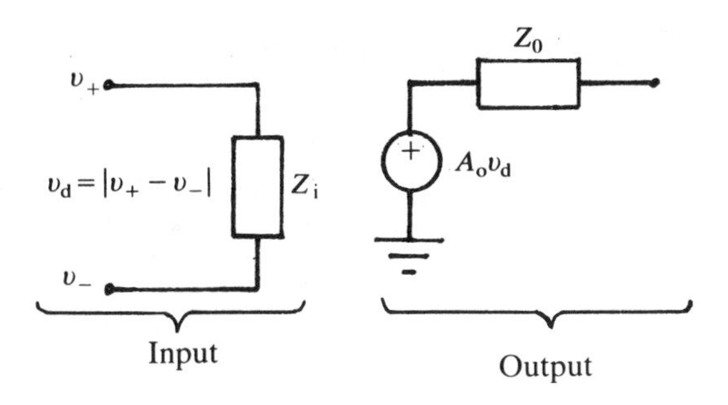

Fig. 8.2.

200 000. The term *open-loop* is necessary because as we shall see, the output can be connected to one of the inputs via a *feedback* loop. In this case the *closed-loop* voltage gain is considered.

One problem arises as a consequence of the high open-loop voltage gain. Suppose that the pd between the inputs of a '741' is 1 mV. The output (relative to earth) will be 200 000 × 1 mV or 200 V. However, as the dc voltage supply is ± 15 V, the output tends to saturate in the manner illustrated in Fig. 8.3. It is only over the relatively small linear region that effective voltage amplification occurs.

Before considering the operation of some basic op-amp circuits, it will be helpful to digress somewhat and investigate in rather more detail the significance of the term *voltage feedback*.

8.2 GENERALIZED VOLTAGE FEEDBACK SYSTEM

Fig. 8.4 is a schematic representation of an amplifier with feedback. The function of the *feedback* system is to attenuate the output υ_o by some fraction of k — the *feedback fraction*. In its simplest form the feedback system could, perhaps, be a potential divider network. The output $k\upsilon_o$ from it is then combined with the input signal υ_i at the adder to give the input e to the amplifier itself. Now $k\upsilon_o$ and υ_i can be combined either in-phase with each other or in anti-phase.

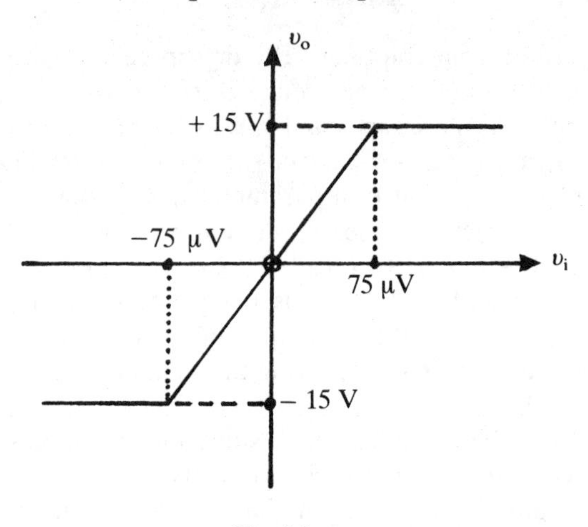

Fig. 8.3.

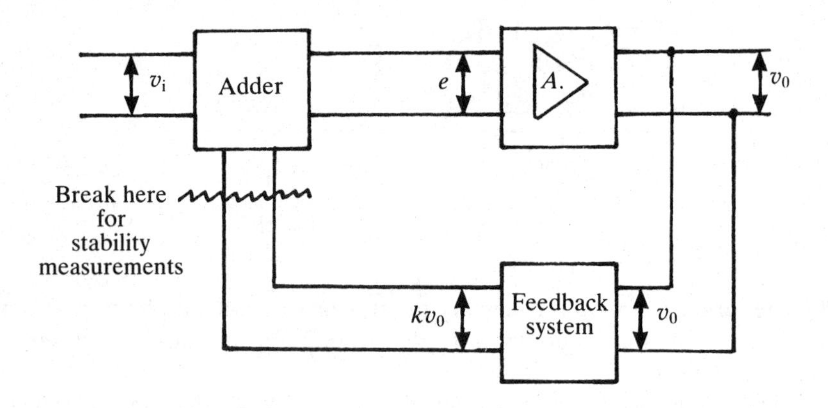

Fig. 8.4.

8.2.1 Positive feedback

Let us firstly consider the in-phase case. We see immediately that

$$e = v_i + kv_o \ , \tag{8.1}$$

but, using the definition of the open-loop voltage gain,

$$v_o = A_o e \ , \tag{8.2}$$

Substituting from (8.1) into (8.2) gives

$$v_o = A_o(v_i + kv_o)$$

from which

$$v_o/v_i = A_o/(1 - kA_o) \quad . \tag{8.3}$$

This is the closed-loop voltage gain A.

What is interesting about (8.3) is that when

$$(1 - kA_o) = 0 \tag{8.4}$$

we have

$$A = \infty \quad .$$

This result implies that even when $v_i = 0$, v_o will still have a finite value; sufficient of the output signal is being fed back to sustain the amplification process. The system exhibits *positive* feedback. It is the basis of oscillator design.

8.2.1.1 Stability

Equation (8.4) marks the onset of instability in an amplifier. The general question of stability in amplifiers can be investigated experimentally by breaking the feedback loop in Fig. 8.4 and measuring the feedback voltage kv_o, which, for convenience, we shall call v_F, produced by the input signal e. The phase shift θ between v_F and e also needs to be determined. These measurements should be repeated at various frequencies.

The results may be displayed on a *Nyquist* diagram. It is essentially a polar diagram. An arrow of length equal to the magnitude of the voltage gain v_F/e is drawn from the origin at an angle which depends on the magnitude and sign of θ. If θ is positive (v_F *leads e*) then the angle is measured anticlockwise from the $\theta = 0$ direction, whereas if θ is negative (v_F *lags* behind *e*) it is measured in a clockwise sense. Two examples are shown in Fig. 8.5.

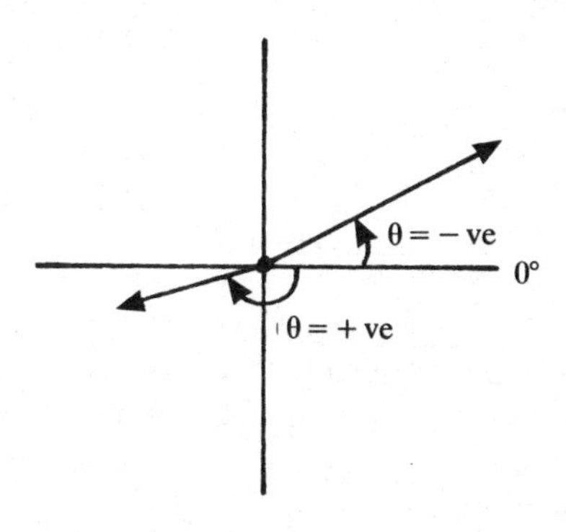

Fig. 8.5.

When the results of all frequencies have been included on the diagram, a curve is then constructed by joining-up the tips of the arrows, see Fig. 8.6. Also included is

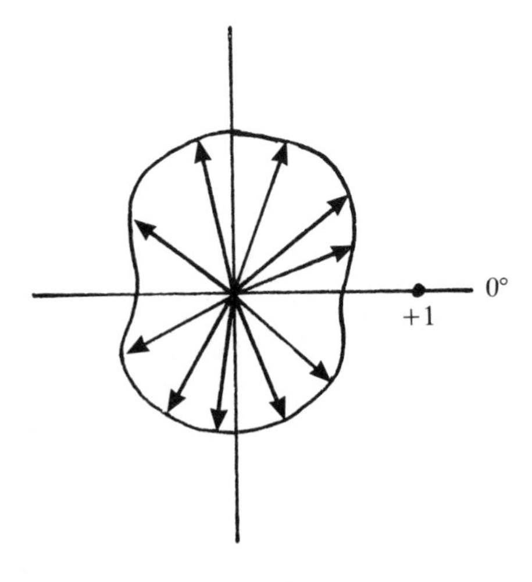

Fig. 8.6.

the $+1$ point. This point is for $v_F/e = 1$ and $\theta = 0$ or 2π (the in-phase condition); it marks the onset of instability. If the locus lies entirely within a circle of radius 1 unit then the amplifier will be stable at all frequencies, as in Fig. 8.7a. However, it is

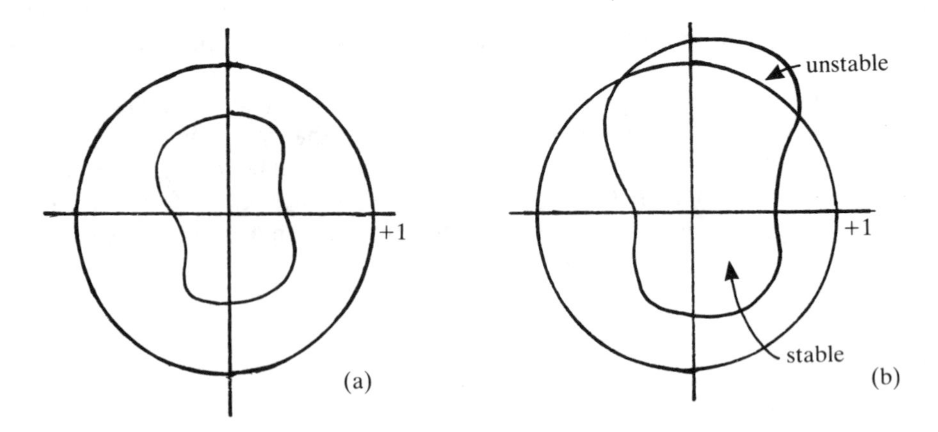

Fig. 8.7.

possible for the locus to lie partly outside this critical circle, as Fig. 8.7b indicates. In this case the amplifier is stable only over a certain range of frequencies.

8.2.2 Negative feedback

Return to Fig. 8.4. Now let the feedback signal be added in antiphase with v_i so that

$$e = v_i - kv_o \ . \tag{8.5}$$

The same algebraic procedure leads to

$$A = A_o/(1 + kA_o) \ . \tag{8.6}$$

An important point to note is that the closed-loop gain A is now always less than the open-loop gain A_o. So the amplifier with negative voltage feedback is always stable. One special case emerges if it is assumed that $A \ll A_o$, for then $A/A_o \rightarrow 0$, which means that $(1 + kA_o) \rightarrow \infty$ and $kA_o \gg 1$. So (8.6) can be rewritten

$$\begin{aligned} A &= A_o/kA_o \\ &= 1/k \ . \end{aligned} \tag{8.7}$$

This result is very interesting because it suggests that the closed-loop voltage gain depends only on the nature of the feedback circuit and not on the amplifier itself. Temperature variations, which can affect the characteristics of the transistors in the amplifier, and thereby the overall voltage gain, are no longer a problem!

Another interesting pay-off from negative feedback is that distortion in the input signal can be substantially reduced. We shall not prove this result, however.

8.3 IDEAL CHARACTERISTICS OF AN OPERATIONAL AMPLIFIER

To assist us in analysing some of the many op-amp circuits we shall assume that the op-amp is an ideal system.

CHARACTERISTICS:

(i) The impedance of the input stage in Fig. 8.2 is infinite;
(ii) The open-loop voltage gain is infinite;
(iii) The impedance of the output stage is zero;
(iv) The bandwidth is infinite.

(i) signifies that NO current actually enters the op-amp. In a '741' op-amp some current does flow ($\sim 0.1\ \mu$A).
(ii) implies that the pd between the two inputs is ZERO.
(iii) implies that the signal generated by the voltage source in the output stage is NOT ATTENUATED. The output impedance of a '741' is about 75 Ω.
(iv) is not really essential for analysing the circuits. However, it is pertinent to mention that the bandwidth of an op-amp without feedback is quite small; A_o is about 200 000 for a '741' up to 10 Hz, but beyond this frequency there is a rapid fall-off in gain. In fact, at 25°C the gain has fallen to 1000 at 1 kHz and is zero at 1 MHz. Bandwidth will be discussed more fully in section 8.6.

Let us see how (i) and (ii) — the golden rules — with the help of KCL work in practice, by analysing some basic op-amp circuits.

8.4 A SELECTION OF OP-AMP CIRCUITS

8.4.1 Inverting amplifier

This circuit, with the dc network omitted, is shown in Fig. 8.8. A signal source is

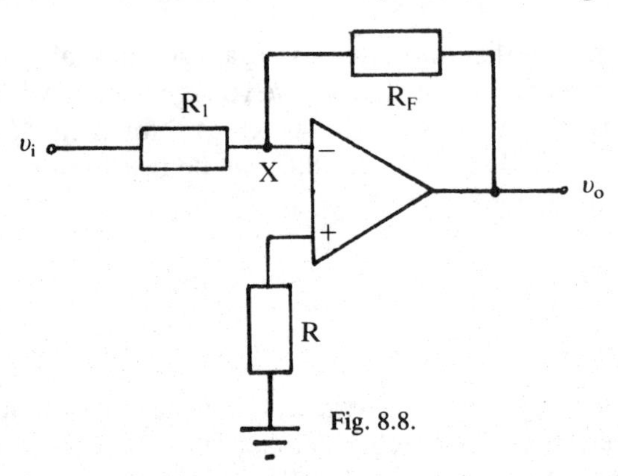

Fig. 8.8.

connected to the inverting (−) input and the feedback loop, with its feedback resistor, connects the output to the − input. From the discussion in 8.1, we must expect to determine that the output signal is in antiphase with the input signal.

STEPS:

(a) As no current enters the non-inverting (+) input its potential υ_+ must be 0 V — RULE (i).

(b) The potential of the inverting input must also be 0 V — RULE (ii).

Although node X is not actually strapped to earth, it is referred to as a *virtual* earth.

(c) Apply KCL at node X to give

$$\upsilon_i/R_i + \upsilon_o/R_F = 0 \tag{8.8}$$

which leads to

$$\upsilon_o/\upsilon_i = -R_F/R_1 \; . \tag{8.9}$$

The right-hand side of (8.9) is the closed-loop voltage gain A. The negative sign verifies that phase inversion has occurred between output and input. A variable feedback resistor allows the circuit designer to alter the gain.

What exactly is happening in this amplifier? υ_o has to do a bit of a juggling act to make υ_- and υ_+ sit at 0 V. If υ_- is initially at some small positive value, for example, then υ_o will go negative so that the potential-divider network can pin the − input at

0V. If v_o goes too far negative, v_- will also go negative which, in turn causes v_o to rise again until v_- finally settles down at 0V.

☆ ☆

AN ASIDE: Op-amps are not perfect devices, even though, in our analyses, they will be assumed to be. The first stage in an op-amp consists of a differential amplifier which has two identical transistors in it. A small base current must, therefore, flow to each transistor for the amplifier to function.

☆ ☆

RULE: Both the inverting and non-inverting inputs must see the same resistance to earth.

If this rule is not obeyed then an *input offset* voltage will develop (see section 8.5.1). In this example, R should equal R_1 in parallel with R_F; that is,

$$R = R_1 R_F/(R_1 + R_F) \ .$$

Even if this procedure is carried out, there is no guarantee that an input offset voltage will not develop. Manufacturing differences between transistors of the same type will give rise to different base currents. However, it is good design practice to minimize the unbalance in the way suggested. It will be assumed in the examples that follow that this procedure has been carried out.

The input impedance of the inverting amplifier is equal to R_1 because v_- is at 0 V. So the circuit designer is able to adjust the input impedance to any desired value.

If $R_1 = R_F$ then $A = -1$. This circuit is termed a *unity-gain* inverter. It does not amplify but can be used as a sign-changer.

8.4.2 Non-inverting amplifier

Fig. 8.9 gives the basic circuit. The source signal is fed to the amplifier via the

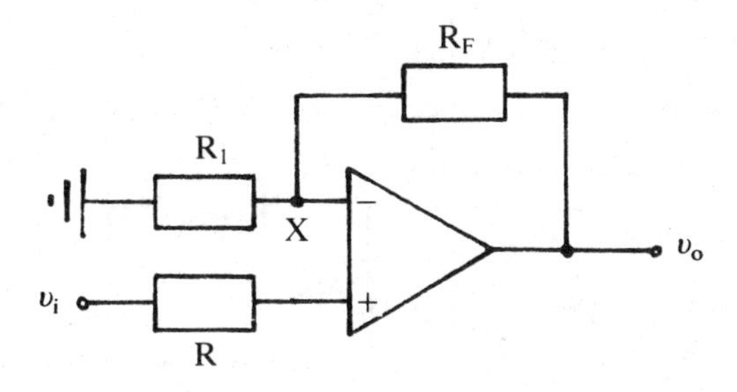

Fig. 8.9.

+ input, whilst the potential of the − input is set by a potential-divider network formed by R_1 and R_F.

By RULE (i)

$$v_+ = v_i \ . \tag{8.10}$$

By RULE (ii)

$$v_- = v_i \ . \tag{8.11}$$

Also, using the potential divider rule,

$$v_- = v_o R_1/(R_1 + R_F) \ , \tag{8.12}$$

therefore, by (8.11)

$$v_i = v_o R_1/(R_1 + R_F) \tag{8.13}$$

and the voltage gain is given by

$$v_o/v_i = 1 + R_F/R_1 \ . \tag{8.14}$$

8.4.2.1 *Voltage follower*

This circuit is quite important because it performs the same function as the emitter-follower; that is, it acts as a buffer amplifier. As we shall see, the voltage amplifier does not amplify but it can be used to match a high-impedance source into a low-impedance load. The relevant circuit is given in Fig. 8.10. The circuit can be directly

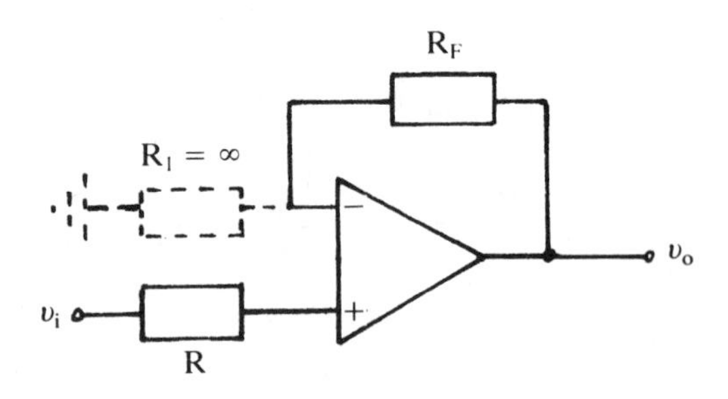

Fig. 8.10.

compared with Fig. 8.9 if an imaginary resistor of infinite value is included. The section shown dashed is equivalent to an open-circuit. So, putting R equal to infinity in (8.14) gives

$$v_o/v_i = 1 \ . \tag{8.15}$$

The input impedance is equal to R.

Worked examples 8.1

Q1. An inverting amplifier is designed with three inputs, v_1, v_2, and v_3, as shown in the diagram. Determine the output voltage. Indicate how the circuit can be modified to act as a *summer*.

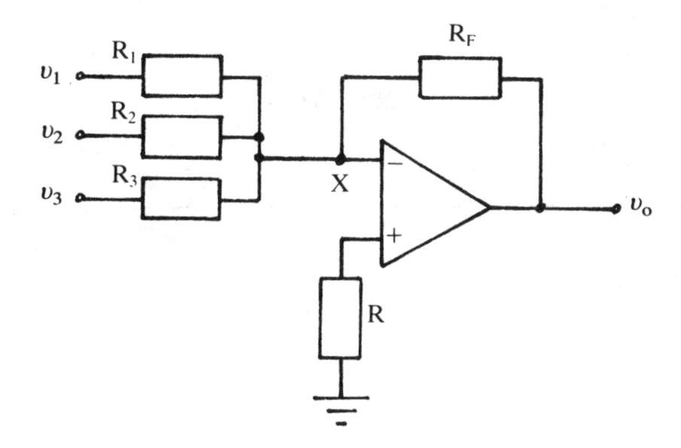

As before,

$$v_- = v_+ = 0 \text{ V}$$

and KCL at node X gives

$$v_1/R_1 + v_2/R_2 + v_3/R_3 + v_o/R_F = 0 \ , \qquad -(1)$$

so

$$v_o = -R_F[v_1/R_1 + v_2/R_2 + v_3/R_3] \ . \qquad -(2)$$

By putting $R_1 = R_2 = R_3 = R_F$, we obtain

$$v_o = -[v_1 + v_2 + v_3] \ . \qquad -(3)$$

Ignoring the negative sign, it can be seen that the output signal is the sum of the three input signals.

A unity-gain inverter (section 8.4.1) can be used to remove the negative sign.

Q2 How can the summer in Q1. be used as a digital-to-analogue converter?

Suppose that the series resistors in the input lines of the summing amplifier have values: $R_F = R_1 = 100$ kΩ, $R_2 = 50$ *k*Ω, and $R_3 = 25$ kΩ. Then substitution into (2) of Q1 gives

$$v_o = -[v_1 + 2v_2 + 4v_3] \ .$$

If, for the sake of argument, the input voltages are either 0 V or 1 V, then Table 8.1

Table 8.1

v_3	v_2	v_1	v_o
0	0	0	0
0	0	1	1
0	1	0	2
0	1	1	3
1	0	0	4
1	0	1	5
1	1	0	6
1	1	1	7

can be set-up. With v_1 as the least significant bit, it is clear that binary data are being converted into analogue data. The resitive ladder network connected to the − input can be extended if a longer binary word is being considered.

Q3. Modifying the inverting amplifier circuit of Fig. 8.8 so that the output is always: (i) 10 times larger than the input; (ii) 10 times smaller than the input.
For (i): (8.9) says that the output is a factor of R_F/R_1 times bigger than the input. So (i) can be satisfied simply by making R_F equal to 10 times the value of R_1. For example, if R_1 is 10 kΩ then R_F must be 100 kΩ. The op-amp, with its feedback resistor, is acting as a *multiplier*. It is straightforward to alter the multiplier — just alter the value of the feedback resistor.

For (ii): the same principle suggests that R_F must be 10 times smaller than R_1. The inverter circuit is acting as a *divider*.

So it is the ratio of the feedback resistor to the input resistor which controls the mathematical operations of multiplication and division.

8.4.3 Miller integrator

The circuit shown in Fig. 8.11 is an extension of the inverting amplifier; a capacitor is

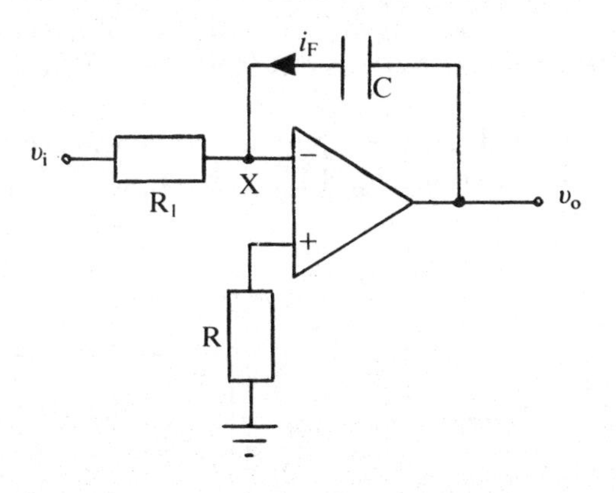

Fig. 8.11.

used in the feedback loop. Rules (i) and (ii) give:

$$v_+ = v_- = 0 \ . \tag{8.16}$$

If the charge on the plates of the capacitor is q and the current in the feedback is i_F then

$$\begin{aligned} i_F &= dq/dt \\ &= d(Cv_o)/dt \\ &= C dv_o/dt \ . \end{aligned}$$

Now KCL at node X gives

$$v_i/R_1 + i_F = 0 \tag{8.17}$$

or

$$v_i/R_1 + C\ dv_o/dt = 0$$

from which

$$v_o = -(1/R_1C) \int_o^t v_i\ dt \ . \tag{8.18}$$

It is assumed that the capacitor is initially uncharged. The product R_1C is the time constant of the circuit. Typically, if C is 1 μF and R_1 is 1 MΩ, then the time constant is 1 s.

The output signal is an integrated version of the input. $1/R_1C$ can be regarded as a scaling factor.

If a dc potential is applied to the − input, the output will decrease linearly to the saturation value, as illustrated in Fig. 8.12. The output is referred to as a *ramp*

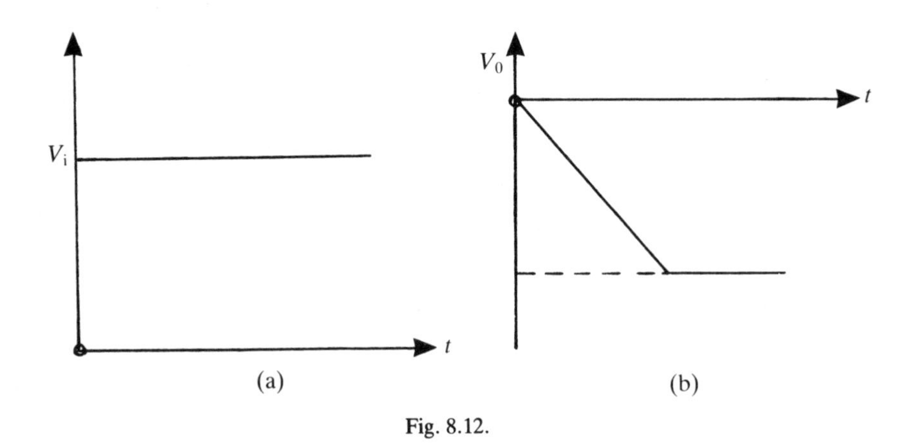

Fig. 8.12.

function. Apart from the negative sign, we also came across the ramp function in Q2 of *Worked examples* 3.4.

The output generated by an input train of square waves consists of a train of triangular waves.

8.4.4 Differentiator

The required circuit is given in Fig. 8.13. The capacitor is in the input line and the

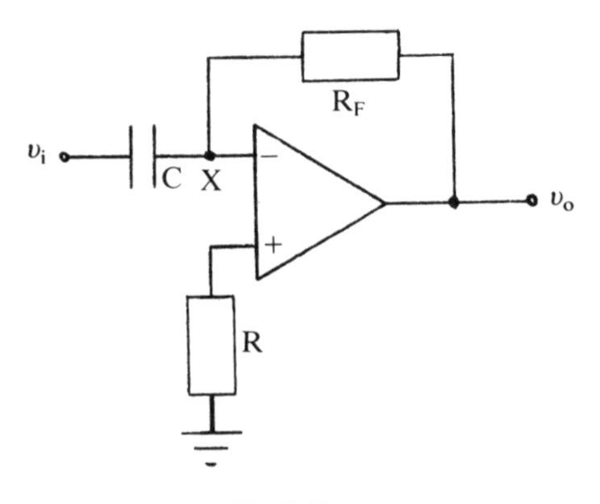

Fig. 8.13.

resistor is in the feedback loop. The voltage gain can be determined by using the same approach as that in section 8.4.3.

Node X is a virtual earth, so by KCL

$$C\,\mathrm{d}v_i/\mathrm{d}t + v_o/R_F = 0 \tag{8.19}$$

or

$$v_o = -CR_F\,\mathrm{d}v_i/\mathrm{d}t\ . \tag{8.20}$$

The output signal is a differentiated form of the input (see Q1 of *Worked examples* 3.4).

8.4.5 Current-to-voltage amplifier

The basic circuit is shown in Fig. 8.14. The − input is connected directly to a current

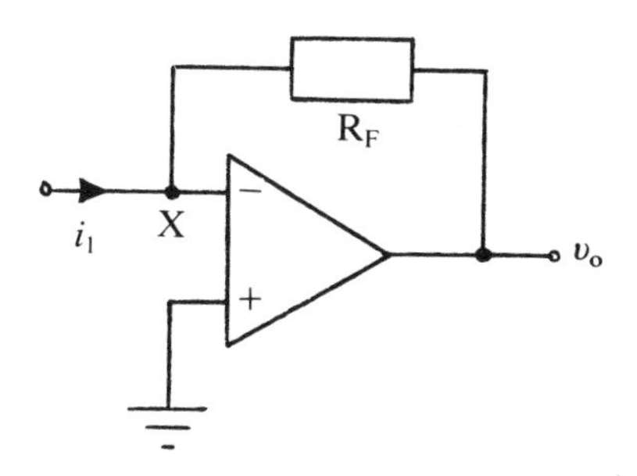

Fig. 8.14.

source supplying current i_1. Node X is a virtual earth, so

$$i_1 = v_o/R_F\ , \tag{8.21}$$

therefore,

$$v_o = -R_F i_i\ . \tag{8.22}$$

Clearly the degree of amplification depends on the value of R_F.

8.4.6 Current-to-current amplifier

As the name suggests, a current fed to the input of the op-amp is to be amplified. The circuit is given in Fig. 8.15. The feedback resistor is connected between the load resistor R_L and the − input. The currents in the various branches are indicated in the figure.

Applying KCL at Y:

$$i_i + i_s = i_o\ . \tag{8.23}$$

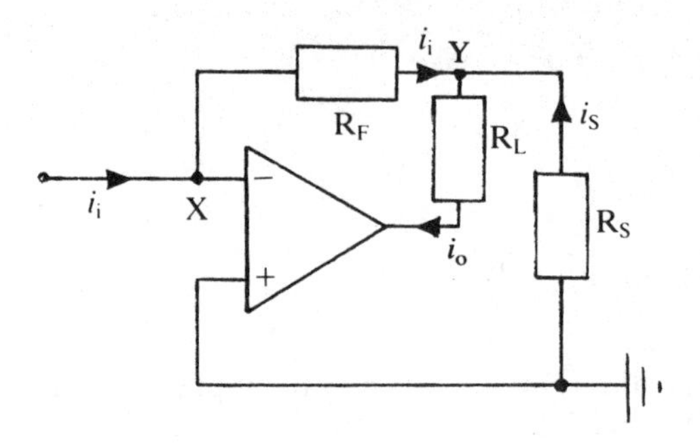

Fig. 8.15.

However, as

$$i_s = -V_Y/R_S$$

(8.23) becomes

$$i_i - V_Y/R_S = i_o \ . \tag{8.24}$$

Further simplification is possible because X is a virtual earth, and

$$V_Y = -i_i R_F \ .$$

Hence (8.24) becomes

$$i_i[1 + R_F/R_S] = i_o$$

or, the current gain

$$i_o/i_i = 1 + R_F/R_S \ . \tag{8.25}$$

8.4.7 Charge-to-charge amplifier

Any reference to charge should bring to mind a capacitor. This is the case with this amplifier in which there is a capacitor in the − input line and in the feedback loop, see Fig. 8.16.

KCL at node X gives

$$dq_1/dt + dq_F/dt = 0 \tag{8.26}$$

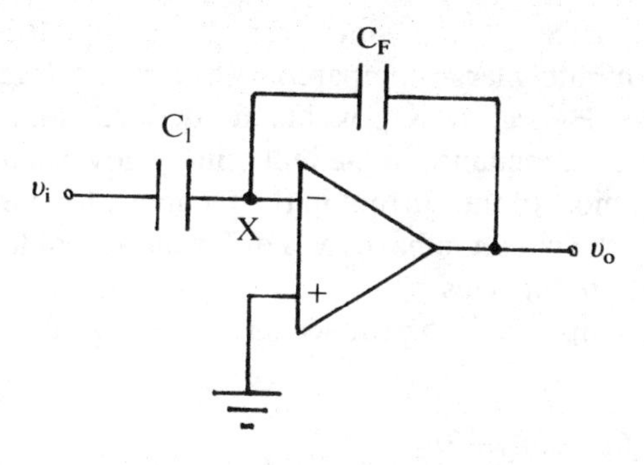

Fig. 8.16.

where q_1 and q_F are the charges on the input and feedback capacitors. Thus

$$q_1 = -q_F$$

or

$$C_1 v_i = -C_F v_o \ .$$

This enables the voltage gain to be determined as

$$v_o/v_i = -C_1/C_F \ . \tag{8.27}$$

Worked examples 8.2
Q1. Analyse the following circuit at low and high frequencies:

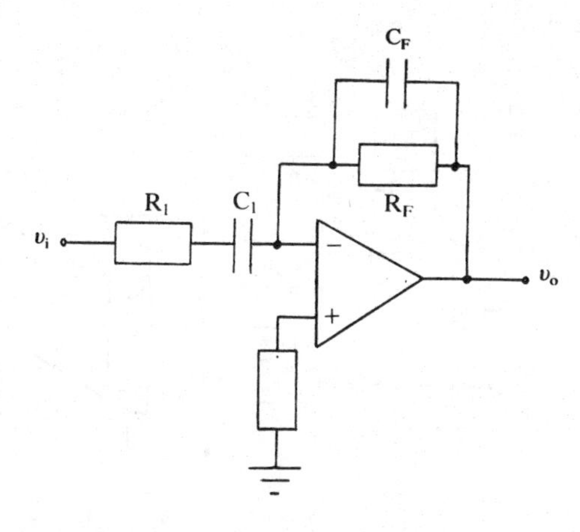

The first point to realize is that the problem is concerned with the relative values of the individual impedances in the input line and in the feedback loop. As the impedance of C_1 is infinitely large compared with R_1 at low frequencies, the op-amp is isolated from dc. Hence, it is possible to deduce immediately that at low frequencies the overall impedance in the input line is governed by the capacitor. By the same reasoning, most of the current in the feedback loop must flow through the resistor. As a result, the circuit behaves as a differentiator at low frequencies and as an integrator at high frequencies.

Q2. Design an op-amp circuit to produce an output signal given by

$$v_o = 8v_1 + 6v_2 - 5v_3 - 3v_4 \ .$$

This expression will be similar to the extension to (8.9) used in Q1 of *Worked examples* 8.1 if unity-gain inverters are placed in the v_1 and v_2 input lines. So let us write (-1) to indicate the need for a unity-gain inverter. Then the above expression becomes

$$v_o = (-1) \times [8v_1 + 6v_2] - [5v_3 + 3v_4] \ .$$

Now compare this with the general expression

$$v_o = (-1) \times R_F[v_1/R_1 + v_2/R_2] - R_F[v_3/R_3 + v_4/R_4] \ .$$

Chose R_F to be 240 kΩ. Then:
$R_F/R_1 = 8$ and $R_1 = 30$ kΩ
$R_F/R_2 = 6$ and $R_2 = 40$ kΩ
$R_F/R_3 = 5$ and $R_3 = 48$ kΩ
$R_F/R_4 = 3$ and $R_4 = 80$ kΩ.

Any reasonable resistance value between 1 kΩ and 1 M Ω is acceptable in order to avoid loading any previous circuitry.

The circuit might look like

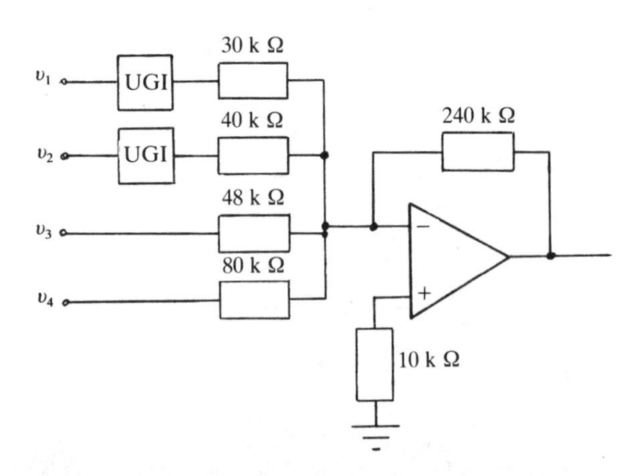

Another possible design is to attach v_3 and v_4 to the $-$ input of the op-amp and v_1 and v_2 to the $+$ input. Then with the same value of feedback resistor, we will have the following circuit.

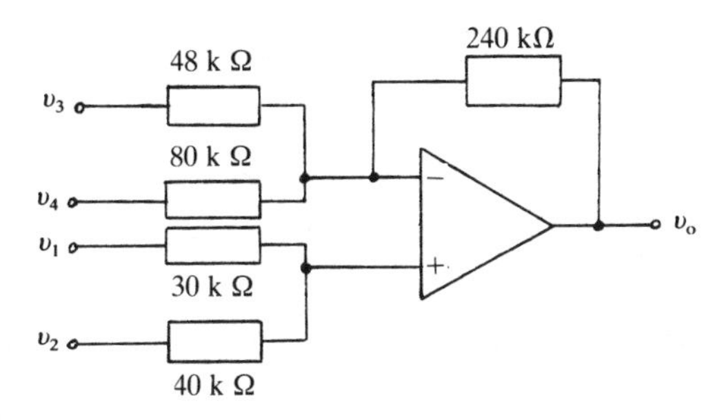

Q3. The input impedance of an ideal op-amp (that is, with no feedback) is infinite. What is the impedance of an op-amp with a feedback resistance R_F.

It is helpful to draw the equivalent circuit of the op-amp by dividing it into its input and output stages, see Fig. 8.2. Then we have

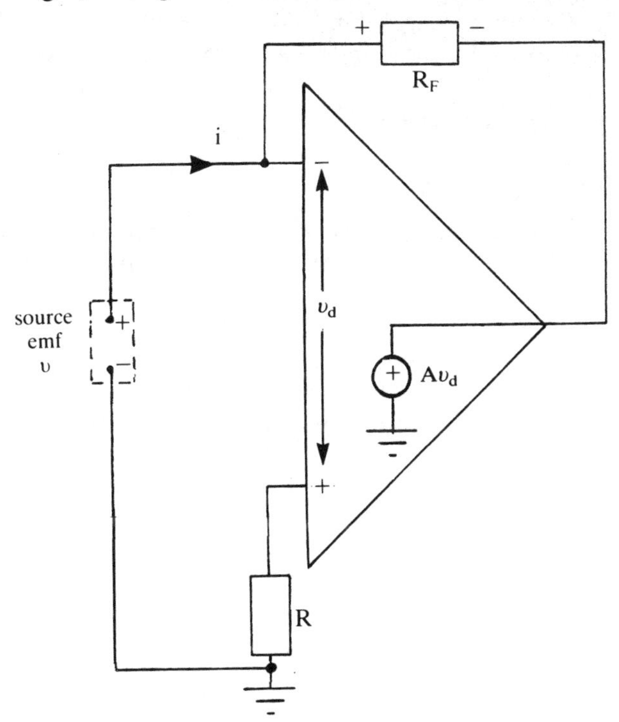

The input impedance is v/i. This is the impedance that the source sees on looking into the op-amp. As no current flows into the op-amp, all the current from the source flows through the feedback loop, the output voltage source, earth, and back to the source.

Applying KVL in an anti-clockwise direction to this loop, we obtain

$$v - Av_d - R_F i = 0 \ . \qquad (1)$$

Also, applying KVL to the input loop,

$$v + v_d = 0 \ . \qquad (2)$$

Substituting $v = -v_d$ from (2) into (1) gives

$$v + Av = R_F i \qquad (3)$$

from which we find that

$$Z_i = v/i = R_F/(A+1) \ . \qquad (4)$$

This expression for the input impedance is sometimes referred to as the *Miller* input impedance. In cases where A is large, but finite, Z_i is R_F/A.

If a resistance R_1 is placed in the − input line, a similar analysis will show that the input impedance is equal to R_1 for large A. However, as stated in section 8.4.1, this result can be obtained without analysis because node X is a virtual earth.

8.5 THE OP-AMP IN REAL LIFE

The characteristic of the *ideal* op-amp have enabled a number of basic circuits to be explored. However, it is proper to mention briefly the ways in which the *real* op-amp differs from its ideal counterpart.

8.5.1 Input-offset voltage

The transistors in the first stage of the op-amp, called the differential amplifier, require some base current, as Fig. 8.17 indicates. If V_d is zero then theoretically V_o

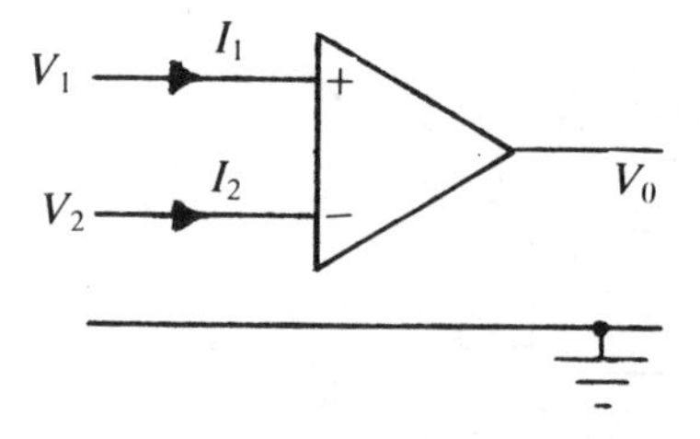

Fig. 8.17.

will be zero also. However, owing to slight manufacturing differences the two transistors in the differential amplifier will not be identical, and V_o will have a small finite value. By varying V_1, say, V_o can be adjusted to zero. Then the difference between V_1 and V_2 when this occurs is called the *input offset* voltage. For a '741' it is about 2 mV.

8.5.2 Input-base current

This is the mean of I_1 and I_2 with V_1 and V_2 held at 0 V. Thus

$$I_B = (I_1 + I_2)/2 \; . \tag{8.28}$$

In the '741' it is about 80 nA.

8.5.3 Input offset current I_{IO}

With V_1 and V_2 both equal to 0 V, the input offset current is the magnitude of the difference between the two base currents; that is,

$$I_{IO} = |1_1 - I_2| \; . \tag{8.29}$$

In the '741' it is about 20 nA.

8.5.4 DC voltage gain

Owing to the possibility of there being a small output voltage generated by the input offset voltage, it is more sensible to define A_o by using the relation

$$A_o = \frac{\text{change in the output voltage}}{\text{change in the input voltage}} \; . \tag{8.30}$$

It is typically greater than 20000. A_o may also be expressed in V mV^{-1} when it is equal to 20.

8.5.5 Common-mode rejection ratio

In theory, an op-amp should not differentiate between input potentials of 0 V and 0.01 V and 6 V and 6.01 V, say. Why? because only the difference is amplified. However, in practice, this is not so. The addition of the constant value of 6 V to each input does affect the output voltage. An output can be generated by strapping together the inputs at 6 V (call it V_{CM}). This, again, occurs because of slight differences between the two transistor networks in the differential amplifier in the input stage of the op-amp. The term *common-mode* is given to denote this case. The common-mode voltage gain A_{CM} is defined as

$$A_{CM} = V_o/C_{CM} \; . \tag{8.31}$$

This is zero for an ideal op-amp, but it should be as small as possible. The common-mode rejection ratio is given by

$$\text{CMRR} = A_o/A_{CM} \; . \tag{8.32}$$

Typically, the CMRR (in decibels) is ~ 100 for a high-quality op-amp.

8.5.6 Slew rate

The output must be able to follow variations in the input, or else the latter will be distorted. The slew rate is defined as the maximum time-rate of change of the output signal; that is,

$$\text{S.R.} = \mathrm{d}v_o/\mathrm{d}t|_{\text{MAX}} \tag{8.33}$$

This is constant for a given op-amp.

The output signal will not be distorted so long as

$$\text{S.R.} \geqslant \mathrm{d}v_i/\mathrm{d}t|_{\text{MAX}} \; . \tag{8.34}$$

Let us suppose that

$$v_i = E_P \sin \omega t \quad .$$

Then

$$\mathrm{d}v_i/\mathrm{d}t = \omega E_P \cos \omega t$$

and, because the maximum value of cos ωt is ± 1,

$$\mathrm{d}v_i/\mathrm{d}t|_{\text{MAX}} = \pm \omega E_P \; .$$

Therefore, to avoid any distortion,

$$\text{S.R.} \geqslant \omega E_P \; . \tag{8.35}$$

This inequality implies that since the slew rate is a constant for a given op-amp, it is not possible to use the op-amp at really high frequencies and at large amplitudes at the same time.

8.6 BANDWIDTH OF THE VOLTAGE GAIN

A Bode plot for an op-amp under open-loop conditions shows that the voltage gain is constant over a relatively small frequency range and then decreases fairly rapidly to zero. Although smaller than the open-loop voltage gain, the closed-loop gain behaves in a similar way but possesses a longer plateau region. Fig. 8.18 depicts the overall frequency response of the op-amp. The bandwidth of the op-amp is usually taken to be the frequency range from 0 to f_i; the frequency f_i, corresponding to the 3 dB point, is called the *corner* frequency.

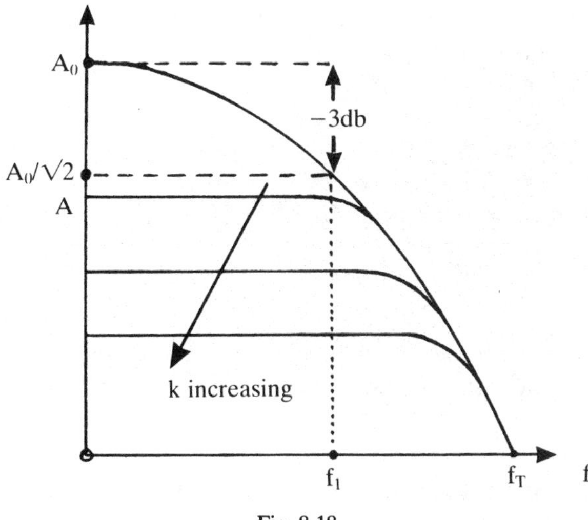

Fig. 8.18.

Why should the voltage gain fall off with frequency? It is necessary to think of those components which have frequency-dependent properties. The capacitor and the inductor should spring to mind. Although it is difficult to imagine an op-amp possessing inductive properties, it should be quite easy to suggest where capacitance occurs. For example, input and output leads might have stray capacitance associated with them, as does, for example, the emitter/base and collector/base regions of bipolar transistors and the gate region of a MOSFET transistor.

Using the equivalent circuit for the op-amp in Fig. 8.19, it can be shown that the 3

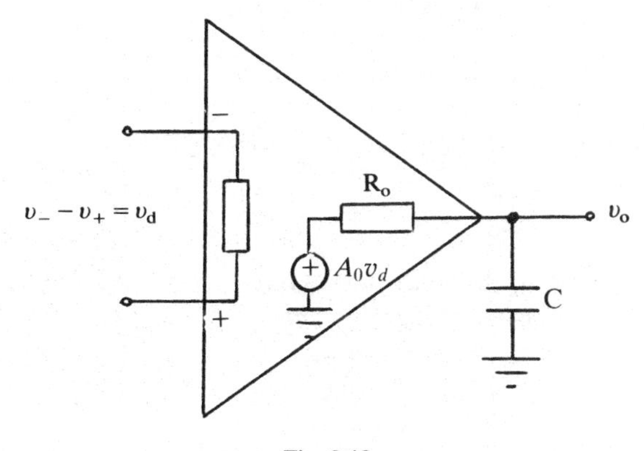

Fig. 8.19.

dB point occurs at a corner frequency given by

$$f_1 = 1/(2\pi RC) \tag{8.36}$$

(see Q1 of *Worked examples* 8.3 below), and that A_o decreases beyond f_1 at 20 dB per decade (or 6 dB per octave). The way that the voltage gain decreases in this region is referred to as the *roll-off*. The frequency at which the gain becomes zero is known as the *transition frequency* f_T. It is equal to the product $A_o f_1$.

The closed-loop gain A behaves in a similar way except that the corner frequency (and therefore the bandwidth of the op-amp) has increased to $f_1(1+kA_o)$. The output signal is in phase with the input at low frequencies but *lags* the input us the frequency is increased. For example, the phase lag is $\pi/4$ at the corner frequency and is $\pi/2$ at extremely high frequencies ($f \gg f_1$).

Worked examples 8.3

Q1. Using the equivalent circuit of the op-amp shown in Fig. 8.19, obtain an expression for the open-loop voltage gain at low and high frequencies. Determine relations for the corner frequency and the phase shift.

To take into account any phase shift between the input and output signals we shall work with complex quantities. So let the complex capacitive reactance be $\mathscr{X}$ (see section 4.6.3). Using the potential divider rule, we can obtain the output at some frequency f as

$$\mathscr{V}_o = [\mathscr{X}/(\mathscr{X}+R_o)] \,.\, A_o \mathscr{V}_d \;. \qquad (1)$$

(i) At low frequencies, $\mathscr{X}$ becomes very large so that if

$$\mathscr{X} \gg R_o$$

then (1) becomes

$$\mathscr{V}_o = A_o \mathscr{X}_d \;. \qquad (2)$$

The complex voltage gain $\mathscr{A}_o$ is identical with A_o. This result also implies that the output signal is *in phase* with the input signal.

(ii) At higher frequencies, $\mathscr{X}$ is comparable with R_o. Let the complex open-loop voltage gain at frequency f be $\mathscr{A}_{OF}$, then as $\mathscr{X}$ is $1/j\omega C$, (1) becomes

$$\mathscr{A}_{OF} = (1/j\omega C)/[R_o + 1/j\omega C].A_o$$

$$\mathscr{A}_{OF} = A_o/[1 + j\omega R_o C]$$

On multiplying numerator and denominator by the complex conjugate $[1 - j\omega R_o C]$ we obtain

$$\mathscr{A}_{OF} = A_o[1 - j\omega R_o C]/[1 + (\omega R_o C)^2] \;. \qquad (3)$$

A_{OF}, the modulus of $\mathcal{A}_{OF}$, is, therefore, given by

$$A_{OF} = A_o\sqrt{\{[1+(\omega R_o C)^2]/[1+(\omega R_o C)^2]^2\}}$$
$$= A_o/\sqrt{[1+(\omega R_o C)^2]} \ . \quad (4)$$

At the 3 dB point,

$$A_{OF} = A_o/\sqrt{2} \ .$$

So, by (4),

$$\omega_1 R_o C = 1$$

and

$$f_1 = 1/(2\pi R_o C) \quad (5)$$

where ω_1 and f_1 refer to the 3dB point.
If $\mathcal{A}_{OF}$ is written in polar form as

$$\mathcal{A}_{OF} = A_{OF} \exp(j\phi)$$

then ϕ indicates that the output *leads* the input. By (3),

$$\tan \phi = -\omega R_o C \ . \quad (6)$$

(6) tells us that at very high frequencies, $\tan \phi \to -\infty$ and $\phi \to 3\pi/2$. So the output leads the input by $3\pi/2$ or lags it by $\pi/2$. This is the largest phase shift possible.

Q2. For a given op-amp, the gain-bandwidth product is approximately constant. Explain the origin of this statement.

Consider the ideal Bode plot illustrated in the following figure, in which the 3 dB point is assumed to be coincident with the end of the plateau region.

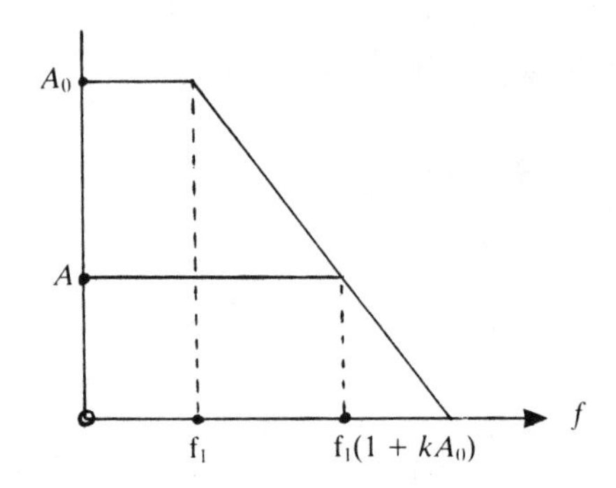

The gain-bandwidth product under open-loop conditions is simply $A_o f_1$. Under closed-loop conditions, as section 8.12 informs us, the corner frequency is $f_1(1 + kA_o)$, and the gain-bandwidth product is

$$A_o/(1 + kA_o) \times f_1(1 + kA_o)$$

which is $A_o f_1$.

This result is applicable only to those amplifiers which have roll-off of 20 dB per decade.

Bibliography

Dance, J. B., *Op-amps* (Newnes Technical Books, 1978).
Gibson, J. R., *Electronic logic circuits* (Edward Arnold, 1983).
Horowitz, P. & Hill, W., *The art of electronics* (CUP, 1982).
Jones, M. H., *A practical introduction to electronic circuits* (CUP, 1982).
Malvino, A. P., *Electronic principles* (Tata McGraw-Hill Publ. Co. Ltd., 1986).
Digital principles and applications (McGraw-Hill Book Co., 1986).
Parr, E. A., *How to use op amps* (Bernard Babani (Publishing) Ltd., 1986).
Stonham, T. J., *Digital Logic Techniques* (van Nostrand Reinhold (UK), 1986).
Waterworth, G., *Work out electronics* (Macmillan Education Ltd., 1988).

Index